international® AIR POWER REVIEW

AIRtime Publishing
United States of America • United Kingdom

international AIR POWER REVIEW

Published quarterly by AIRtime Publishing Inc.
120 East Avenue, Norwalk, CT 06851
Tel (203) 838-7979 • Fax (203) 838-7344

Created for AIRtime Publishing by Aeromedia Communications Ltd

Softbound Edition ISSN 1473-9917 / ISBN 1-880588-85-4
Hardcover Deluxe Casebound Edition ISBN 1-880588-82-X

Origination by Universal Graphics, Singapore
Printed in Singapore by KHL Printing

International Air Power Review is published quarterly in two editions (Softbound and Deluxe Casebound) and is available by subscription or as single volumes.
Please see details opposite.

Acknowledgments
The author would like to express his gratitude to the many people who provided help and hospitality during the production of the Viggen article, including: at F21 – Peter Eriksson, Christopher Hellström, Stefan Helsing, Dick Henriksson, Boris Jonsson, Björn Lindroth, Joergen Olsson, and all the personnel of 1 Div. and 1 Komp.; at F4 – Peter Wiklund; at FMV – Jonas Beronius, Pia Ericson, Ulf Hugo, Monica Liljegren, Ulf Lindström, Nils Lundqvist and Sievert Ohlander; at Saab – Vanje Brännström, Bernt-Ove Hedman, Ida Holmberg, Håkan Lindell, Anna Samuelsson, Berndt Weimer and Heidi Wendt; at Ericsson Microwave – Göran Engström, Dan Mårtensson and Evakarin Nilsson; at Volvo Aero – Fredrik Fryklund; thanks also to Daniel Karlsson and Peter Liander

The author of the Javelin article would like to thank Mrs Valerie Adams, Brian Carroll, Bill Hunt (MoD), Tim Kershaw and John Lewer of the Jet Age Museum, Roger Lindsay, Ian White, the National Archices and the RAF Museum. Thanks also to Naomi Capps (Qinetiq), Chris Esplin-Jones (Cobham plc), Tom Smith (Rolls-Royce Heritage) and Mike Stroud for their help with photos

Contact and Ordering Information (hours: 9 am-5 pm EST, Mon-Fri)
addresses, telephone and fax numbers

International Air Power Review, P.O. Box 5074, Westport, CT 06881, USA
Tel (203) 838-7979 • Fax (203) 838-7344
Toll free within USA and Canada: 1 800 359-3003
Toll free from Australia (13-15 hours ahead) : 0011 800 7573-7573
Toll free from New Zealand (17 hours ahead): 00 800 7573-7573
Toll free from Japan (14 hours ahead): 001 800 7573-7573

International Air Power Review, Postbus 3946, 4800 DX Breda, Holland
Toll free (to our US East Coast office) from the United Kingdom, Belgium, Denmark, France, Germany, Holland, Ireland, Italy, Luxembourg, Norway, Portugal, Sweden and Switzerland (5-6 hours ahead): 00 800 7573-7573
Toll free from Finland (6 hours ahead): 990 800 7573-7573

website
www.airtimepublishing.com

e-mails
airpower@airtimepublishing.com
inquiries@airtimepublishing.com

Subscription & Back Volume Rates

One-year subscription (4 quarterly volumes), inclusive of ship. & hdlg./post. & pack.:

Softbound Edition
USA $59.95, UK £48, Europe EUR 88, Canada Cdn $99,
Rest of World US $79 (surface) or US $99 (air)

Deluxe Casebound Edition
USA $79.95, UK £68, Europe EUR 120, Canada Cdn $132,
Rest of World US $99 (surface) or US $119 (air)

Two-year subscription (8 quarterly volumes), inclusive of ship. & hdlg./post. & pack.:

Softbound Edition
USA $112, UK £92, Europe EUR 169, Canada Cdn $187,
Rest of World US $148 (surface) or US $188 (air)

Deluxe Casebound Edition
USA $149, UK £130, Europe EUR 232, Canada Cdn $246,
Rest of World US $187 (surface) or US $227 (air)

Single/back volumes by mail (each):

Softbound Edition
US $16, UK £10.95, Europe EUR 18.50, Cdn $25.50
(plus ship. & hdlg./post. & pack.)

Deluxe Casebound Edition
US $20, UK £13.50, Europe EUR 22, Cdn $31
(plus ship. & hdlg./post. & pack.)

Prices are subject to change without notice. Canadian residents please add GST. Connecticut residents please add sales tax.

Ship. & Hdlg./Post. & Pack. Rates (for back volume and non-subscription orders)

	USA	*UK*	*Europe*	*Canada*	*ROW (surface)*	*ROW (air)*
1 item	$4.50	£4	EUR 8	Cdn $7.50	US $8	US $16
2 items	$6.50	£6	EUR 11.50	Cdn $11	US $12	US $27
3 items	$8.50	£8	EUR 14.50	Cdn $14	US $16	US $36
4 items	$10	£10	EUR 17.50	Cdn $16.50	US $19	US $46
5 items	$11.50	£12	EUR 20.50	Cdn $19	US $23	US $52
6 or more	$13	£13	EUR 23.50	Cdn $21.50	US $25	US $59

Volume Fourteen

CONTENTS

Major Features planned for Volume Fifteen
Focus Aircraft: HALE/MALE UAVs, **Warplane Classic:** IAI Kfir, **Variant Briefing:** Mikoyan MiG-23/27 Part 2, **Air Power Analysis:** Russia Part 4, **Type Analysis:** Vought F7U Cutlass, **Air Combat:** MAG-13 in Iraqi Freedom, **Technical Briefing:** *Charles de Gaulle*, **Photo-feature:** F/A-18 Super Hornet

PROGRAMME UPDATE

Aerial Common Sensor (ACS)

On 2 August 2004 the US Army announced the selection of Lockheed Martin Space Systems as the winner of its Aerial Common Sensor (ACS) programme, and awarded it a $879 million contract. Although the programme could be worth as much as $6 billion, the initial award allows the project to begin the system development demonstration (SDD) phase. The 66-month long phase includes the production of the first five ACS aircraft. Testing will begin in 2006 and the system will subsequently enter low-rate initial production (LRIP) in 2007, with full-rate production following in 2009.

Tasked with battlefield intelligence, surveillance, reconnaissance (ISR) and target identification missions, the new aircraft will replace the army's RC-7 Airborne Reconnaissance Low Multifunction (ARL-M) and RC-12 Guardrail/ Common Sensor (GR/CS), as well as the US Navy's EP-3E Aries II.

Like its predecessors, the ACS is composed of manned airborne collection platforms, ground-based facilities, communications sub-systems and a satellite remote capability. The programme's major equipment has been broken down into three components that include the airborne platform (APS), airborne mission equipment (AMES) and ground exploitation station (GESS) sub-systems. ACS differs from the earlier GR/CS in that the aircraft, or APS, will carry onboard system operators. The Embraer (Empresa Brasileira de Aeronautica SA) ERJ-145 airframe will serve as the APS. Embraer has already developed ISR models for the air arms of Brazil, Greece and Mexico. The two services could respectively purchase as many as 38 and 19 aircraft.

The AMES features a mission-tailored sensor suite that includes signals intelligence (SIGINT) and imagery intelligence (IMINT) equipment. The SIGINT suite comprises communications (COMINT), electronic (ELINT) and measurement and signatures intelligence (MASINT) systems. A variety of sensor payloads is planned, including synthetic aperture radar/moving-target indicator (SAR/MTI), foliage penetrating (FOPEN) radar, electro-optical/infrared (EO/IR) and hyper/multi-spectral imaging (HSI/MSI).

The GESS will utilise the Distributed Common Ground Station-Army (DCGS-A), which supports remote data processing. As mentioned earlier, the ACS operators provide an 'on-board' capability for processing, exploiting and disseminating data, which will be transmitted between the aircraft and the ground element via the Common Data Link (CDL).

Planning for the ACS began in the mid-1990s, and in January 2000 the Army announced its intention to enter into the concept exploration (CE) phase. The US Army Communications-Electronics Command (CECOM) selected three teams to conduct evaluations, and subsequently chose Lockheed Martin and Northrop Grumman to conduct the Component Advanced Development (CAD) phase. Each team was initially awarded $6 million in April 2002.

PROJECT DEVELOPMENT

India

New Medium Combat Aircraft

Despite protracted delays with its indigenous Tejas Light Combat Aircraft (LCA) programme, which is currently planned to equip five squadrons to replace current IAF Jaguars and Mirage 2000s between 2009-15, the Bangalore-based Aeronautical Development Agency (ADA) has announced long-term design and R&D plans for a fifth-generation deep-penetration strike aircraft. ADA Project Director (LCA) Kota Harinarayana said earlier this year that the newly-proposed stealthy medium-range multi-mission combat aircraft (MRMMCA) would be a natural progression from the LCA. Discussions were already progressing with the IAF on the MRMMCA's basic operating requirements, for development over 15-20 years.

As currently envisaged, it would use technology and expertise gained by ADA from developing the multi-role LCA, and would be open to joint development and co-production, for which agreement was announced with Russia earlier this year. The single-seat tailless MRMMCA would be powered by two indigenous Kaveri turbofans, and utilise smart structures and antenna, plus advanced integrated modular avionics. These would be based on shared processing architecture, with only two or three on-board computers, compared with nearly 28 in the LCA.

Meanwhile, Tejas flight development is continuing with the first two GE F404-powered technology demonstrators and the first prototype (PV-1), all flying by mid-2004, having completed more than 200 sorties. PV-2 was expected to follow in August, plus three more similar prototypes, before the first of eight limited series production (LSP) LCAs now on order from HAL becomes airborne in 2006-07. Flight tests totalling about 3,200 hours are planned to achieve full certification.

HAL chairman N.R. Mohanty said earlier this year that Rs4,000 crore ($879.22 million) IAF orders were expected soon for another 40 LCAs, including eight two-seat combat/ trainer versions which, like their predecessors, would be F404-powered. Some 200 subsequent single-seat and 40 two-seat LCA versions planned for IAF procurement would utilise indigenous GTRE Kaveri turbofans, now undergoing flight development.

When the RAF's 'Red Arrows' display team visited Blackpool for an air show, the opportunity was taken to stage this formation with four RAF Typhoon T.Mk 1s from nearby BAE Systems Warton. The quartet represents the RAF's first two units: No. 17(R) Sqn and No. 29(R) Sqn, the operational evaluation and conversion units.

HJT-36 engine change planned

In a surprise decision in May, the Indian government placed a $200 million contract in Russia for development of the new 21.55-kN (4,845-lb) thrust AL-55 turbofan designed by NPO Saturn for production of about 250 by the Ufa Engine Building Association (UMPO), and installation in HAL's new HJT-36 intermediate jet-trainer (IJTs). The first of two IJT prototypes began its flight development from Bangalore on 7 March 2003, followed by the second on 26 March 2004, powered by a 14.12-kN (3,175-lb) thrust SNECMA/Turboméca Larzac 04-H20 turbofan. SNECMA had also offered uprated Larzac U4-03 versions, developing 16.66 kN (3,745 lb) with full-authority digital engine control (FADEC), for the IAF's initial 16 pre-production HJT-36s.

The first of these nearing completion will also be Larzac-powered, since the AL-55 is at an early stage of development, with only component runs to date. The IAF and Naval Aviation plan to begin replacing some 250 HAL HJT-16 Kiran II jet trainers by 211 HJT-36s from 2005, although initial AL-55 production deliveries were not scheduled before 2006.

International

Russia resumes An-70 funding

Renewed impetus for joint development and production of the Russo/Ukrainian Antonov An-70 four-propfan heavy-lift transport was expected from recent summit agreements between the two partners. Since its first flight on 16 December 1994, An-70 development by the joint Medium Transport Aircraft International Consortium (MTAIC), of 1993 origin, is slowly nearing completion. The programme has included 345 stalls in 45 different configurations of the An-70's high-lift devices, and at various power settings of its ZMKB Progress/Zaporozhye D-27/CB-27 propfans.

Loads of up to 35.3 tonnes total weight have been parachuted on four platforms from altitudes between 500-4000 m (1,640-13,123 ft), in completely automatic modes. Flights have also been completed with a normal maximum cargo load of 35

The Navy's ACS will replace the EP-3E in the Sigint-gathering role, but will also have an important maritime and littoral/overland reconnaissance function. The greater emphasis placed by the Navy on Sigint means that its version will probably have two more operators.

Lockheed Martin claims that 70 percent of the ERJ-145 will be built in the US and final assembly will be carried out at a new facility at Cecil Commerce Center in Jacksonville, Florida. Once completed, the aircraft will be flown to the L-3 Communications facility in Greenville, Texas, where the mission equipment will be installed. An advanced version of the Rolls-Royce AE3007 turbofan engine, installed in the commercial versions of the ERJ-145, will power the ACS. Designed to operate at an altitude of 37,000 ft (11278 m), the aircraft will carry a crew of six comprising two pilots and four systems operators. A navalised variant of the ACS would likely be equipped with two additional system operator stations. It will be capable of deploying worldwide within 72 hours and can be sustained in the field for a period of 30 days. The Northrop Grumman team had proposed using the General Dynamics Gulfstream G550 as its aerial platform.

The first army examples of the as-yet undesignated aircraft will be delivered in 2009 and fielding will be complete by 2017. The army aircraft will be assigned to five aerial exploitation battalions in 'systems' that each comprise seven aircraft. Two examples will be used for training and an additional aircraft will support follow-on testing and development. The US Navy had been following the development of the ACS and joined the programme in June 2003. It has a requirement for 14-19 examples and its first aircraft could be purchased in 2008 and fielded in 2012. The latter service has not yet determined how many aircraft it will actually order.

Tom Kaminski

Lockheed Martin ACS Team

Lockheed Martin – prime contractor and lead systems integrator
Embraer – ERJ-145 airframe
L-3 Communications – aircraft integration
Harris Corp – communications equipment
General Dynamics – modelling and training
Argon Engineering – signals intelligence sensors

This is how the Army's ACS is expected to look when it appears in 2006. The configuration draws heavily on that developed for the Brazilian air force's R-99B (EMB-145RS), particularly in the area of the large ventral fairings. Like the R-99B, the ACS will have two large ventral fins to restore longitudinal stability.

Embraer ERJ-145 specification

Overall length	29.87 m (98 ft 0 in)
Wingspan	20.04 m (65 ft 9 in)
Height	6.75 m (22 ft 2 in)
Cabin width	2.11 m (6 ft 11 in)
Cabin height	1.82 m (6 ft 0 in)
Powerplant	two Rolls-Royce Allison AE 3007A turbofans each rated at 33.04 kN (7,426 lb st) thrust
Maximum ramp weight	19300 kg (42,549 lb)
Maximum take-off weight	20600 kg (45,415 lb)
Maximum landing weight	18700 kg (41,226 lb)
Maximum payload	5500 kg (12,125 lb)
Maximum cruise speed	833 km/h (450 kt)
Service ceiling	11278 m (37,000 ft)

tonnes, and with 47-tonne overloads. The An-70 recently carried a 20-tonne payload over a 3000-km (1,619-nm) range, after a take-off run of only 610 m (2,000 ft).

Under the original MTAIC agreement, Russian air force An-70 requirements totalled 164 aircraft, from local airframe construction by the Polyot manufacturers' association, and Salyut/Stupino groups for the D-27 propfans. In Ukraine, similar respective tasks are being undertaken by the Aviant Kiev State plant and Motor Sich joint-stock company, for 65 An-70s. Aviant is already building five An-70s, but Russian progress with the programme, including a $49 million contribution, had been virtually halted through funding shortages.

At a 16 June Russo/Ukraine three-day summit meeting, however, General Vladimir Mikhailov, C-in-C of the Russian Federation air force (RFAF), and Ukraine's Lt Gen. Yaroslav Skalko signed agreements to continue An-70 trials, to expedite service entry "in a short time". These views were endorsed in July at a Crimea meeting of the Russian and Ukrainian prime ministers, which included agreement to continue the An-70 as a joint programme, and resumption of financial support from Moscow. Antonov also claims that the Czech Republic has ordered three An-70s, apparently through Russian trade-debt offsets.

Tranche 2 order delayed

Concern was being expressed in July, particularly by Germany, that delays in a go-ahead decision, then due, for production commitments to 236 multi-role Tranche 2 Eurofighters and 510 EJ200 turbofans, could add up to Euro 2 billion ($2.423 billion) to already inflated programme costs. A Tranche 2 Eurofighter go-ahead was originally scheduled for four-country signature by December 2003. Its non-implementation was attributed in Germany, where parliamentary approval for 68 Tranche 2 Eurofighters has been secured, to UK difficulties between the Ministry of Defence and BAE Systems in accommodating the required funding in the MoD's budget economies.

With Eurofighter's other partners, comprising Italy and Spain, German industry sources claimed that the MoD and BAE had still not agreed how to pay for cost overruns in Tranche 1 air-defence Typhoon procurement. Their differences are believed to centre on the role-extension costs of accelerating Tranche 2 ground-attack capabilities.

Newly appointed Eurofighter chief executive Aloysius Rauen had warned that in the absence of a UK Tranche 2 funding commitment by July, it would be difficult to prevent a gap in the supply chain. Programme continuity and protection requirements were essential, he added, otherwise a production break could present "very difficult" problems. While acknowledging these possible risks, BAE chief executive Mike Turner said he would wait until MoD contract terms met company requirements.

There are additional concerns, however, that Tranche 2 Typhoon production launch delays could affect tight delivery schedules accompanying Austria's order for 18 Eurofighters from this batch, due to start in May 2007. Meanwhile, the RAF is seeking a limited air-to-surface capability from integration of Litening targeting pods in some of its 55 initial Tranche 1 Eurofighter T.1/F.2s, to deliver Raytheon GBU-10/16 Enhanced Paveway II laser/GPS-guided bombs, after 2007 operational clearance.

Tiger service clearance

After delays of about 18 months, type certification and qualification for the UH (Unterstützungs-hubschrauber) German battle-support version of the Eurocopter Tiger combat helicopter variant, achieved in August from the Bundesamt für Wehrtechnik und Beschaffung (BWB), followed similar French DGA (Délégation Générale pour l'Armement) clearance last spring. These were essential preliminaries for initial deliveries of some of the first eight French and five German helicopters to the joint Tiger training centre at Le Luc, in southern France, from October 2004.

Both the UH-Tiger and French attack/escort Tiger HAP will complete further certification and qualification trials in 2005, when expected functionality improvements and additional equipment beyond the original development contract will be incorporated. Service entry for Germany's Heeresflieger (army aviation) Tigers is planned for 1 April 2006 at Fritzlar air base. Apart from 80 Tigers each ordered by France and Germany, Australia is taking delivery of 22, and Spain 24.

Italy

M-346 starts flight development

Aermacchi experimental chief test pilot Olinto Cecconello made a successful initial 55-minute flight on 15 July of the first of three prototype M-346 advanced/lead-in fighter trainers now being built. He said that the M-346 confirmed its high thrust/weight ratio and outstanding field performance, taking off in 400 m (1,312 ft) and landing in 520 m (1,706 ft). Three more test flights followed by the month-end.

Ground-running of the M-346's two 28-kN (6,300-lb) Honeywell ITEC F124-200 turbofans started in April at Aermacchi's Venegono flight test centre, allowing taxiing trials to start two weeks later.

As a 10-15 per cent scaled-down version of the Yak/AEM-130 proof-of-concept demonstrator, and 1000 kg

On 6 May 2004 Boeing rolled out the first production MH-47G special operations Chinook for the US Army. The aircraft was rebuilt from a CH-47D.

The first Bombardier Sentinel R.Mk 1 for the RAF flew on 26 May with its mission system installed by L-3 Integrated Systems. The aircraft will be flown by No. 5 Squadron from RAF Waddington.

(2,204 lb) lighter, the M-346 has been further modified to meet western structural, flight-control, maintenance, supportability and production standards. Reducing the canopy depth and cross-sectional area for slimmer profiles became possible because of the smaller and lighter Martin-Baker/SICAMB Mk 16L zero-zero ejection-seats. Replacement of the original chined forward-fuselage by a circular-section nose allows radar installation for weapons training or combat roles. Eight underwing and one centre-fuselage hardpoints will accommodate up to 3000 kg (6,614 lb) of external stores.

Claimed M-346 performance is Mach 0.95 in level flight, and sustained angle of attack capabilities up to 35°. As the only current twin-turbofan advanced trainer, apart from its Yak-130 progenitor, the M-346's high dry-thrust margin allows sustained turns of some 14°/sec at a constant 5-6 *g* and Mach 0.88 at 4572 m (15,000 ft).

The M-346 is one of four advanced trainers being evaluated from a Euro7.9 million ($9.7 million) feasibility study for the 12-nation Advanced European Jet Pilot Training (AEJPT) or Eurotraining programme. With Dassault Aviation, Spain's EADS CASA, EADS Military Aircraft in Germany and Saab, Aermacchi was a signatory to the international collaboration agreement in July 2002, for feasibility studies of a joint European integrated training system. This includes all flying and ground-based training for advanced and lead-in phases, to meet European Air Staff Target specifications.

In a crowded advanced/lead-in-fighter trainer, and light attack export field, the M-346 faces strong competition from the Aero L-159B, BAE Hawk 100 series, a proposed single-turbofan Dassault project, EADS Military Aircraft Mako AT/LCA, Korea Aerospace Industries/Lockheed Martin T-50 Golden Eagle, a proposed down-rated JAS 39 Gripen, and the Yak-130 itself.

Apart from Eurotraining interest, the M-346 is tailor-made for the Italian air force's emerging requirement to replace some 85 MB-339A advanced trainers. These have operated successfully since 1979, and upgraded MB-339CD versions are still being produced.

United States

Raptor update

The USAF's F/A-22 Combined Test Force recently completed electromagnetic compatibility (EMC) and electromagnetic interference (EMI) testing of the Raptor's communications, navigation and identification (CNI) systems in the Benefield Anechoic Facility at Edwards AFB, California. The tests verified that the aircraft's systems are not affected by outside electronic emissions and that the aircraft is not generating unintentional emissions.

EA-18G assembly begins

Northrop Grumman began assembling the first bulkhead components for the initial Boeing EA-18G electronic attack aircraft on 1 July 2004. Boeing will deliver two EA-18Gs to the US Navy under the programme's system development and demonstration (SDD) phase. Northrop Grumman serves as the principal sub-contractor to Boeing as part of the Super Hornet programme and builds the entire centre and aft fuselage. The EA-18G will eventually replace the Navy's fleet of EA-6Bs and will carry out a variety of missions including surveillance and electronic jamming of enemy threat radars and communications systems. The aircraft's electronic attack suite will be based on the Increased Capability (ICAP) III system that is currently being tested and fielded by the Prowler. The SDD phase will span a five-year period and includes laboratory, ground and flight-testing. The EA-18G is expected to achieve initial operational capability (IOC) in 2009 and plans call for the procurement of 90 examples.

During August 2004 the Canadian Forces leased a General Atomics Mariner in support of its Atlantic Littoral ISR Experiment (ALIX). The Mariner deployed to CFB Goose Bay, Labrador, from where it carried out three 23-hour surveillance missions over the Arctic Ocean, Gulf of St Lawrence and the Grand Banks. The missions simulated a domestic emergency, a peace support operation and a domestic security mission. Imagery collected by the UAV's EO/IR cameras and its maritime radar was passed, in real-time, to Canadian warships and a command centre in Ottawa. The leased UAV was assigned the Canadian designation CU-163.

UH-60M progress

Sikorsky and the US Army have completed 325 flight test hours with the initial pair of UH-60M prototypes at Sikorsky's Flight Development Center in West Palm Beach, Florida. UH-60M flight testing began in September 2003 and overall performance, structural, handling quality and propulsion testing is now 95 percent complete. The new model has operated at a maximum weight of 24,500 lb (11113 kg) in the self-deploy configuration, and external loads of 9,000 lb (4082 kg) have been carried to an altitude of 20,000 ft (6096 m). Night Vision Goggle (NVG) compatibility was also concluded recently. While UH-60M no. 1 is primarily dedicated to airframe testing, the no. 2 aircraft is supporting structural, performance and avionics tests. Flight tests are currently focused on the aircraft's new glass cockpit displays, the embedded GPS inertial navigation system, and verifying the T700-GE-701D engines and the Improved Hover Infra-Red Suppression System (IHIRSS) compatibility with the new airframe. The programme currently involves eight aircraft, including a single UH-60A, five UH-60Ls and one UH-60Q that are being modified to the new configuration and a single new production UH-60M airframe. Low-rate initial production (LRIP) will begin in 2005.

Latest Knighthawk tests

Sikorsky Aircraft is conducting tests of an MH-60S helicopter that is equipped with the Northrop Grumman Airborne Laser Mine Detection system (ALMDS). Initial testing of the Airborne Mine Countermeasures (AMCM) system comprised a captive carry test that evaluated the interface between the ALMDS, and the Carriage, Stream, Tow, and Recovery System (CSTRS). The latter system provides the interface between the helicopter and various AMCM systems. The CSTRS is equipped with a BRU-14 bomb rack when carrying the ALMDS. It is, however, replaced by a towing system that allows the carriage, and stream and recovery of towed systems. Once the contractor completes the captive carry testing the aircraft will undergo additional testing with the US Navy at NAS Patuxent River, Maryland, and in Panama City, Florida. The AMCM version of the MH-60S helicopter first flew in July 2003. Besides the CSTRS, the AMCM derivative is equipped with a removable operator's station in the cabin, and additional internal fuel. The AMCM variant will achieve initial operational capability (IOC) during 2005.

Sikorsky acquires Schweizer

Sikorsky Aircraft recently acquired Schweizer Aircraft, which although most famous as a glider manufacturer, now primarily builds light helicopters and reconnaissance aircraft. Its lineup includes the air vehicle for the Northrop Grumman RQ-8 Fire Scout VTUAV that has been purchased by both the US Navy and Army. Schweizer, which has been a Sikorsky sub-contractor for more than 25 years, is headquartered in Elmira, New York, and is the United States' oldest privately held aircraft manufacturer. It will be operated as a wholly owned subsidiary of Sikorsky, which is itself a division of United Technologies.

JSF production progress

Assembly of the forward fuselage and wing for the initial F-35A Joint Strike Fighter began recently at Lockheed Martin's facility in Fort Worth, Texas. BAE Systems also began assembling the first aft fuselage at its facility in Samlesbury, England. The events mean that each of the aircraft's four major sub-assemblies is now in production. Construction of the centre fuselage began in May at Northrop Grumman's facility in Palmdale, California. All four sub-assemblies will be mated in the spring of 2005 and the aircraft will be completed later that year. The first flight for the conventional take-off and landing (CTOL) configured F-35A is due to take place in 2006.

Northrop Grumman has begun integration range testing of the fighter's AN/APG-81 fire control radar at its Baltimore, Maryland, facility. The radar will undergo approximately 6-8 months laboratory testing in preparation for the initial development flight-testing aboard the company's BAC One-Eleven test aircraft in early 2005. Both air and surface modes will be tested aboard the BAC 1-11 prior to delivery of the first radar to Lockheed Martin's Mission Systems Integration Lab (MSIL), in late 2005. Once installed in the MSIL, the radar will be integrated with the F-35's mission systems.

USMC seeks tiltrotor gunship

The USMC's deputy commandant for aviation recently approached Bell Helicopter and asked the contractor to investigate the development of an armed variant of its commercial BA609 tiltrotor that could serve as an escort for the MV-22B tiltrotor transport.

Bell has already been working on several ideas for tiltrotor escorts that included a BA609 derivative and plans to make a presentation to the USMC. Its concepts include a newly designed attack tiltrotor, a hybrid that combines the V-22's wing and rotors with a newly designed gunship fuselage. The most advanced concept is based on a new type of tiltrotor aircraft that could fold its rotors in flight and use a jet engine.

Developed by Bell and AgustaWestland, the BA609 first flew in March 2003 and is currently undergoing flight-testing. The USMC believes that flight-testing of an armed variant could begin as early as 2015.

Sova-Attack unveiled

On 20 August 2004 the ATL consortium unveiled a new privately-funded Mi-24 'Hind' upgrade – the Sova-Attack – at the TEREM Branch Letetz maintenance facility in Sofia, Bulgaria. The ATL group includes ATE of South Africa, TEREM and the trade company Liconex. Foreign partners include Thales from France and the South African companies Aerosud, Cumulus, Kentron and IST. The Sova-Attack is aimed at a number of potential customers, principally Bulgaria. ATL is also working on a privately-funded Mi-17 upgrade (named Sova-Support). The Sova-Attack was expected to fly in October 2004, with the Sova-Support following in late 2004 or early 2005.

The Sova-Attack demonstrator is an ex-Bulgarian Mi-24D which was acquired in early 2004. It has been upgraded to a partial ATE Super Hind Mk II standard, its core system being based on that developed for the Denel Rooivalk helicopter and already proven in service on ATE-upgraded Algerian 'Hinds'. New equipment in the Sova-Attack includes a Kentron Cumulus Argos 550 nose turret with TV, FLIR, auto-tracker and laser rangefinder/designator, and a Kentron F2 20-mm cannon. The wing has been upgraded to carry two four-round launchers for the Kentron ZT35 Ingwe laser beam-riding anti-tank missile. The new navigation system incorporates digital Doppler and embedded GPS, while the cockpits have been overhauled with colour displays, HUD (front only) and NVG compatibility. A programmable Thales/Vinten chaff/flare dispenser is fitted, although the Lipa IR jammer is retained and there are no plans to fit new RWR or missile approach warning systems. These could be integrated on customer request.

The Sova-Attack demonstrator is seen during its roll-out at the TEREM works. The aircraft is similar in configuration to the Mi-24s upgraded by ATE for Algeria. The large box on the side of the forward fuselage feeds ammunition to the F2 cannon in the nose. Note the Ingwe missile launchers.

Bulgaria's Mi-24D/V fleet is nearing the end of its original service life, and structural life extensions are required for service beyond 2005/06. In May 2004 the Bulgarian air force requested a minimum upgrade of an initial batch of six Mi-24Ds as an interim solution, with a longer-term upgrade being applied to 12 aircraft. The interim Mi-24Ds would then be sold to offset the cost of the 12-aircraft major upgrade. The upgraded aircraft could serve until 2010/12.

Tenders are expected to be requested in late 2004. As well as ATL, BAE Systems, Elbit, IAI Lahav and Sagem are bidding for the contract. Russian proposals from Mil, Rostvertol and Ulan-Ude have reportedly been rejected as the Russian equipment does not meet a number of important NATO criteria.

Alexander Mladenov

Upgrades and Modifications

Australia

New Hornet pods and weapons

Australia's Defence Department has approved a plan to equip the RAAF's F/A-18 fighters with an updated precision target designation system. The Hornets are currently equipped with the AN/AAS-38 Nite Hawk targeting system; however the dated equipment will be replaced by one of three systems that are currently being assessed. The equipment includes the Raytheon AN/ASQ-228 advanced targeting forward looking infrared (ATFLIR) pod, the Lockheed Martin Pantera (Sniper XR advanced targeting) pod, and the AN/AAQ-28 Litening AT pod, which is manufactured by Northrop Grumman and Rafael. Selection is planned for 2005 and the first squadron is expected to be equipped with the new system by early 2007.

Australia will also purchase Enhanced Paveway II (EP2) and Joint Direct Attack Munition (JDAM) kits that will provide the Hornets with an all-weather precision bombing capability at a cost of around $35.4 million. The GPS-guided kits will equip bombs that are already in the RAAF inventory. The initial integration of the bomb kits will be complete by 2008.

New missile capability planned

Australia has announced plans to equip its F/A-18 fighters and AP-3C maritime patrol aircraft with long-range missiles. Three weapons are currently being considered and include the AGM-154 joint air-to-surface stand-off missile (JASSM) built by Lockheed Martin, a variant of the KEPD 350 precision-attack cruise missile manufactured by Taurus Systems, and the AGM-84K SLAM-ER built by Boeing. Australia will select its new missile in 2005 and the weapons will be introduced into service between 2007 and 2009.

Canada

Updated Aurora flies

Trials of the first of two CP-140 Auroras equipped with new avionics was carried out in Halifax, Nova Scotia recently. Canada's fleet of 18 Aurora maritime patrol aircraft is being equipped with Navigation and Flight Instruments Project developed by CMC Electronics as part of the larger Aurora Incremental Modernization Project (AIMP). This is the second of four blocks associated with the AIMP. The modifications were performed by IMP Aerospace in Halifax. Flight-testing is being carried out by the contractors and CAF personnel and will be completed in the autumn of 2004.

Egypt

More FMS contracts

Continuing US military aid to Egypt included notification to Congress on 6 July by the Defence Security Cooperation Agency (DSCA) of a possible $108 million Foreign Military Sale contract to upgrade three CH-47C Chinook medium-lift helicopters to the newer CH-47D configuration. In addition to associated equipment and services, Egypt's FMS package will include eight Honeywell T55-L-712 turboshaft engines, spare and repair parts, support equipment, publications and technical data, communications equipment, maintenance, personnel training and training equipment. Boeing Helicopter will be prime contractor for this programme, which will apparently supplement some 14 or so EAF army support Meridionali/Boeing CH-47C/Ds operated since 1981.

On 7 July, a further DSCA notification concerned a proposed FMS contract, worth up to $94 million, to supply Egypt with three ex-USAF C-130H Hercules, plus associated support equipment, training, and services. Prime contractor will be Lockheed Martin Aeronautics Company to refurbish the C-130s before their EAF transfer, to supplement some 22 C-130H/H-30s.

Japan

First Hawkeye 2000 conversion

Progress with the Japanese Air Self Defence Force upgrade programme for its 13 Northrop Grumman E-2C Hawkeye AEW&C aircraft was reported on 14 July with the initial flight of the first to incorporate Hawkeye 2000 enhancements. These were installed by a Japanese Hawkeye industry team which includes Kawasaki Heavy Industries (KHI), Toshiba, and Yamada Corporation. Northrop Grumman's Integrated Systems sector, the US Navy's E-2C prime contractor, undertook the engineering and kit production for the modification programme, which Kawasaki Heavy Industries is completing at its Gifu facility in Japan.
The JASDF so far plans to upgrade nine of its E-2Cs to Hawkeye 2000 standards, which include new mission computers and displays, and replacement of earlier AN/APS-138 radar with the Lockheed Martin AN/APS-145, for significant enhancements in operational capability and reliability.

New Zealand

P-3 upgrade programme revived

The RNZAF's long-deferred Project Sirius follow-on upgrades for its six Lockheed P-3K Orion maritime patrol and surveillance aircraft were finally revived in August by selection of US L-3 Communications Integrated Systems division as preferred contractor. Four companies received requests for tender last year, for the revised $NZ300 million ($189 million) two-phase Project Guardian programme. The NZ Defence Ministry then invited L-3 to submit a best and final offer later this year.

Apart from ASW and surface attack capabilities, multi-role and inter-operability enhancements are envisaged in the revived P-3K radar, digital communication, data-management, electronic sensors, and navigation systems upgrade requirements for No. 5 Squadron Orions at Whenuapai air base. As outlined in the NZ Defence Long Term Development Plan (LTDP), the upgrade will follow contract nego-

Aermacchi plans to fly an upgraded development of the company's S-211 basic jet trainer. Redesignated M-311, the new trainer features digital avionics, HOTAS, a strengthened airframe and undercarriage, and an uprated P&WC JT15D-5C turbofan. It will complete Aermacchi's range of military training aircraft from the primary SF-260, advanced/lead-in fighter MB-339CD and new M-346 next-generation successor. The S-211 was developed by SIAI-Marchetti (later Agusta) and was entered in the US JPATS competition. This is the former JPATS demonstrator at Farnborough in 2004.

Both UH-1Y (above) and AH-1Z have begun armament trials with Patuxent River-based HX-21. Above is a 'Yankee' firing rockets. The AH-1Z is being fitted with a new exhaust system (left) to reduce IR signature.

tiations ending in October, to launch a two-year design, development and test programme.

L-3 IS will provide the mission system design, integration, intercom and data management systems, plus their installation and test. New Zealand-based Safe Air Ltd and Beca Applied Technologies will work closely with L-3 throughout the programme, and provide long-term support for the updated P-3s, due to reach full operational status between 2006-08.

Last November, an RNZAF P-3K reached a milestone 20,000 flight hours while flying a support mission in the Gulf region. Major Project Kestrel wing and tail structural improvements incorporated earlier in the New Zealand's Orion fleet should allow their continued operation for at least another 12 years.

Two RNZAF Boeing 757-200s delivered from early May 2003 in a passenger-only configuration to replace veteran Boeing 727s, are completing local modifications to add a cargo capability for tactical and strategic transport roles. The 757s are supplementing five C-130H Hercules operated by No. 40 Squadron, at Whenuapai, for virtually world-wide transport support.

Romania

Soim upgrade progress

Romanian air force service entry was recently achieved of the first four of 25 indigenous Avioane Craiova Soim advanced jet-trainers upgraded by the manufacturers and Israel's Elbit Systems as IAR-99Cs, with F-16-type digital NATO-compatible avionics. The aircraft were delivered from mid-2003 with modular mission and air data computers, GPS/INS, and twin multifunction cockpit and head-up displays, developed jointly with Aerostar SA for Romania's MiG-21/29 Lancer/Sniper upgrade programmes. They will allow the IAR-99C to fulfil lead-in fighter training (LIFT) and light ground-attack roles, with both eastern and western weapons.

The prototype IAR-99C started flight development on 22 May 1997, but service acceptance trials were completed only in 2003. An improved defensive aids suite is also being installed, plus on-board combat training systems, including simulated radar, RWR and weapons operation. Avioane is completing another four IAR-99Cs at Craiova, and at least 16 more upgrades are scheduled by late 2005. Joint export marketing is also being explored by Avioane and Elbit.

Russia

Tanker plans released

The Russian Air force has decided to bolster its air refuelling fleet by equipping a number of its Ilyushin Il-76MD transports as convertible tanker/transports. Once modified, the aircraft will be redesignated as Il-76MDK. The service currently operates around 40 convertible Il-78s and dedicated Il-78M tankers, and the addition of the Il-76MDK will double the size of its tanker fleet by 2015.

United States

H-1 upgrade progress

The Bell UH-1Y fired weapons for the first time at Fort A.P. Hill, Virginia, recently. Bell Helicopter and USMC test pilots from the H-1 Integrated Test Team conducted the test. Under this phase of testing approximately 400 70-mm (2.75-in) rockets, 12,800 rounds of machine-gun ammunition, 136 flares, 104 chaff canisters and 104 decoys were fired over the course of some 40 flights. The AH-1Z undertook its weapons testing at the Army's Yuma Proving Ground in Arizona, where it demonstrated the capability to fire 2.75-in rockets and its M197 20-mm Gatling gun using the Thales Top Owl helmet-mounted sight and display system. As part of the testing the AH-1Z will also fire Hellfire anti-tank and AIM-9 air-to-air missiles.

The initial AH-1Z was recently equipped with an updated exhaust system designed to reduce the aircraft's vulnerability to heat-seeking missiles. The integrated engine exhaust management system (IEEMS) reduces the aircraft's infrared signature by diverting the engine's hot exhaust gases out and away from the aircraft's tailboom. The aircraft returned to NAS Patuxent River following a short flight test programme at Bell's Amarillo, Texas, facility. The system will also be installed on the UH-1Y, which is currently being modified, as well as the current fleet of AH-1Ws. Flight testing of the AH-1Z and UH-1Y will continue at Pax River and a modified AH-1W is scheduled to undergo flight tests beginning in October 2004. The first operational system will be fielded in March 2005. Bell Helicopter has received a $7.2 million contract that covers the production of 24 turned exhaust kits for AH-1W helicopters.

Orion delivered

The US Navy recently took delivery of the first P-3C to be refurbished by L-3 Communications under the Enhanced Special Structural Inspection (ESSI). L-3 is refurbishing 16 P-3Cs under a $48 million contract that will add up to 5,000 hours of service life to the airframe.

Yemen

MiG-29 upgrade plans

Some of 16 YAF MiG-29s and four two-seat MiG-29UB combat trainers acquired from Moldova and Russia from 1994 are to be upgraded by RSK MiG to multi-role MiG-29SMT standards. Main changes include an RPKB Ramenskoye digital avionics system, similar to that developed for China's Su-30MKKs, and an uprated Phazotron NIIR N019M or N010 Zhuk-M fire-control radar, plus greater internal fuel capacity.

PROCUREMENT AND DELIVERIES

Algeria

Trainer plans and MiG-29s

Algeria has expressed interest in purchasing up to 80 MiG-AT advanced jet trainers from Russia, plus Ilyushin Il-103 basic trainers. The trainer is being developed jointly by Russia's RSK-MiG and a consortium of French companies that includes Snecma and Thales. The deal could be worth as much as $1 billion.

Earlier reports of the planned supply by Russia to the Algerian air force (AQJJ) of up to 49 upgraded MiG-29s, reported variously as M1/M2s, or more probably less advanced MiG-29SMTs, were confirmed in Moscow in June. A proposed $1.8 billion contract was expected to include weapons, spares and support equipment, to supplement 24 'Fulcrum-Cs' delivered in 1999. Russian reports claimed that the MiG-29s would follow 193 combat aircraft supplied to Algeria earlier.

Deliveries of a number of Mi-171Sh transport helicopters have also been completed to Algeria by the Ulan-Ude Aviation Plant (UUAZ), according to its General Director Leonid Belykh. He said that Algeria had received its Mi-171Sh helicopters through the Rosoboronexport state arms agency, and additional deliveries had been completed by UUAZ to Iran. Active discussions were also under way with Iran, to create a service centre for maintenance of UUAZ helicopters delivered earlier. Belykh also said that UUAZ has begun working on a contract for delivery of Mi-171Sh helicopters to Yemen, plus two for Malaysian police and military use.

Australia

NH90 selection announced

Government plans were announced on 31 August to acquire 12 NH Industries NH90 transport helicopters costing some $A1 billion ($709 million), via Australian Aerospace, for Army Aviation Corps (AAAC) troop-lift roles, to meet long-standing Project Air 9000 Phase 2 requirements. Designated MRH-90s, Australia's new aircraft will be customised for shipboard operations, with extra corrosion protection, folding rotor blades and other enhancements. These include a rear vehicle loading ramp, flexible cabin configuration, full fly-by-wire flight-controls, and a digital cockpit

Subject to satisfactory contract negotiations, they will equip an additional helicopter squadron, releasing an AAAC Sikorsky S-70A-9 squadron to provide dedicated east coast Australian Special Forces support. MRH-90 deliveries for the new AAAC squadron at Townsville are planned between 2007-08. Australia will become the 11th NH90 customer, from 357 firm orders, plus 86 options additionally from France, Germany, Finland, Greece, Italy, Netherlands, Norway, Oman, Portugal, and Sweden.

Brazil

FAB considers Impalas

Negotiations have recently been reported with South Africa for possible FAB acquisition of surplus SAAF Atlas Impala versions of the Aermacchi MB.326 advanced jet-trainer. These are required to replace Brazil's 30 or so remaining EMBRAER-built EMB-326GBs (T/AT-26s) from 166 originally built under licence, and operated since 1972 on drug-interdiction as well as advanced and lead-in fighter training roles. Most have now reached their authorised fatigue-lives, and the FAB has allocated some $8 million to acquire 22 SAAF Impalas, with options for 23 more, plus 48 Rolls-Royce Viper turbojets, spares and support equipment.

Chile

Air force procurement progress

Flight development is planned to start before the year-end at Edwards AFB of the first of six GE F110-129-powered Block 50 F-16Cs, and four two-seat F-16Ds, ordered under a $320 million May 2003 FMS contract. Equipped to operate with Rafael Python 4 agile AAMs, in conjunction with a helmet-mounted sight, their delivery will start after completion of the development programme, from February 2006. The FACh needs 24 F-16s, but funding is still awaited for the additional 14.

Recent FACh procurement has also included six US Army surplus Bell UH-1Hs, and four used Bell 412s requiring overhaul. Four Bell 412s are also being sought for Chilean naval aviation (CANAC), while the army has a requirement for up to 20 Mil Mi-17s.

China

Upgraded Su-30s delivered

Senior Russian industrial sources have revealed that Sukhoi deliveries to China in 2003 comprised the first of 24 KnAAPO-built upgraded multi-role Su-30MK2 versions of the MKK, with more advanced glass-cockpit avionics from the Ramenskoye Design Company (RPKB), as a key member of the 17-Russian company Technocomplex group. The Su-30MK2 are reported to have followed earlier KnAAPO supplies to China of 76

Raytheon is demonstrating an updated version of the T-6 to customers. The T-6B (right) introduces a Flight Visions Inc. FV-4000 Modular Missions Display Processor, stores management system SparrowHawk HUD and colour displays.

Su-30MKKs, as well as 80 or so Su-27SK air defence fighters and two-seat Su-27UBK combat trainers, within overall procurement contracts of up to $7 billion. Su-30MK2 upgrades include NIIP Tikhomirov N001VEP multi-mode radar and Kh-31 missile. The MK2 fleet is believed to be earmarked for service with the PLANAF (navy).

Colombia

Renewed light attack plans

Government plans were announced in June to buy 24 new turboprop aircraft to replace the FAC's ageing fleet of Cessna A-37B and Rockwell OV-10D attack aircraft, for drug trafficking interdiction, and use against paramilitary rebel forces. Some $234 million had originally been allocated for this programme, suspended in November 2002, for which evaluations of leading international contenders have now been renewed.

Nine of 16 Mil Mi-17-1V 'Hip-H' transport helicopters acquired by Colombian army aviation (EdC) in 1997-02 will undergo major overhauls and minor systems upgrades from a new $18 million contract. This was negotiated recently at government levels, for completion by Russia's Kazan Helicopters.

Ongoing US aid to Colombia is planned to include three Basler Turbo BT-47/DC-3 turboprop conversions, for operation by the National Police Air Service. They will supplement nine similar but heavily-armed AC-47 turboprop conversions, operated by the Colombian air force (FAC) as gunships in drug-interdiction roles.

New transport

The USAF has awarded Raytheon a $29.4 million contract that covers the purchase of a single used B737-74V Boeing Business Jet that will be delivered to Colombia.

Denmark

EH101 and C-130J re-orders

New RDAF equipment announced in the DKr19 billion ($3 billion) June defence budget included follow-on procurement of four EH101 utility helicopters and a fourth Lockheed Martin C-130J tactical transport, both from options in earlier contracts. Westland Helicopters exhibited the first of 14 RDAF EH101s ordered in December 2001 at the July Farnborough air show, for imminent delivery, although the additional four, costing a reported DKr1.6 billion ($260 million), are for Danish naval operation. Like its predecessors, the RDAF's fourth Super Hercules will be an extended-fuselage C-130J-30. Following recent initial deliveries, RDAF C-130Js will be operational by the year-end.

AIM-9X procurement planned

A Letter of Offer and Acceptance (LOA) was formally executed in June between the Danish government and the US Navy, for delivery of Raytheon AIM-9X Sidewinder advanced agile AAMs for RDAF Lockheed Martin F-16s. Employed with a helmet-mounted sight, the AIM-9X is a fifth-generation short-range missile, which achieved USAF initial operational capability in November 2003. An 18-year AIM-9X production run is planned, involving more than 10,000 missiles for the USAF, Navy and Marine Corps, and another 5,000 missiles for the international market. Denmark is the fourth AIM-9X export customer, after South Korea, Poland and Switzerland.

Egypt

Tankers approved

The US State Department has given its approval for the Egyptian government to acquire KC-135 tankers. The subject aircraft will likely be former USAF KC-135As stored at AMARC in Arizona and to be refurbished before delivery.

France

C-130J lease considered

Discussions were reported early in 2004 by French Defence Ministry officials and Lockheed Martin concerning the possible lease of C-130Js, for interim service pending initial deliveries of AdA's planned 50 Airbus A400M military transports. The first of these is not expected until late 2009, but retirement of AdA's 66 Transall C.160s, including nine equipped as air-refuelling tankers, will start next year.

Germany

Dutch P-3s acquired

Negotiations between the German and Dutch Defence Ministers, Dr Peter Struck and Henk Kamp, started last autumn, concerning possible FRG acquisition of eight Netherlands' Naval Aviation (MLD) Lockheed P-3C II.5 Orion maritime patrol aircraft, were satisfactorily concluded on 13 July. Agreement was then reached on the German purchase of eight aircraft, costing Euro295 million ($357.35 million), including all logistic elements, spare parts, documentation, ground servicing and testing equipment, plus a simulator and training of German armed forces flight and technical personnel. Lockheed Martin is continuing the Capability Upkeep Programme (CUP) upgrades at its Greenville, South Carolina, plant of 10 MLD P-3Cs by March 2006, the first of which was redelivered to its Valkenburg base on 7 July. Portugal has been negotiating to buy the remaining five Dutch P-3Cs, through a reported Euro72 million ($87.2 million) contract. The ex-Dutch P-3Cs will begin replacing German Marineflieger Breguet Atlantic maritime patrol aircraft at Nordholz next year, for completion by 2006.

India

Helicopters delivered

Rosoboronexport and Kamov recently delivered the last of nine Ka-31 helicopters to the Indian navy. The Ka-31 is reportedly equipped with the E-801M Oko (Eye) airborne electronic warfare radar and is capable of detecting airborne targets at low altitudes over water.

IN aircraft-carrier plans

Two Indian government contracts were recently signed with the Italian Fincantieri group, for design and construction assistance with the IN's new 38,000-tonne aircraft-carrier or Air Defence Ship (ADS). With a similar displacement and propulsion system, based on four General Electric LM 2500 gas-turbines, to the Italian navy's recently-launched *Cavour* aircraft-carrier, the ADS will be developed in Italy by a joint Indo-Italian team, for construction at Cochin.

The ADS will supplement the 46,300-tonne 1978-built long-stored Russian aircraft-carrier *Admiral Gorshkov*, bought by the Indian government in January from a Rs6,900 crore ($1.491 billion) contract, of which Rs449.29 crore ($97.1 million) has so far been released. After a major $666.6 million refurbishment and modernisation, for completion by August 2008, this will operate 12 single-seat MiG-29K navalised multi-role fighters and four two-seat MiG-29KUB combat trainers from a

Having made its first flight on 26 July 2004, the first AgustaWestland Super Lynx 300 for the Royal Thai Navy was handed over at Yeovil on 9 September. The Thai Lynxes will remain in the UK for pilot and technician training, before being delivered to Thailand in early 2005.

$640 million package, with 30 options, plus $30 million for IN personnel training. MiG-29K delivery is scheduled between 2007-2012.

Ireland

Helicopter selection imminent

Defence Ministry evaluations started in August of competitive tenders received for six twin-turboshaft helicopters for Air Corps operation. Two light utility military helicopters are required mainly for military pilot/aircrew training, and general-purpose roles. Another four, plus two options, capable of carrying eight fully-equipped troops or equivalent cargo, plus two crew, are required for operational and utility roles. A military and civilian evaluation team was expected to complete examination of tenders from AgustaWestland, Eurocopter and Sikorsky by October, for year-end selection and contract signature.

Italy

C-130 transfers approved

Lockheed Martin is refurbishing two of 14 AMI Lockheed C-130H tactical transports recently replaced by C-130J/J-30s, from a $13 million firm fixed-price US government contract, prior to their transfer to the Colombian and Romanian air forces. Contract negotiations were completed in June, via USAF's HQ Warner Robins Air Logistics Center in Georgia, for delivery from September.

Japan

F-2 programme problems

Cost escalation and shortfalls in planned production output by Mitsubishi of 130 multi-role F-2 developments of the F-16, plus emerging fatigue-cracks in their composite wings, have threatened the future continuation of this joint close-support fighter programme. By mid-2004, Mitsubishi had delivered 45 F-2s to the JASDF, which received funding for only five this year. Their unit price has also increased from $73 million to $109 million, which virtually matches JASDF F-15J costs. Nevertheless, Lockheed Martin received a $130 million contract from Mitsubishi early this year to supply aft fuselages, 80 per cent of all port wing-boxes, and other components for another six F-2s.

AH-64D procurement progress

Further progress with the little-publicised Japanese Defence Agency (JDA) procurement of some 60 Boeing AH-64D Longbow Apache all-weather attack-helicopters came in July, with a report from Longbow International (LI). As a Lockheed Martin and Northrop Grumman joint-venture, LI delivered the first two of 23 APG-78 Longbow fire-control radar units in July to Mitsubishi Electric Corporation (MELCO), from a Japanese government contract.

Jordan

Helicopter strength increased

Agreement was announced in July between the US Government and Sikorsky Aircraft for procurement of another eight UH-60L Black Hawk helicopters for the Royal Jordanian air force (RJAF), increasing its overall total to 13. The new UH-60Ls will enter RJAF service in 2006, for special operations, armed support, utility, SAR, and VIP transport roles.

Delivery was also completed this summer by Eurocopter of the ninth and last EC 635 military versions of the EC 135 (formerly the MBB BO108), which include five for special operations with the RJAF, and four for the police. The EC 635 has provision for armament, including GIAT 20-mm NC621 cannon, HMP 400 machine-guns, and 12-round FZ 237 70-mm (2.75-in) rocket-launchers.

This wooden mock-up shows how the Sikorsky CH-148 Cyclone might appear in Canadian Forces colours. The H-92 (S-92) was announced as the winner of the Maritime Helicopter Project on 23 July 2004, having beaten the AW Cormorant (EH101) and Lockheed Martin/Eurocopter NH90. The order covers 28 aircraft.

Malaysia

Agustas ordered

Malaysia has purchased 11 A 109 helicopters from AgustaWestland and will accept its initial example in November 2005. The remainder will be delivered in monthly instalments. The Malaysian Army will operate the aircraft in the gunship role. Malaysia is also reportedly interested in purchasing the larger AB 139 utility helicopter for use in the SAR role.

Transport helicopters delivered

The Ulan-Ude aircraft manufacturing plant in Russia has delivered four Mi-171Sh utility helicopters to Malaysia. A total of 10 aircraft will eventually be delivered to the Royal Malaysian Air Force. The manufacturer also delivered examples to the Malaysian Police in June 2004.

Nepal

Helicopters delivered

The Royal Nepalese Army recently took delivery of two advanced light helicopters (ALH) from India. Known as the Dhruv, the ALH is manufactured by Hindustan Aeronautics (HAL).

Oman

NH90s ordered

RAFO rotary-wing reinforcement will result from a 24 July order signed in Muscat by the Omani Defence Affairs Minister and NH Industries (NHI) for 20 NH90 tactical transport helicopters (TTHs), with enhanced RTM 322-01/9 powerplants. As a wholly-owned subsidiary of EADS International, Eurocopter has a 62.5 per cent shareholding in NHI, followed by Agusta with 32 per cent and Stork Fokker with 5.5 per cent. The contract includes a comprehensive support package, plus training aids and mission preparation stations. RAFO NH90 service entry is planned by 2008.

Pakistan

More Mirages acquired

Expanded support of the Dassault Mirage III/5 for long-term continued PAF service is planned from current Pakistani acquisition from the Libyan government of some 40 Mirage 5D/DEs and 10 two-seat Mirage 5DD combat trainers, which have been retired for about a decade. The PAF currently operates about 100 Mirage III/5 series from nearly 200 purchased from Dassault and other sources since 1967, and is buying the 50 LARAF Mirage and all supporting spares, including 150 overhauled SNECMA Atar 9 turbojets, for cannibalisation.

PAF transport expansion

In a two-stage programme announced on 2 September, although first discussed in 2002, Lockheed Martin is being awarded a $8.4 million firm fixed-price FMS contract to buy seven ex-RAAF C-130E Hercules aircraft for the Pakistan air force (PAF) by December 2004. Lockheed Martin will then undertake Programmed Depot Maintenance, modification and upgrade of six of these C-130Es by October 2005. They will supplement 16 C-130B/E/L-100s operated by the PAF since 1963, when not grounded from spares shortages.

The PAF tactical transport force is being further expanded by four twin-turboprop CN-235M-220s from Indonesian Aerospace (IAe), through a $49 million contract. The PAF received two CN-235s by September, with the remaining two due by the year-end.

Bell 412 helicopters delivered

In its new status as a US ally, the Pakistan government is receiving 26 Bell 412EP medium twin-turboshaft utility helicopters and associated equipment, from a $230+ million Bell Helicopter Textron contract placed earlier this year. The first nine Bell 412s arrived in Pakistan in June, for an unspecified government agency, with the remainder due by mid-2005. Already operating about nine Agusta-Bell 205A-1s and Bell UH-1Hs, Pakistan Army Aviation may be the most likely Bell 412 recipient, primarily for personnel transport, emergency medical service and disaster relief roles.

Philippines

Medium transport helicopter

The Philippine Navy has announced plans to purchase a single medium transport helicopter. The Bell 412, already in service with the Philippines Air Force, is understood to be the preferred choice. The air arm is also seeking government approval to refurbish seven BO105 helicopters that are currently in storage. A single BO105 is currently operational.

South Korea

Military spending increases

Defence Ministry budget requests announced in June and totalling KRW21.5 trillion ($18.485 billion) for the coming year, are 13.4 per cent higher than the current year, to strengthen RoK capabilities against North Korean aggression. The increase accompanies security concerns over

Reagan hosts South American carrier aircraft

USS* Ronald Reagan *sails with NAe* São Paulo *off the coast of Brazil (above). The US carrier had Air Wing Five aboard. During joint operations Brazil's Skyhawks made a number of 'touch and goes' (right).

When the US Navy's newest carrier, USS *Ronald Reagan* (CVN 76) deployed from Norfolk to its new base of San Diego in June 2004, the naval aviators of Brazil and Argentina grabbed the opportunity to undertake a series of approaches and bolters on the vessel. First, on 8 June, while the carrier was sailing along the coast of Rio de Janeiro state, it was joined by a Brazilian Navy submarine and by the carrier NAe (Navío Aeródromo) *São Paulo* (A-12) with its wing of six A-4KU (local designation A-1) Skyhawks of VF-1. The Skyhawks performed bolters on a US carrier for the first time, as part of the Passex exercise. *Reagan* subsequently entered Rio de Janeiro's harbour.

On 17 June, as *Reagan* headed south, two Argentine Super Etendards of the 2ª Escuadrilla Aeronaval de Caza y Ataque (EA32) and three Grumman S-2T Turbo Trackers of the Escuadrilla Aeronaval Antisubmarina (EA2S) performed a series of bolters. Alouette IIIs and S-61D Sea Kings of Escuadrilla Aeronaval de Helicópteros 1 and 2 (EAH1 and 2) operated to and from the ship carrying personnel.

USS *Ronald Reagan* was travelling together with the 'Ticonderoga'-class cruiser USS *Thomas S. Gates*. On 20/21 June the carrier passed through the Straits of Magellan, the first nuclear-powered carrier to do so.

Santiago Rivas and Juan Carlos Cicalesi

Argentina's naval aviation has not had a carrier of its own for many years, and uses every opportunity to practise carrier operations. US Navy personnel were impressed by the skill of the Argentine pilots, despite their lack of carrier experience.

US plans to withdraw about a third of its 37,000 troops from the RoK by the end of next year. Seoul is expediting its defence expansion plans, since Washington is pushing its global realignment faster than expected. The proposed 2005 budget represents 2.9 per cent of GDP, however, or only 0.1 per cent more than in 2003-04. It includes KRW7.3 trillion for arms and equipment; KRW14.1 trillion for current outlays; and KRW5.6 trillion for 181 projects listed under 'combat and readiness'.

These include the acquisition of AWACS airborne early-warning systems; joint command control, communication, computer and intelligence systems (C4I), as well as the Boeing F-15K acquisition programme, K1A1 tanks, and unmanned patrol aircraft (UAVs). KRW72 billion has also been allocated for 17 new combat projects, including GPS-guided missiles, launching military spy satellites and aerial defence systems, and the Patriot ABM system, through the SAM-X programme.

The state-run Korea Institute of Defence Analysis said the RoK needed to raise its spending to 3.2 per cent of GDP, as an additional KRW209 trillion would be needed by 2020 for its current equipment needs. In June, Boeing announced its submission of the AEW&C B737 to the RoKAF, for its E-X airborne surveillance and command and control programme. A few weeks later, Korean Aerospace Industries (KAI) delivered the 140th and last licence-built RoKAF Lockheed Martin KF-16, from a 1991 KRW6.6 trillion ($5.5 billion) contract.

Sri Lanka

More An-32s ordered

Recent equipment orders for the Sri Lankan air force have included a contract for four Antonov An-32 high-altitude twin-turboprop tactical transports. Delivered from the Aviant factory in Ukraine, these will supplement three or four SrLAF An-32s remaining airworthy from seven delivered from 1995, and will be accompanied by five-year delivery of spares to expand the fleet's serviceability.

Sudan

MiG-29 deliveries completed

In July, the MiG Russian Aircraft Building Corporation (RSK MiG) announced completion of deliveries to Sudan of 10 MiG-29SEh upgraded fighters and two two-seat MiG-29UB combat trainers, with spares and support equipment, from a $370 million December 2001 contract. The first two MiG-29 jets, incorporating export upgrades, including Phazotron N019 radar allowing the launch of Vympel RVV-AE MRAAMs, arrived in Sudan last December, reportedly ahead of schedule. Two more arrived in January, with the remainder following within a few months.

US sources claim that, in addition to the new MiG-29s, Sudan's air force operates nine Northrop F-5E/Fs, 27 Chinese Shenyang F-5s and F-6s, CAC F-7Bs and three Russian MiG-23B combat aircraft.

Switzerland

New defence funding details

Federal Council funding of SF647 million ($511.8 million) approved in May for the 2004 Armaments Programme, included SF109 million ($86.224 million) for acquisition of two EADS-CASA C-295M twin-turboprop tactical transport aircraft. These will rectify the long-standing SAFAAC lack of transports, to undertake international humanitarian relief, peace support and crisis management missions. The 2003 defence budget restriction programme calls for spending reductions of SF670 million ($530 million) more between 2004-07. Nominal 2004 defence funding is actually 19 per cent lower than in 1990, representing a real-terms reduction of 36 per cent. Equipment procurement expenditures have been reduced by 57 per cent over the same period.

Other priorities will include digital data-link system integration into the recently-formed Florako high-performance national air-space surveillance, defence and air operations management organisation, costing SF268 million ($212 million). Datalinks will also be installed in SAFAAC Boeing F/A-18C/D Hornet fighters, within their current upgrade programme, funded in the 2001 and 2003 Armament Plans.

Further SAFAAC F/A-18C/D enhancements are planned from Swiss government signature of a US Navy Letter of Offer and Acceptance (LOA) for procurement of Raytheon's advanced AIM-9X Sidewinder high off boresight agile AAM.

Taiwan

E-2 Hawkeye expansion

Earlier this year, Northrop Grumman Systems, at Bethpage, New York, received an $18.167 million firm-fixed-price Foreign Military Sales contract from Naval Air Systems Command, at Patuxent River, Maryland, for procurement of one E-2T Hawkeye 2000 AEW tactics trainer for the Taiwanese government. The RoCAF has operated four E-2T Hawkeyes since 1995, and took delivery on 17 August of the first of two upgraded Hawkeye 2000s from a previous order. The second was due in October.

Thailand

US UH-1 transfers

The last 10 of 30 refurbished ex-US Army Bell UH-1H utility helicopters were formally handed over to the Royal Thai Army (RTA) at its Aviation Centre in Lopburi Province on 29 June. US-operated between 1967-1990,

All change at Laage – on 4 August 2004 Jagdgeschwader 73 'Steinhoff' dispatched its last nine MiG-29s to Bydgoszcz for service with the Polish air force. The first of a total of 22 was transferred in 2003. Above is one of the final nine MiGs to be operated by JG 73, a MiG-29GT which originally served with the East German air force's JFG 3 at Preschen as '179 Black'. One MiG-29 (29+03) has been retained in Germany for museum display. Meanwhile, JG 73 has been receiving aircraft from the first production batch of Typhoon two-seaters, an example of which is seen below wearing '30+' type designator codes.

the UH-1s were delivered to Thailand in three equal batches, after Stateside refurbishment 'like new', from a $30 million late-2001 FMS contract. This also included spare parts and training.

Turkey

Programme problems

After several years of international evaluations, Turkey's Defence Procurement Executive Committee announced the cancellation earlier this year of its $3 billion attack helicopter and UAV programmes. In 2000, the Bell AH-1Z was selected for the $2 billion attack helicopter programme, from local co-production and assembly of at least 50 of 145 for the Turkish armed forces. Contracts scheduled by 2002 failed to materialise, however, because of other funding priorities, technology transfer, and additional complications. Selection between the US General Atomics and Israel's Elbit Systems submissions for Turkey's UAV requirements has also been deferred.

Delays have further been reported with progress of the THK's Peace Eagle acquisition of four Boeing 737-700 AEW&C aircraft, for which a $1.6 billion contract was signed in June 2002. Initial deliveries were scheduled to Turkey in 2007, but development of Northrop Grumman's new multi-role electronically-scanned array (MESA) dorsally-mounted radar has reportedly still to achieve its specification requirements.

United Kingdom

WAH-64D deliveries completed

In a landmark ceremony held at the July Farnborough air show, the UK Army Air Corps' 67th and last Apache WAH-64D AH.Mk 1 attack-helicopter was formally delivered by Westland Helicopters to Defence Procurement Minister Lord Bach, on behalf of the Defence Procurement Agency. Although the AAC WAH-64s are not yet operational, Westland-led Team Apache, which includes Boeing, Lockheed Martin, Longbow International, BAE Systems, RTM and Aerosystems International, claimed to have successfully enhanced the basic US AH-64's capability. This included integration of RTM 332 turboshafts for improved performance, BAE's HIDAS defensive aid system for better survivability, and a CRV7 rocket system for improved accuracy and target effect.

Clearance has additionally been achieved of the WAH-64's armament systems and UK-specific software, which recently received MoD Military Aircraft Release approval. AAC Apache squadrons will be the cornerstone of the UK's new Joint Helicopter Command, within the Allied Rapid Reaction Corps.

United States

USAF tanker proposal

The USAF continues to wrestle with its plan to acquire 100 KC-767A tankers from Boeing and retire its oldest KC-135s. In response to the service's ongoing analysis of alternatives, International Air Systems (IAS) recently submitted a proposal under which it would provide the USAF with 100 new aerial refuelling tanker/transport aircraft for a period of 10 years at a cost of $1.8-2.2 billion. Its plan includes 30,000 hours of flight time as well as logistics and maintenance support. IAS would provide the refuelling and airlift services with new B737 and B777 aircraft acquired with private financing. Under the plan 28 tankers would be turned over to USAF units, which would operate and maintain the aircraft. A further 35 aircraft would be operated by IAS in support of military refuelling operations, while the final group of 37 aircraft would be operated as commercial carriers until required for tanker operations. When not flying missions in support of the military the aircraft would be used for contract passenger and cargo services. IAS and the USAF would share the profit from these services. The USAF's original plan called for it to lease 100 KC-767As over a period of six years and purchase each aircraft outright at the end of the lease, at a total cost of $29.8 billion. Under a subsequent plan it would have leased the initial 20 examples but purchased 80 aircraft at a cost of $23.5 billion. The US Department of Defense has deferred a decision on the USAF's tanker recapitalisation programme until a Mobility Capability Study is completed in late 2004.

Upgraded Global Hawk flies

The first Lot 2 RQ-4A Global Hawk was delivered to Edwards AFB, California, recently. Known as AF-3, the air vehicle is the first to be equipped with upgrades developed as a result of combat experience over Afghanistan and Iraq. Lot 2 includes four RQ-4As along with a single launch, recovery element (LRE) and mission control element (MCE).

Longbow milestone

The US Army recently accepted the 500th AH-64D Apache Longbow helicopter, from Boeing at its Mesa, Arizona, facility. The helicopter is actually the 383rd example delivered to the army, which will eventually take delivery of 501 AH-64Ds. Longbows have also been ordered by the air arms of Egypt, Greece, Israel, Kuwait, the Netherlands, Singapore and the United Kingdom. The 500th aircraft was built to the latest so-called Block II configuration. Boeing is developing additional improvements for the Longbow Apache that include a new composite rotor blade, the joint tactical radio system, and improved power and drive systems. The systems will be incorporated in later block updates. The contractor also recently celebrated the delivery of the last of 22 AH-64D maintenance trainers to the US Army Aviation Logistic School at Fort Eustis, Virginia.

Production MH-47G and CH-47F

Boeing recently delivered the first production CH-47F and MH-47G to the US Army. The initial aircraft, which was delivered two months ahead of schedule, is the first of seven helicopters included in Lot 1. The remaining aircraft in this lot, which will all be built to MH-47G special operations configuration, will be delivered by March 2005. A further 30 MH-47Gs will be remanufactured before the next CH-47F delivery occurs in 2006. This first CH-47F will initially support flight-testing and demonstrations.

The initial MH-47G, which was remanufactured from an earlier CH-47D, is one of three being modernised under the army's reconstitution programme and will replace MH-47Es lost in Afghanistan and Iraq. Plans call for Boeing to remanufacture and deliver 58 additional MH-47Gs by 2011. The helicopter features a number of upgrades over earlier models, including more efficient engines, improved avionics, reduced interior vibration and improved air transportability. In addition, it is equipped with an integrated digital Common Avionics Architecture System (CAAS) that will enable crews to integrate global communications, navigation, survivability and situational awareness capabilities. The service currently operates a fleet of 34 Special Operations Chinooks, comprising 23 MH-47Es and 11 MH-47Ds.

Predators ordered

General Atomics Aeronautical Systems was recently awarded a $26.4 million contract covering the purchase of seven MQ-1L Predator air vehicles, six ground data terminals and test equipment. The equipment will be delivered to the USAF's 57th Wing at Indian Springs AFAF in Nevada.

Navy transport plans

The US Naval Air Systems Command has announced plans to order at least seven Boeing C-40A transports between 2005 and 2009. A single example was requested in the 2005 Defense Bill.

Seahawks ordered

Lockheed Martin has been awarded a $14.8 million advance acquisition contract associated with the delivery of eight low-rate initial production (LRIP) III MH-60R helicopters. The helicopters are included in the FY2005 Defense Bill. Sikorsky Aircraft received a $12.3 million contract that related to the actual production of the same helicopters.

Yemen

Ka-52 order reports

According to recent Russian newspaper reports, the RSK MiG group is in the final stages of negotiating a $150 million contract to provide 12 Kamov Ka-52 two-seat attack helicopters to Yemen. MiG General Director Valery Toryanin was quoted as saying that the contract would be concluded "in the immediate future" which, if true, would be the first production order for this type. Initial Russian Federation air force (RFAF) orders have reportedly been placed, however, for a dozen or so Kamov Ka-50 versions.

Beauvechain Trainer Meet

To coincide with the 25th anniversary of Belgian Alpha Jet operations, a Trainer Meet was organised at Beauvechain (Bevekom) air base on 26 June 2004. The Alpha Jets will have moved to Cazaux in France by the end of the year, and the Fouga Magister will be retired. The Belgian helicopter wing is due to move in, although the SF.260 trainers will remain. The Meet attracted a number of trainers from European air forces. The local base displayed an Alpha Jet in an unusual scheme (above), and a Magister in the colours of the famous 'Red Devils' display team (right). Among the notable visitors were a recently-delivered Pilatus PC-9/M from the Irish Air Corps (below), and a recently upgraded SF-5M from the Spanish air force's Ala 23 at Talavera (below right). Among others, the Czech air force sent an L-159, while the French air force provided a special scheme Mirage F1B.

Air Arm Review

Australia

Role changes and retirements

On 26 June the AAAC formally withdrew its eight remaining Bell UH-1Hs from long-term service as Bushranger gunships, with provision for side-mounted Miniguns and rocket-pods. These had been developed by the RAAF for support operations in Vietnam with No. 9 Squadron gunship flight, formed on 23 April 1969, and operated until 1986. All battlefield helicopters were then transferred to the AAAC, which will start replacing them in December with the first of 22 next-generation Eurocopter AS 665 Tiger dedicated armed reconnaissance helicopters on order.

Eighteen Tigers will be assembled by Australian Aerospace. The Bushranger UH-1Hs will continue operating without armament in utility and troop-lift ('slick') roles with the AAAC until at least 2007.

On 30 June the RAAF also withdrew its last five twin RR Dart turboprop-powered HS 748-228 navigation trainers from their East Sale air base, in Victoria. These were the last of eight Hawker Siddeley 748s, acquired by the RAAF, plus two equipped for RAN electronic warfare training, from 1967.

Belgium

Radical air component changes

Redesignation of the air force (FAB) as the Belgian Defence Air Component accompanies other recent major changes and reorganisation, which include incorporation on 1 March of the army's Light Aviation Group as the BDAC Helicopter Wing. Reductions are also being implemented of the FAB's active upgraded Lockheed Martin F-16A/B MLU combat aircraft from 90 to 72 by the year-end, many already being in Weelde air base storage. These are being offered for sale, and the BDAC's active F-16 inventory will be further cut to 60 by 2015.

Similar reductions are also planned in the BDAC Helicopter Wing, which will comprise four squadrons operating from Beauvechain, initially with 44 multi-role Agusta A 109HA/HOs, plus a further squadron with 28 Aérospatiale SA 318C Alouette-Astazou IIs. Twelve A 109s will be withdrawn in the next few years, together with the ageing Alouettes, and the army's 10 Britten-Norman BN-2A Islanders. Currently used for maritime patrol and liaison, the BN-2A light-twins are being replaced by 18 army-operated SONACA/IAI B-Hunter UAVs.

Canada

Labrador retired

424 Transport & Rescue Squadron (TRS) commemorated the retirement of the CH-113 Labrador after 41 years of service on 26 June 2004, at CFB Trenton, Ontario. The squadron formally retired the 'Lab' during July when it completed the transition to the CH-149 Cormorant.

India

IAF crash statistics quoted

A total of 157 Indian air force pilots and 35 civilians had been killed in IAF MiG-21 crashes since 1971, Defence Minister Pranab Mukherjee informed the Rajya Sabha in mid-July. These crashes represented financial losses amounting to Rs17.68 billion ($385 million), he added, but did not include earlier accidents involving IAF MiG-21s operated since 1964. With current upgrades of 125 MiG-21bis by HAL to MiG-21-93 Bison standard, he said the type was the backbone of the IAF's combat fleet, although earlier versions were currently being retired. Mukherjee claimed that there had recently been a definite decline in the number of IAF MiG-21 aircraft accidents, which other sources indicate form by far the largest proportion of over 470 IAF crashes recorded since the 1970s. These further include over 56 IAF MiG-23s.

The IAF has also crashed 38 of its 146 SEPECAT Jaguar strike-fighters, in which 12 pilots were killed, since their 1979 service entry. Seven of the IAF Jaguar losses occurred in the past year, including four (two of which collided), this spring.

MiG-21 retirements

The Indian Air Force has announced plans to retire nearly 70 MiG-21 fighters during 2005 as they reach the end of their design service lives. Built between 1966 and 1970, the fighters are used as trainers at Tejpur and Bagdogra AFS.

Trainer deliveries announced

Delivery of the first Hawk trainers to the Indian Air Force, under its Advanced Jet Trainer (AJT) programme, will take place in September 2007. The last of the 24 trainers built by BAE Systems will follow in February 2008. Hindustan Aeronautics Limited (HAL) will manufacture subsequent aircraft under license between 2008 and 2010.

P-3s purchase likely

Defence officials have reportedly reached an agreement under which the US will provide India with eight upgraded P-3B maritime patrol aircraft (MPA). The aircraft would likely

This CF-188A from 425 Tactical Fighter Squadron (3 Wing, Bagotville) was given special markings for the 2004 display season to commemorate the 80th anniversary of the RCAF. In previous seasons there have been 'eastern' and 'western' display aircraft, but budgetary cuts forced the adoption of just one aircraft for 2004. Eighty Canadian Hornets are undergoing the CF-18M Avionics Upgrade Program.

In late June 2004 USS* Enterprise *and a partially equipped Air Wing One (without Tomcats, Super Hornets or USMC Hornets) took part in a JMC maritime exercise off Scotland. On board were this special-scheme F/A-18C of VFA-82 (above), and the S-3s of VS-32. The Viking (left) is carrying an AGM-84H SLAM-ER missile on the port pylon, balanced by the associated datalink pod.

receive an avionics upgrade prior to delivery. India has been actively seeking a new MPA since two of its five Il-38s were lost in a mid-air collision in October 2002.

Italy

New carrier launched

The Italian Navy will double its STOVL aircraft-carrier strength from the launch on 21 July of the 27,100-tonne *Cavour*, which Fincantieri started building in mid-2001 from a Euro1.4 billion ($1.7 billion) November 2000 order, for 2007 delivery from its Genoa shipyard. Four General Electric-Avio gas-turbines generating 88 MW (118,121 shp) will confer a sustained speed of 52 km/h (28 kt), and a 12972-km (7,000-nm) range at 30 km/h (16 kt).

With a length of 244 m (800 ft), and a 180-m (590-ft) flight-deck with a 12° bow ramp, *Cavour* will operate up to eight Italian navy aircraft, including some of 15 Boeing AV-8B Harriers, future Lockheed Martin Joint Strike Fighters, or 12 EH 101, NH90 or SH-3D helicopters. It will also accommodate up to 1,300 people, including the ship's 451 complement, 203 aircrew and up to 465 amphibious task force personnel.

Russia

RFAF awaits Yak-130 deliveries

Deliveries of the first of an initial 10 Yak-130 advanced jet trainers are due to start to the RFAF from 2006, following completion of state flight-tests in 2005, according to OAO Yakovlev OKB general director and Irkut corporation chairman Oleg Demchenko. From November, the first production Yak-130, which started flight development on 28 April, powered by two 24.5-kN (5,512-lb) ZMKB Progress/Salyut AI-222-25 turbofans, is being joined by a second aircraft from Sokol's Nizhni Novgorod factory.

Demchenko said the RFAF wanted up to 250 Yak-130s by 2015. Possible buyers included Algeria, where the Yak-130 is being marketed alongside the Su-30MK, and India. China was excluded from potential current Yak-130 markets, added Demchenko, since Sukhoi had been working for two years with the Hongdu Aviation Industry Group (HAIG) on joint development of the twin-turbofan supersonic L-15 advanced trainer/light ground-attack aircraft. "We are also developing a concept design of this type for Russia," he added.

South Africa

Rooivalk shortcomings aired

Parliamentary criticism of Denel's AH-2A Rooivalk attack helicopter programme was expressed earlier this year, including slow delivery rates and achievement of SAAF operational status, plus an apparent lack of export interest. No. 16 Squadron had received 10 of 12 SAAF Rooivalk attack helicopters on order at Bloemspruit by the beginning of this year, from current annual allocations of R200 million ($32 million).

Funding increases of nearly 30 per cent had been recommended in the next fiscal year, however, to include retrofit installation of planned upgraded Rooivalk systems, and complete operational trials programmes. The increases would also cover army APC vehicle upgrades, and new portable SAM systems.

Systems upgrades of the SAAF's nine Lockheed C-130Bs were also criticised for falling behind schedule, plus the fact that only two Hercules were maintained in regular service. Three ex-USN C-130Fs are now being offered for disposal from long-term storage, through Armscor, together with two cargo Boeing 707s, 18 Dassault Mirage F1s, 12 Dassault Cheetahs (Mirage IIIs), 10 Aérospatiale/Denel Oryx (Puma) transport helicopters, and 12 Aermacchi/Atlas MB.326 Impala Mk 1s. Also for sale are quantities of 30-mm DEFA cannon ammunition, 68-mm (2.67-in) SNEB rockets, and 250-kg (551-lb) bombs.

United States

Army Hueys transferred

The US Army transferred six UH-1H helicopters to the USAF in a ceremony held at Fort Rucker's Lowe Army Heliport, Alabama, recently. The helicopters were the first of 40 to be transferred to the USAF, at a rate of six per month through October 2004. The service will use 24 of the Hueys in support of its Specialized Undergraduate Pilot Training – Helicopter (SUPT-H) programme at Fort Rucker, and plans to update the aircraft to Huey II configuration. Initially 12 aircraft will be updated to the later configuration, while four helicopters will be used as a source of spare parts. The USAF's training programme is conducted in two three-month phases, or blocks, by the 23rd Flying Training Squadron. Although the initial Block I training had previously been conducted using procedures and syllabus associated with the US Army's Rotary Wing Qualification Course (RWQC), both blocks will be now be conducted using USAF procedures.

KC-130J approved for service

The USMC recently released the KC-130J's operational evaluation report, which recommends that the aircraft be approved for full fleet introduction and operational service. The report, which was prepared by the US Navy's Operational Test & Evaluation Force, is based upon operational testing that assessed the effectiveness and suitability of the KC-130J for several critical operational requirements. Subsequent tests to certify the aircraft's missile defence systems began in May at NAS Patuxent River, Maryland, and at NAWS China Lake, California. The aircraft will achieve initial operational capability with VMGR-252 at MCAS Cherry Point, North Carolina in September 2004.

Viking joins NASA

NASA's Glenn Research Center at Lewis Field in Cleveland, Ohio, recently took delivery of an S-3B Viking. Assigned the civil registration N601NA, the former anti-submarine warfare aircraft will support the centre's Icing Branch and will eventually replace a de Havilland Canada DHC-6 that is currently in service. It is scheduled to perform research flights

On 30 August 2004 the Swedish helicopter wing undertook its last flight with the Agusta-Bell AB 206 JetRanger. Sweden received 34 JetRangers – 22 HKP 6As for the army, 11 HKP 6Bs for the navy and a single HKP 6C used by the FMV. The HKP 6Bs were withdrawn in 2003, leaving a few HKP 6As flying with the 1st Helicopter Squadron at Boden. This aircraft is a navy HKP 6B, seen before Swedish helicopters were grouped in one wing.

NATO Tiger Meet 2004

The Luftwaffe's Aufklärungsgeschwader 51 hosted the 40th NATO Tiger Meet at Schleswig-Jagel between 30 August and 5 September 2004. The event drew another fine array of tiger-decorated aircraft (and tiger-striped cars), including an immaculate Mi-24V from the Czech air force's 231 vrl (left) and Mirage 2000C from EC 1/12 (above). From Spain came 142 Escuadrón/Ala 14 with four Mirage F1Ms (below left). This unit is called the 'Jaguares'.

The Tiger Association's latest official member is the Swiss Air Force's Fliegerstaffel 11, which sent five F-18s to NTM 2004, including this specially painted example.

beginning in November 2006. The S-3B was last assigned to VX-1 at NAS Patuxent River, Maryland, and carried BuNo. 160607.

Army Aviation changes

The 1st Infantry Division (Mechanized) will bid farewell to the AH-64A helicopter when the 1-1st AVN is deactivated in May 2005. Its personnel will transfer to Fort Rucker, Alabama, where they will transition to the AH-64D. The unit, which is currently stationed at Ansbach AHP, Germany, will be reactivated at Fort Hood, Texas, in November 2005 and will return to Germany in June 2006 after completing the Longbow transition. In related news, the 2-6th Cavalry – which recently completed its Longbow transition – will be assigned as the 1st ID(M)'s second attack helicopter battalion and will be redesignated as 3-1st AVN.

The last four UH-1H helicopters assigned to the Joint Readiness Training Center at Fort Polk, Louisiana, departed Polk Army Air Field recently. The aircraft, which had been assigned to the JRTC Flight Detachment, were transferred to the Army Aviation Technical Test Center (ATTC) at Fort Rucker, Alabama.

The US Army deactivated the 229th Aviation Regiment, which had been stationed at Fort Bragg, North Carolina. The unit, which reported to the 18th Aviation Brigade, was known as the 'Flying Tigers'. Two attack helicopter battalions assigned to the regiment and stationed at Simmons Army Airfield have being reassigned. The 1st attack battalion (1-229th Aviation), which operates the AH-64D, joined the 3rd Infantry Division (Mechanized) as part of its redesigned aviation brigade, and was renamed the 3-3rd AVN. The 2nd battalion (2-229th), which is equipped with the AH-64A, was transferred to the 82nd Airborne Division at Fort Bragg and became the 3-82nd AVN.

Although the Nebraska Army National Guard had previously announced plans to station up to 24 AH-64As at a new aviation support facility, at Central Nebraska Regional Airport in Grand Island, these plans have been revised. Rather than equipping with Apaches, the Army will station 10 light- and heavy-lift helicopters at the facility, comprising four OH-58s and six CH-47Ds.

The 2nd Battalion, 6th Cavalry Brigade (2-6th Cavalry) was recently certified as the US Army's 10th AH-64D Apache Longbow battalion. Based in Illesheim, Germany, the battalion completed eight months of training at Fort Rucker, Alabama, and Fort Hood, Texas, before being certified as combat-ready.

The 1-6th Cavalry at Camp Eagle Army Heliport, Wonju, Korea, recently transferred its AH-64As stateside and its personnel have relocated to stateside facilities where they are undergoing transition to the AH-64D.

Navy squadron news

VMFA-321 'Hell's Angels' was deactivated on 30 September 2004 at NAF Washington/Andrews AFB, Maryland. The squadron had operated the F/A-18A as part of the 4th Marine Air Wing's MAG-49.

As part of its ongoing tactical aviation integration (TACAIR) with the USMC, the US Navy has announced that VFA-82 'Marauders' will be disestablished. Although a formal timeline was not included, it will likely occur before the end of 2005. Equipped with the F/A-18C, the squadron is one of two Navy units stationed at MCAS Beaufort, South Carolina.

VFA-27 'Maces' returned to NAF Atsugi, Japan, with 12 F/A-18E Super Hornets and rejoined Carrier Air Wing (CVW)-5 during September 2004. The squadron had spent the past several months at NAS Lemoore, California, where it completed its transition from the earlier F/A-18C.

Several units that provided local search and rescue (SAR) at US naval air stations were recently deactivated. The last HH-1N assigned to the Air Operations Department at NAS Brunswick, Maine, was transferred to NAS Fallon, Nevada, following a farewell ceremony. The aircraft had been assigned to the station's search and rescue department. The three HH-1N search and rescue helicopters operated by the air operations department's search and rescue division at NAS Lemoore, California, were recently transferred to other facilities. Two went to the USMC at Camp Pendleton, California, where they are now used for training. The final aircraft was reassigned to the Naval Air Warfare Center at NAWS China Lake, California. Two SAR HH-1Ns assigned to Naval Base Guantanamo Bay, Cuba, have also been removed. The aircraft were also transferred to MCAS Camp Pendleton and assigned to HMT-303 in support of training efforts.

VFC-13 'Saints' will establish a detachment at NAS Key West, Florida, once it takes delivery of eight former Swiss Air Force F-5E fighters.

Navy OSA fleet reorganised

The US Navy has completed a reorganisation of its CONUS-based operational support airlift (OSA) fleet. As a result, its UC-12B fleet has been consolidated at two bases comprising Chambers Field/Naval Station Norfolk, Virginia, on the east coast and NAS North Island, California, on the west coast. UC-12Bs that had been previously assigned to NAS Jacksonville and NAS Key West, Florida, NAS Whidbey Island, Washington, NAS Fallon, Nevada, and NAS Lemoore and NAF El Centro, California, have been transferred to the USMC at MCAS Cherry Point, and MCAS New River, North Carolina, MCAS Yuma, Arizona, and MCAS Miramar, California, or to training duties at NAS Corpus Christi, Texas. The transfers began in April 2004 and were completed in September 2004. The UC-12s had been operated by air operations departments assigned to the air stations.

USAF changes

The 305th Air Mobility Wing (AMW) at McGuire AFB, New Jersey, took delivery of its first C-17A Globemaster III on 24 September 2004. Named *Spirit of New Jersey*, the aircraft is assigned to the wing's 6th Airlift Squadron but will also be operated by the Air Force Reserve Command's 514th AMW/732nd AS, an associate unit also stationed at McGuire. The two wings, whose personnel began training for the new aircraft in March 2003, will also share maintenance responsibilities for 13 C-17As.

The USAF has begun constructing new facilities at Hickam AFB, Hawaii, that will support the deployment of eight C-17As. The first C-17A will arrive at Hickam in January 2006 and deliveries will continue at a rate of one aircraft each month. The $30 million construction project includes the addition of new operations, maintenance and simulator facilities. Three new hangars and a new combat landing strip are part of another $190 million project. The location for the new 3,500- to 5,000-ft (1067- to 1524-m) runway has not yet been determined, but four sites are under consideration. The C-17As will replace C-130Hs currently assigned to the Hawaii ANG's 154th Wing.

Project Aquila

F-5E Tiger II for Austria

In February 2004, the question of how to protect Austrian airspace was at the top of the political agenda in Austria. While Austria had selected the Eurofighter Typhoon as its new main combat aircraft, the aircraft are not due for delivery until mid-2007. The reason for this is not a delay on the part of the Eurofighter consortium – though that continues to be the reason given by the country's left-wing political opposition – but Austria's political situation which, between the time the Typhoon decision was taken in July 2002 and the signing of the contract, involved early elections and negotiations over the ensuing coalition. Furthermore, the contract has come under fire in parliament, and although it was finally passed in August 2003 it was then far too late for the delivery of the originally-planned Tranche 1 Typhoons in 2005. Consequently, Austria has become the first customer for Tranche 2 – even before the final capability details of this variant have been hammered out.

Meanwhile, the Austrian Air Force's 23 existing Saab J 35OE Drakens are the last to be operated by any country. The fact that none has been lost in an accident since delivery in 1986/7 is considered testimony to the force's maintenance and piloting skills. At the same time, it was acknowledged that it was impossible for the Draken's service life to be prolonged until the arrival of the Eurofighter.

However, the Swedish lobby in the Austrian Air Force and the country's defence ministry is (historically, at least) strong, and following negotiations with Swedish Industry (Telub), the MoD was seeking a further (Swedish) solution to bridge the 'fighter gap'. Simultaneously, a bridging lease of six Tranche 1 Eurofighters from the air arms of the four nations responsible for developing the aircraft – Germany, Italy, Spain and the UK – was also being pursued. Neither option progressed too well: the Swedish side was influenced by Austria's failure to choose the 'logical' Gripen, and only the German Luftwaffe clearly signalled any willingness to divert any Eurofighters to Austria.

On a cold February night in a remote Swiss guesthouse, Austria's Defence Minister Günther Platter had an informal meeting with his Swiss counterpart Samuel Schmid. The Austrian mentioned the 'deadlock' over his interim solution and left with a solid offer of help by Samuel Schmid an hour later – in the shape of used F-5s. That exit-strategy quickly succeeded and by 9 March 2004 an agreement had been signed. Austria will get 12 single-seat F-5Es to operate until 2008. The pure-leasing costs were were given as Euro14 million per year by the Swiss agency 'Armasuisse', while the Austrian MoD gave a maximum level of Euro75 million

When the first four F-5s for Austria were delivered, they were welcomed by two J 35OE Drakens. The elderly Swedish fighter has become increasingly difficult and costly to support, and will be retired in 2005.

Austria's F-5s will operate primarily from Graz-Thalerhof as the other main fighter base – Zeltweg – is being rebuilt to house the OLk's Typhoons when they arrive in 2007. The F-5s will fly purely on air defence duties, for which they carry the AIM-9P Sidewinder.

in total for four years to operate the F-5s. This includes all repair work other than line maintenance done by the Swiss company RUAG, spare parts and even the calculated fuel costs for a package of 5,000 flying hours. This equates to little more than 100 hours per year for 12 pilots for four years.

One last obstacle to be overcome was the US administration in the Pentagon and the US Congress, because the aircraft were initially acquired through a US Foreign Military Sale (FMS). Many in Austria wondered if the US might delay the deal because the Austrians had not selected the F-16 Block 52. As it transpired, they need not have worried, as the deal was sanctioned and the transfer was secured in Congress by early June 2004.

As the Swiss authorities had awaited this confirmation before letting the Austrians touch the Tigers, it was as soon as the following day that the first four Austrian pilots arrived in Dübendorf near Zürich to start their training, accompanied by 30 Austrian technicians. Lt Col Georg Gappmeier, Captain (Cpt) Guenther Taschler, Cpt Joerg Sandhofer and Cpt Werner Kriebitz immediately began F-5 familiarisation under the supervision of the Swiss Air Force's chief flight instructor, Lt Col Peter Starkl. The training programme was initiated on the F-5F and included emergency procedures, landing run with no flaps, single-engined handling, IFR flying and outer-envelope handling, including spins.

Training was carried out in four-week stages: theory and type familiarisation taking place during the first week, with a theoretical introduction to the radar and training flights (including landings and take-offs) in the F-5F in the second, four solo flights (all on 24 June), plus tactical training and 1 vs. 1 air combat in the F-5E during the third week, and formation leader training and 2 vs. 2 air-combat in the final week. Most of the flying involved in what is referred to as Project Aquila was done in the airspace around Santis, Davos, Chur and St Moritz. According to Lt Col Starkl, "the Austrians' skills and levels are absolutely 'top' – but I expected nothing else."

According to Austrian technicians at Graz, only minor modifications in the communications (VHF/UHF combination) had still to be installed locally by RUAG. When the four pilots delivered the first four Tigers to Austria two days before the official hand-over, only the pair of Drakens on alert and a photo-ship Saab 105 welcomed them into Graz-Thalerhof. Draken operations will now be scaled down and concentrated at Zeltweg, while large-scale construction work will get underway there in readiness for the arrival of the Typhoon.

Thanks to the delivery of the Project Aquila F-5Es, the Austrian authorities appear – officially, at least – to be relaxed in the light of the mid-2004 debate over the delayed signature of Typhoon Tranche 2 production, particularly the subject of whether to incorporate improved air-to-ground functions. While such features were described by the Austrian AF's commander, Lt Gen. Wolf, as "nice to have but not a priority", spokesmen for Defence Minister Platter underlined their confidence that the Austrian contract for delivery from mid-2007 on will be fulfilled, and that there will be no interruption in Typhoon production.

Georg Mader

Indian deployment to Alaska

Cope Thunder 2004

In July 2004 the Indian Air Force undertook an historic deployment to Alaska to participate in the USAF's Cope Thunder exercise. The deployment notched up a number of 'firsts: first overseas IAF deployment of fighters using tankers, first use of the Il-78 'Midas' tanker outside India, and the first multi-national exercise for IAF fighters. It was also the IAF's longest ever deployment, entailing a round trip of almost 20000 km (12,428 miles).

Cope Thunder was created in 1976 to train multi-service pilots from the US in realistic air combat training, with a variety of roles and aircraft types. It runs several times a year and takes place at Eielson AFB and Elmendorf AFB, in airspace made up of three heavily instrumented weapons ranges, 17 permanent military operating areas and two restricted areas, totalling more than 66,000 square miles (170927 km²) of Alaskan and Canadian airspace. The 353rd Combat Training Squadron is responsible for organising, planning and executing the training at Cope Thunder, and oversees the ranges which are equipped with 28 threat simulators and 235 targets. Up to 80 aircraft can operate and be tracked at a time during the exercise. In 1998 Cooperative Cope Thunder was created to focus on Pacific Rim international participation. The exercise featured multi-national airlift operations combined with interdiction, personnel recovery, close air support and counter-air missions. In Cooperative Cope Thunder 2003 participants

Some of the IAF contingent is seen parked at North Bay in Ontario during its deployment to Alaska. Behind the Jaguars are the two Il-78MKI tankers. All of the 10 aircraft in the group were freshly painted for their North American adventure.

Lockheed Martin U-2

Renewed life for the Dragon Lady

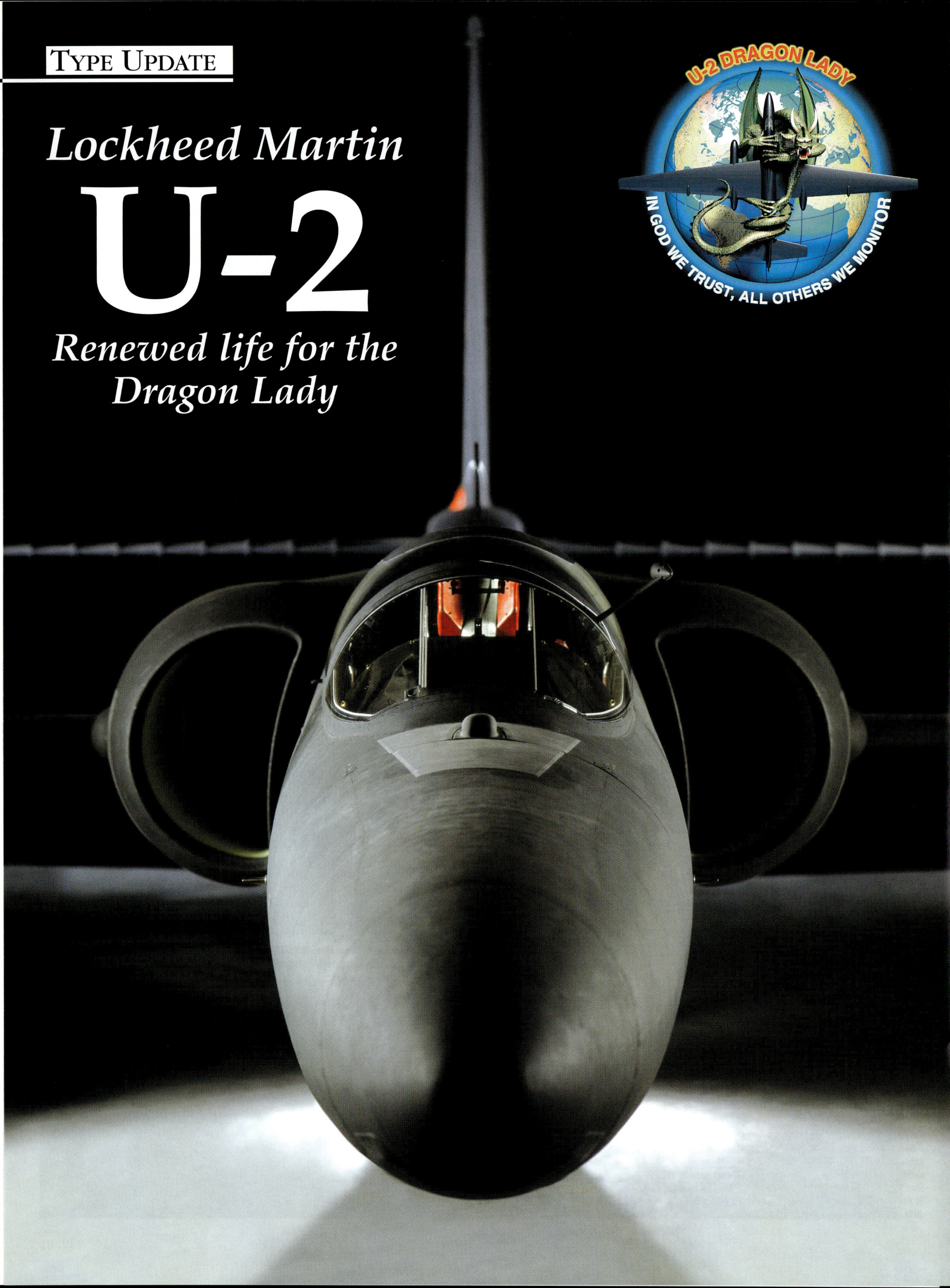

In demand as much as ever, the U-2R remains as one of the cornerstones of the USAF's reconnaissance effort. Since its major role in Operation Desert Storm in 1991, the type has added to its combat record with significant contributions to the wars in former Yugoslavia, Afghanistan and Iraq. It has also undergone a much-needed upgrade programme to maintain its usefulness in an era of ever more sophisticated intelligence-gathering. Here the development and actions of the U-2 in the last few years are reviewed.

***Above:** 80-1067 is a test aircraft permanently assigned to the Warner-Robins Air Logistics Center flight test detachment at Palmdale. It wears a red Skunk Works badge on the fin. Here it is seen fitted with the ETP/Senior Spur pod. The flight test team also has another U-2 assigned – 68-10336.*

***Left:** Another view of 80-1067 on the Palmdale ramp shows it fitted with a non-standard satcoms-type antenna on the spine. It is also fitted with an instrumented boom in the port superpod for accurate air data measurements.*

In September 1996, the southern 'No-fly' zone over Iraq, which had been established by coalition forces after the Gulf War, was extended to the 33rd parallel. Iraq objected, of course, and its air defence system stepped up harassment of US and British fighters. Despite sanctions, Iraq had managed to rebuild and refine its air defence system, including SA-2/3/6/8 SAMs and fibre-optic links between radar sites and control rooms. In Operation Desert Strike, the US launched cruise missiles against SAM sites and command bunkers.

U-2s flying from Taif were operating in the zone, helping to identify Iraqi air defence developments. They were protected by co-ordinating the sorties with RC-135 Rivet Joint aircraft, which could broadcast threats detected by their wider frequency-range ELINT and COMINT sensors. U-2s were also flying north of the zone, including over Baghdad itself, on the separate Olive Branch missions for the UNSCOM. By early 1996, some 280 such missions had been flown, each one notified to the Iraqis two days in advance in respect of entry and exit points, but not the course to be flown once inside the country. The imagery from these flights helped the UNSCOM inspectors track movements of banned weapons from suspected sites.

However, as tensions mounted over the increasingly intrusive inspections, the Iraqi regime stepped up its objections to the U-2 flights. This culminated in overt threats to shoot them down in late October 1997. During that same month, the U-2s moved to join the rest of US airpower in Saudi Arabia at Al Kharj near Riyadh, also known as Prince Sultan Air Base (PSAB). There, they eventually became part of the 363rd Expeditionary Air Wing, as the 363rd Expeditionary Reconnaissance Squadron. Despite Iraq's objections, the Olive Branch flights continued at an average weekly interval, until Iraq stopped co-operating at all with UNSCOM in August 1998. Operation Desert Fox, a four-day bombing campaign, followed in December 1998.

Operation Allied Force

Meanwhile, the tragedy of the Balkans continued to play out, and the U-2 continued to observe. In 1998, Serb paramilitary forces stepped up a campaign of repression in the semi-autonomous Yugoslav province of Kosovo. NATO countries pressured President Milosevic into accepting photo-reconnaissance flights over the province. The first U-2 mission was flown on 16 October by the 99th ERS, temporarily based at Aviano while the runway was resurfaced at Istres, France.

But the situation on the ground deteriorated, and on 24 March 1999, NATO launched Operation Allied Force to evict Serb forces from Kosovo. During the 78-day campaign, the 99th ERS flew 189 missions and provided over 1,300 hours of collection time. By flying out of NAS Sigonella, Italy, instead of Istres, the U-2s were closer to the action and could therefore increase their time-on-station within a standard nine-hour mission. With cloud covering the region often, and with much mountainous territory to be covered, images from the high-flying U-2's ASARS radar sensor proved vital in identifying targets.

Right: One of the 9th RW's four 'two-headers' taxis for a training mission from Beale. It was originally built as a TR-1B, but was then redesignated U-2RT. After re-engining with the F118 engine it became a U-2ST and, following another rationalisation of fleet designations, became a TU-2S.

Left: The end of the 1991 Gulf War did not signal the end of U-2 operations over Iraq. The Taif detachment became the 4402nd RS(P), and continued to fly over Iraq to monitor compliance with UN resolutions. Here a U-2R returns to its desert base after one such Olive Branch mission in December 1993. Monitoring overflights of Iraq continued until 1998, resuming again in February 2003 just days before the US and UK launched Operation Iraqi Freedom. Having moved to Al Kharj in 1997, the U-2 detachment maintained a watch on Iraq from safe airspace throughout the break in UN overflights, becoming the 363rd Expeditionary Reconnaissance Squadron.

While Predator UAVs and JSTARS grabbed most of the publicity, "the U-2 was the backbone of our ISR architecture," noted USAFE's Director of Operations, Major General William Hobbins, when Operation Allied Force was all over. "We never dropped a bomb on a target without having the U-2 take a look at it," he continued. Although Serbian air defences were never completely eliminated, 39 radars and 28 aircraft were destroyed after being identified by U-2 missions. Moreover, this conflict demonstrated a significant closing of the 'sensor-to-shooter' cycle, which had become a key aim in US air combat doctrine.

Rapid targeting

From the mid-1990s, Beale-based U-2s and other US reconnaissance platforms had participated in various exercises to develop the Rapid Targeting System (RTS) and Real-Time Information to the Cockpit (RTIC). By using powerful computer hardware and software, imagery and SIGINT data from multiple sources, including satellites, was matched or 'fused' to provide target locations with enough accuracy for precision attack. Selected imagery frames were annotated and transmitted via JTIDS links from the ground to an E-3 AWACS and on to an F-15E, allowing the data to cue the Strike Eagle's APG-70 radar.

During the Kosovo conflict, the U-2 provided 66 RTS packages to strike aircraft. Moreover, because it could carry SIGINT as well as imaging sensors, the U-2 proved to be a 'one-stop shop' for providing the targeting data. On one U-2 mission, the Senior Ruby ELINT sensor operator, housed thousands of miles away in the ground station at Beale AFB, detected a mobile SA-6 radar. He quickly took another direction-finding 'cut', and passed the information to his colleagues nearby who were controlling the ASARS sensor. The U-2 pilot was ordered to reverse track, so that imagery could be obtained. This imagery was then passed to the CAOC in Vicenza, Italy, and from there to an F-15E which struck the radar. Total elapsed time was 42 minutes.

After the Serbs were evicted from Kosovo, the U-2 continued to monitor the Balkans from Sigonella. In June 2000, the 1000th Sentinel Torch mission was flown. Each one of them had involved the relay of sensor data all the way back to Beale for interpretation by DGS-2 (Deployable Ground Station-2).

Six aircraft and the personnel of OL-FR pose for a group photo in 1997, as the last U-2Rs were replaced by the U-2S. At the time the Istres-based detachment was busy covering events in former Yugoslavia. Two years later the detachment had become the 99th Expeditionary Reconnaissance Squadron, and was fully employed in providing quality targeting imagery and electronic data for attack aircraft in Operation Allied Force. Wartime missions were flown from Aviano AB, or later NAS Sigonella, in Italy, which the U-2s were using as 'bolt-holes' while the Istres runway was under repair.

Collier Trophy award

The U-2 has been breaking records ever since it first flew in 1955 – but mostly in secret. In late 1998, Lockheed Martin, the USAF and NASA decided to demonstrate that the aircraft was still very much alive and, indeed, rejuvenated with the new F118 engine. Unfortunately, the USAF would not agree to declassify the maximum altitude, so the three new payload-to-altitude world records set in November and December 1998 were still not a true reflection of the aircraft's performance. Still, they helped the U-2 win the 1998 Collier Trophy – the top aeronautical award in the US. At a time when the U-2 programme was still fighting for development funds against a new generation of UAVs, the publicity was useful.

In fact, although some USAF leaders had fallen in love with the Global Hawk and now described the U-2 as "a sunset system", support for new U-2 funding initiatives was growing where it really mattered. The users in CENTCOM, EUCOM and PACOM were not willing to forego the U-2's proven performance for projections of future UAV performance. The purse-holders in Congress took a similar view. By the end of the 1990s, two major upgrades to the airframe had been approved, as well as significant improvements to the U-2's three most important sensor systems.

Power-EMI upgrade

To take advantage of the increased electrical generating capability of the F118, a new 'power distribution backbone' was designed. Using new copper wiring and the latest fibre-optic cables, the new system replaced bundles of old wiring which had been dedicated to specific sensors. This saved some weight, and made it much easier to swap sensors on the airframe. The total generator power was increased to 45 kVA, of which about 36 kVA was available to the sensors. Moreover, electro-magnetic interference

NASA ER-2 developments

In 1996 the USAF reclaimed 80-1069, the U-2R which had been loaned to NASA nine years earlier. The remaining pair received the new GE F118 powerplant, although their designation remained ER-2. The military also subsequently reclaimed the STARLink system.

NASA's Airborne Science programme moved to the Dryden Flight Research Center at Edwards AFB in 1997. From here, the ER-2s were as busy as ever, with experiments related to Earth resources, celestial observations, atmospheric chemistry and dynamics, oceanic processes, satellite calibration and data validation, and so on. By now, NASA's ER-2s and predecessor U-2Cs had flown over 4,000 missions. Some experiments were co-ordinated with NASA's DC-8, such as the 1998 deployment to Patrick AFB, Florida, when an ER-2 flew over Hurricane Bonnie at 65,000 ft (19812 m) while the DC-8 flew into the storm.

The most-flown ER-2 sensor was now the AVIRIS hyper-spectral scanner, but data derived from the wet-film RC-10 camera was still in demand. A modified multi-spectral scanner built by Daedalus Enterprises and NASA – the MODIS (Moderate Resolution Imaging Spectrometer) – was also added to the ER-2's repertoire. This records 50 channels of 16-bit data in the visible, near infrared, mid-infrared and thermal portions of the spectrum.

Back in 1987, measurements from an ER-2 deployed to Punta Arenas in southern Chile had confirmed the extent of ozone layer depletion over the South Pole. Monitoring of this serious phenomenon at both Poles had since become a key task for the aircraft. In early January 2000, NASA 809 deployed with the DC-8 to very cold, very dark Kiruna, Sweden, for more ozone layer research inside the Arctic Circle. The SOLVE (SAGE III Ozone Loss and Validation Experiment) campaign would require flights to the North Pole and, uniquely, the first flight by a U-2-type aircraft over Russia since the downing of Gary Powers 40 years earlier.

NASA pilot Dee Porter flew the first mission to the North Pole in total darkness. For navigation in this hostile environment, a backup GPS/INS system named PARS – Platform Attitude Heading Reference – was added to the ER-2, supplementing the primary one which integrates GPS with the Litton LTN-92 INS. Porter was able to maintain communication with Kiruna via the commercial Iridium system, using blade antennas on the ER-2.

Porter also flew the first mission into Russia, on 27 January – a six-hour straight line southeast past Moscow to a point near the Ukraine border, and back. Russian scientists helped gain the necessary approvals, but Porter had to enter at strictly-defined waypoints. Some 17 different experiments were performed during the flight. A second, eight-hour flight on 5 February was more complicated, involving an outbound leg to Spitsbergen before turning to cross the Barents Sea and fly over northern Russia. A team of 250 scientists on the ground at Kiruna was devising new flight paths throughout the deployment, since the ozone-depletion vortex is dynamic, moving south to east. In a second series of flights during March 2000, NASA pilot Jan Nystrom also flew into Russia.

***Above:** One of the ER-2s is configured to use the STARLink (similar to Senior Spur) to transmit data across global ranges via satellite link.*

***Below:** ER-2s occasionally deploy away from the Dryden Facility (Edwards AFB) to various parts of the globe as part of NASA's ongoing research. Here one is seen at Kiruna in Sweden in January 2000, as part of operations within the Arctic Circle.*

RAMP cockpit

The third incarnation of the U-2R/S cockpit was introduced by the RAMP upgrade, in which virtually the entire cockpit was replaced with new instrumentation. Gone was the circular driftsight which had characterised the U-2's cockpit for so many years, and in came three large colour MFDs (multi-function displays). One is situated centrally in the head-level position, and is the primary display for aircraft flight and navigation information. In this view it is configured with a split screen, the upper portion displaying flight data such as attitude and speed, while the lower half displays bearing and other navigational data. The lower left screen is configured with a map display, while the right-hand screen has engine and aircraft system displays. Again it can be configured with various 'windows', here displaying engine, oil, cabin pressure and fuel information. Data entry for the nav/comms system is accomplished using the data entry panel and small screen above the left-hand MFD.

At Beale the U-2s are housed in wide barns. Here a U-2S and TU-2S are being prepared for the day's training programme.

This U-2S – assigned to the Lockheed test team at Palmdale – is fitted with a SYERS camera, the aperture being clearly visible in the lower side of the nose. The whole nose section rotates to enable the SYERS to operate in oblique or vertical modes. It also carries the dorsally-mounted Senior Spur satellite uplink pod (now known as the Extended Tether Program – ETP) which allows SYERS or ASARS-2 imagery to be transmitted in near real time across global distances. The original four ETPs are being augmented by a further five systems.

(EMI) from the aircraft's power system was much reduced through better shielding and grounding. This was of particular value to the U-2's SIGINT systems, which could now 'listen' from a much quieter airborne environment.

In a six-year effort costing $140 million, the Power-EMI upgrade would be performed as each U-2 entered depot maintenance at Palmdale. At the same time, two other improvements would be incorporated: a new single-piece windscreen with a better anti-icing system, and replacement of the wing-mounted Trimble GPS with the USAF's standard Garmin model, in a new installation behind the Q-bay. The resulting aircraft would be designated U-2S Block 10 (the original re-engined U-2S was now retrospectively designated Block 0). Conversions began in 2000, as 80-1078 and 80-1095 entered depot maintenance. The first deployment took place in September 2002, to the 5th RS in Korea.

Improved SIGINT

To coincide with the Block 10 deployment, upgrades to all three U-2 sensor systems were planned. Relatively little money had been spent on the U-2's SIGINT systems during the 1990s, partly because they were supposed to be replaced by the Joint SIGINT Avionics Family (JSAF), a common ELINT and COMINT system for the RC-135, EP-3 and U-2. But JSAF proved technically difficult, and was all but abandoned by 2001. Meanwhile, black boxes were proliferating on the U-2, and the systems could not easily be adapted to new threat frequencies. Moreover, complications such as that found in the Iraqi air defence system, where Soviet missiles were netted to French and Chinese radars, could defeat the automated detection techniques used by the U-2's Senior Ruby ELINT sensor.

Raytheon did some work to rationalise the Spear (RS-6B) and Ruby (RA-1/2) systems, and added fibre-optic links between the superpods, resulting in a configuration designated Remote Avionics System 1 (RAS-1), or Senior Glass. With adaptation to the satellite uplink and the addition of some digital receivers, the designation became RAS-1A. In the late 1990s a major reconfiguration with circuit cards and RF converters produced a more flexible system which could be reprogrammed remotely through software. This was designated RAS-1R.

Improved imaging

Ever since it was introduced in the early 1980s, ASARS-2 has been the leading radar reconnaissance sensor in the US inventory. Subsequent refinements included electronic scanning of the antenna (in azimuth) and the addition of a Moving Target Indicator (MTI) mode, although only enough MTI upgrades were purchased to equip the three aircraft which patrol the North Korean border from the 5th RS base at Osan.

In June 1996, work began on the ASARS-2 Improvement Program, or AIP. Using commercially-available hardware, Hughes (subsequently Raytheon) replaced the Receiver/Exciter and Controller boxes with a combined REC, and added an onboard processor for the first time (in the Q-bay). A new Asynchronous Transfer Mode (ATM) connected the radar boxes to each other and to the datalink. The pay-off is dramatic improvements in both resolution and coverage. Three-foot (1-m) resolution is now achieved in the best search mode (versus 10-ft/3-m previously), and 1-ft (0.3-m) resolution is possible in 'spot' mode. The area covered by these modes is increased by 1500 percent and 900 percent, respectively, yet thanks to the powerful new onboard processor, radar data can be produced as quickly. The improved geolocation accuracy makes precision targeting much easier. The range of ASARS is still classified, but possibly in excess of 200 miles (320 km). Flight tests of the upgraded radar were completed in late 1998, and the first AIP sensor was delivered in September 2000. This time, MTI modes will be included in all the systems.

If ASARS/AIP is the Rolls-Royce of radar systems, then SYERS-2 must be the Cadillac of cameras. SYERS was originally a dual-band system, offering visible and infrared modes and very long range, thanks to the big telescope housed in the U-2's nose. Five of them were built by Itek, which was subsequently sold to Hughes, itself later bought by Raytheon. They produced excellent imagery in good weather, but frequent cloud cover and haze over Europe and the Korean peninsula restricted the number of days on which SYERS could usefully be used there, especially if it meant replacing the ASARS.

Thought was given to redesigning SYERS so that it could fit in the Q-bay, thereby leaving the U-2 nose free for ASARS carriage, and allowing both sensors to be carried on the same mission, but there was not enough room. Instead, funding was approved for a multi-spectral upgrade, which has added five new collection wavebands, making seven (three visible, two short-wave IR and two medium-wave IR). The use of colour in the visible bands should improve the identification of targets, while short-wave IR will cut through haze and medium-wave IR will improve night-time resolution. The combination of bands will help defeat the increasingly sophisticated camouflage and concealment techniques which are out there. As with ASARS, the upgrade also offers much greater area coverage, and improved geolocation. (SYERS is now owned and supported by Goodrich, which bought Raytheon's 'high-end' EO systems business in 2001).

In order to cope with the increased data produced by these improved imaging sensors, the U-2's line-of-sight link to the MOBSTR relay station was upgraded to a capacity of nearly 300 mb/sec, and redesignated as the Interoperable Airborne Data Link II (IADL-II). A second line-of-sight system can also be carried, for communication with mobile imagery exploitation systems in the field (the Dual Data Link or DDL).

Is there any role left for traditional, wet-film cameras? Hardly, it seems. They do not provide timely data in today's fast-paced military scenarios, and the chemicals required for processing cannot so easily be disposed of under modern environmental legislation. The U-2's excellent Optical Bar Camera (OBC), a 30-in (76.2-cm) lens panoramic system, is being retained for the moment, but another panoramic camera and two framing camera systems were placed in storage as the new decade began. These were, respectively, the IRIS III, the very-high-resolution H-camera and the broad-coverage F489 for mapping. Also, the 35-mm tracker camera, which had once been used to establish 'ground truth' for both operational and training sorties, was retired because student navigation training sorties could now be evaluated from a GPS read-out.

80-1082 cruises over the Lockheed Martin Skunk Works plant at Palmdale. The majority of U-2 upgrade work is performed here, in the facility where the U-2 line was reopened to produce the FY80 batch.

Glass Dragon

As the high-tech sensor upgrades proceeded, the contrast with the profoundly low-tech U-2 cockpit became acute. Apart from sensor control panels, very little had changed since the U-2R first flew in 1967, and the difficulty of sourcing spares for those old-fashioned instruments was

Above: The RAMP cockpit upgrade was first applied to 68-10336, a permanent test aircraft which acted as prototype. The first 'production' conversion was 80-1082, which was handed over to the Air Force in a ceremony at Palmdale in April 2002.

Left: In common with other USAF types, a block system has been introduced for the U-2S to differentiate between various upgrade states. RAMP aircraft are now known as U-2S Block 20, whereas aircraft which retain the 'legacy' cockpit are either Block 0 (basic) or Block 10 (with the Power-EMI upgrade). A Block 30 upgrade, with better EW defences, is now being defined.

With the RAMP upgrade the U-2's cockpit avionics architecture will allow new systems to be integrated with ease. Pilots have welcomed the clear situational displays, which present more information than those of the 'legacy' cockpit, and which are far easier to interpret.

growing. In October 1998, therefore, Lockheed Martin was awarded a $93 million contract for the U-2 Reconnaissance Avionics Maintainability Program (RAMP) – a new 'glass cockpit'. Three 6 x 8-in (15.2 x 20.3-cm) flat-panel multi-function displays (MFDs) would dominate the new layout, plus an upfront display and control panel which would conveniently group all the vital switches together (in the old cockpit, vital controls such as radio and navigation were housed in awkward-to-reach locations in the lower side panels).

The new system would use a standard MIL1553B databus: a new main avionics processor would convert analogue signals from the airframe to digital. Although off-the-shelf hardware was used (the MFDs came from the C-130J upgrade, for instance), everything had to be adapted to the rigours of high-altitude flight. The touch controls were spaced so that a pilot wearing the bulky gloves of the pressure suit could easily operate them. Operational pilots from the 9th RW made significant inputs to the design of both hardware and software for the RAMP.

New information

The result was much-improved situational awareness. Information which pilots of other aircraft took for granted was now available to U-2 drivers for the first time: ground speed and wind vector were now displayed throughout the flight; the fuel panel accurately reflected the actual system layout; flight plan tracks and waypoints could be displayed on the moving map. An angle-of-attack indicator with aural as well as visual warnings was another first – useful for those final few feet before landing on the bicycle gear, when the Dragon Lady must be persuaded to quit flying by touching down close to the stall speed.

The new cockpit design retained the familiar large control yoke, and the sunshade, and even the rear-view mirror, but another unique U-2 feature was missing. There was no room for the bulky, hooded driftsight display which used to dominate the main panel. Generations of U-2 drivers had scanned the driftsight optics to answer such vital questions as: Are there fighters below me? Am I contrailing? Is the landing gear really down and locked? Goodrich provided a small electro-optic camera which could fit in the driftsight bubble below the fuselage, with its images displayed on one of the MFDs. This neat solution has yet to be funded for production, however.

68-10336, one of two aircraft permanently assigned to U-2 Flight Test at Beale, was the RAMP prototype. The first flight was in December 2000. 80-1082 was the first production modification, and was redelivered to Beale in April 2002. The RAMP aircraft are designated U-2S Block 20.

None of the aforementioned upgrades had reached the field, however, when the U-2's next call to action came. Less than one month after the 9-11 outrage in 2001, they were flying over Afghanistan on Operation Enduring Freedom (OEF). This time, the missions were flown from Al Dhafra air base near Abu Dhabi in the UAE, after Saudi Arabia refused the use of its bases for OEF. The two aircraft assigned to the 363rd ERS remained at PSAB, while additional U-2s were flown from the new base. This operation was designated the 9th ERS, alternatively known as Det 4. Ironically, the hangar at Al Dhafra was shared with the Global Hawk UAV, which was making its first operational deployment.

Although the new base was closer to Afghanistan, the geography of the new conflict meant that data from the U-2's sensors could not first be downlinked to a ground station (eg MOBSTR) before relay to the US by satellite for processing and analysis. Instead, the U-2 had to operate 'off-tether', using its own satellite relay system. Fortunately, after a period in the 1990s when such systems were under-funded, new versions had been procured from L-3 Communications. The original Senior Span had limited bandwidth and proved suitable only for the relay of SIGINT. Now, though, four Senior Spur systems were in service. They provided a full bandwidth link to a classified US satellite architecture, so that ASARS or SYERS imagery could be transmitted across continents. Spur is now referred to as the Extended Tether Program (ETP). An additional five ETPs are being procured.

The OEF commitment once again stretched U-2 resources to their limit: pilots, maintainers, airframes, sensors. In late 2001, the operation at Sigonella was closed in favour of consolidation with the detachment at Akrotiri, Cyprus. This RAF base has housed U-2s ever since 1974; it is now the home of the three aircraft 'belonging' to the 99th ERS, plus a maintenance operation staffed by Lockheed Martin. Here, aircraft can be inspected every 400 hours as required, without ferrying back to the US. (The periodic depot maintenance is every 3,400 hours, and is performed by Lockheed Martin at Site 2, Palmdale).

Operation Iraqi Freedom

On 17 February 2003, U-2 flights over Iraq in support of United Nations weapons inspections resumed. But the flights were short-lived, since the US – supported by the UK – lost patience with the weapons inspection process and launched Operation Iraqi Freedom less than a month later. Once again, U-2s played a vital – if unheralded – part in providing the ISR picture. No fewer than 10 aircraft were deployed to Det 4 at Al Dhafra and, surprisingly enough, the Saudis permitted the two aircraft at PSAB to participate in OIF. The three aircraft at Akrotiri were also involved. As a result, the 15 U-2s were able to provide constant coverage of Iraq, with as many as three aircraft on-task at one time. The value of a manned, high-flying aircraft for battlefield communications relay was demonstrated frequently: U-2 pilots were able to co-ordinate the rescue of seven US Army prisoners-of-war; the search for a pilot of a downed F-14; and the fire support of Special Forces which had been cornered by Iraqi forces.

At the end of this conflict the US announced that it would be withdrawing all aircraft from Saudi Arabia. Consequently, the 363rd ERS at PSAB was closed down in early May. Det 4 at Al Dhafra was expected to remain open, with about four aircraft.

Looking to the future

In 2001, the USAF announced that the 9th RW would operate the Global Hawk UAVs, which had just been approved for low-rate production. A new squadron (the 12th RS) is forming alongside the U-2 units. Over the next few years, therefore, the 9th RW will be actively seeking answers to the big question: can the Global Hawk replace the U-2?

The 'mobile' follows a RAS-1R-equipped U-2S as it launches from Al Kharj for a mission on 11 March. Note the pogo wheel bouncing off the runway after it has been jettisoned. At the time the 363rd Expeditionary Reconnaissance Squadron was engaged in preparation for Operation Iraqi Freedom, which began on the night of 19/20 March.

There is no hurry to answer that question, however. There is plenty of life left in the U-2 airframe, although a fatal crash in Korea in January 2003 reduced the operational fleet to 30 aircraft (it was the first write-off for seven years). In 1999 and again in 2001, Lockheed Martin pitched new U-2 production at the Pentagon, including a new U-2U unmanned version. No contract was awarded, but the fact remains that over $1.4 billion has been spent on U-2 upgrades since 1990. Moreover, there are plenty more improvements from which the Dragon Lady could benefit.

The potential of the improved ASARS radar has yet to be fully exploited: the quality of radar data now streaming from the U-2 allows for advanced manipulation techniques such as coherent change detection and automatic target recognition/cueing. More wavebands could be added to SYERS. A long-mooted hyperspectral sensor could also be added. Some of these developments would make the U-2 even more useful in supporting counter-narcotics or special forces operations.

The aircraft's defensive electronic warfare systems need more attention. Considering just how often U-2 pilots have been sent in harm's way in recent years, it is surprising that EW upgrades have lagged other developments. In July 2001, an aircraft flying over southern Iraq apparently came close to being shot down, when an SA-2 exploded below and behind.

The first task will be to upgrade the current Systems 29F for warning and 56-1 for jamming with a high-speed signal processor for quicker identification of threat radars, and to integrate the EW system displays into the RAMP. This will form part of the Block 30 upgrade. Then, a long-standing plan to fit the ITT ALQ-211 system, which was developed for the US Army's AH-64D helicopters, could be implemented. The BAE Systems (ex Lockheed Martin/Sanders) ALE-55 fibre-optic towed decoy (FOTD) could then be added. (The FOTD has already been test-flown, mounted in the GPS housing on the right wing).

The logistics tail and expense associated with the U-2's unique JP-TS (Thermally Stabilized) fuel could be eliminated by modifying the aircraft's fuel system. Heat would be transferred from the engine oil circuit to the outboard fuel tanks (always the coldest) so that standard JP-8 could be used instead. Alternatively, additives could reduce the freezing-point of JP-8.

L-3 Communications is now funded for a significant upgrade to the U-2's datalinks. A common airborne modem assembly will replace the single-purpose boxes that were dedicated to IADL, DDL and ETP. New nine-inch antennas will be fitted in the Q-bay and tail. In the near future, this new system, designated DDL II, will provide multiple communications options: two line-of-sight (LOS) links; or one LOS plus one ETP (eg SATCOM) link; or one LOS plus an air-to-air (ATA) link; or the ETP plus the ATA.

This kind of flexibility is the key to the future of the U-2. It is not yet clear whether UAVs can match the operational utility of this evergreen reconnaissance system. Having outlasted its supposed replacement – the SR-71 Blackbird – who is to say that the Dragon Lady will not pull the same trick again?

Chris Pocock

***Above:** A 363rd ERS U-2S launches from Al Kharj at the start of an Operation Iraqi Freedom mission on 11 April 2003. It is configured with ASARS-2 nose and the RAS-1A (formerly known as Senior Glass) Sigint suite, which combines Elint and Comint functions.*

***Left:** This 'family portrait' from April 2003 shows the 363rd ERS at its Al Kharj (PSAB) base). U-2Ss also operated from Al Dhafra during the conflicts in Afghanistan and Iraq, the UAE base also hosting the RQ-4 Global Hawk detachment. U-2s from Akrotiri in Cyprus also flew missions over Iraq under the mission name Olive Harvest.*

80-1086 (with ASARS-2 and RAS-1A) of the 363rd ERS rests at Al Kharj, Saudi Arabia, between Operation Iraqi Freedom missions on 11 April 2003. At that time the U-2 fleet stood at 31 aircraft, with 29 in the Primary Aircraft Inventory (plus four TU-2s). Four aircraft survive from the original FY68 batch.

Turkey's Phantoms

With eight squadrons of F-4s in the inventory, the Turkish Air Force (Türk Hava Kuvvetleri, THK) is now the largest Phantom operator in the world. The importance of the type is underlined by the fact that it will be operated in various roles well into the next decade, and 54 F-4s have undergone the extensive F-4E 2020 modernisation programme to increase their effectiveness in the fighter-bomber role. Updates are also being applied to some of the RF-4Es to improve their reconnaissance capability.

Right: The F-4E 2020 upgrade programme is now complete, having produced 54 highly capable aircraft for the two fighter-bomber squadrons.

Left: Turkish Phantoms saw considerable action in the mid-1990s during the counter-insurgency effort against the PKK (Kurdish separatists). This 171 Filo F-4E is seen in 1994, armed with Mk 84 bombs. Note the European One camouflage, which later gave way to the standard THK three-tone 'jungle' scheme.

Below: An F-4E 2020 from 171 Filo prepares to launch on a night training sortie. The aircraft carries an SUU-20 practice bomb dispenser.

In Turkey the Phantom story started on 30 August 1974, when the first two of 40 factory-fresh FY73 F-4Es were delivered to the Turkish Air Force under the Peace Diamond III programme. This first batch of Phantoms was used to re-equip 111 Filo 'Panter' (Panther) and 112 Filo 'Seytan' (Devil), part of 1 AJÜ (Ana Jet Üs = Wing) at Eskişehir, which had previously operated the F-100 Super Sabre. All Phantoms were painted in the USAF South East Asia-style (jungle) camouflage. When Turkey was about to receive the first aircraft of the follow-on batch of 32 F-4Es and eight RF-4Es in 1977, a small conversion unit – named Şimşek Kita (Kita = flight) – was formed at Eskişehir. Future Phantom pilots and navigators destined to fly the F-4 were to be trained in this unit.

Still under Peace Diamond III, the second batch of new-build Phantoms was transferred from St Louis to Eskişehir during 1977-1978. These FY77 Phantoms remained at Eskişehir

while the FY73 F-4Es transferred to 171 Filo 'Korsar' (Pirate) and 172 Filo 'Şahin' (Falcon) at Erhaç (7 AJÜ) in 1978. This enabled the squadrons to retire their fleet of F-100s. Eight RF-4Es were assigned to 113 Filo 'Isik' (Light), which started operating these 'Recce Rhinos' together with the last surviving RF-84F Thunderflashes. The last RF-84F finally retired in 1980.

In 1981, a third squadron was formed under 7 AJU, 173 Filo named 'Şafak' (Dawn). This squadron received 15 former USAF F-4Es: all aircraft were delivered that same year. Most of the Phantoms wore the 'European One' lizard scheme, which was later replaced by the standard THK 'jungle' camouflage scheme after overhaul at 1st HIBM (Hava Ikmat Bakim Merkezi = Air Supply and Maintenance Centre) at Eskişehir. This new squadron became a fighter-bomber unit within the 2nd Tactical Air Force of the THK. A second batch of 15 former USAF Phantoms, all FY66/67 F-4Es, was delivered between 1984 and 1985 as attrition replacements.

In the beginning of 1987 the Turkish government signed an agreement with the United States for economical and military aid worth US $312 million, as well as a long-term credit of US $178 million. The third part of the agreement covered the delivery of 40 ex-USAF F-4E Phantoms, worth US$300 million. The first four aircraft were delivered on 18 June 1987. The remaining 36 aircraft were ferried to Turkey over a six-month period in eight batches of four aircraft, one batch of three and a single delivery flight. The last four F-4Es were officially delivered to Turkey on 23 November 1987. All Phantoms were drawn from the 337th TFS, Seymour Johnson AFB, and AMARC.

After the delivery of the first 20 aircraft, 131 Filo 'Edjer' at Konya retired its last Super Sabres and began conversion to the 'new' Phantom. This squadron was to become the new Phantom conversion unit, which obviously led to the disbandment of the Şimşek Kita at Eskişehir. The Konya-based sister squadron

172 Filo (below) is one of two air defence Phantom units. The commanding officer is Major Korkmaz (right), the pilot with the most flying hours in the THK. Over 3,000 of his 3,500-plus hours have been in the F-4E, and include around 360 logged as combat time with 171 Filo. After an instructor tour with 131 Filo, Korkmaz took command of 172 Filo in 2001.

Above: Training for the Phantom fleet is conducted at Konya by 131 Filo. As well as type conversion, weapons instruction is also provided. For air-to-air gunnery training the Phantom can be equipped with a towed target.

Below: Operating alongside 131 Filo at Konya is the STANEVAL and advanced tactics development unit, 132 Filo. This aircraft taxis from a hardened aircraft shelter with a live AIM-9P Sidewinder.

132 Filo 'Hançer' (Dagger) was disbanded after the retirement of its Super Sabres on 1 November 1987, ending the F-100 era in the Turkish Air Force. At Konya 131 Filo had at its disposal nearly 40 Phantoms, all of them 'European One' (lizard)-camouflaged. These aircraft were slowly repainted in the usual 'jungle' colours, the task being completed in 1995. In 1991, 132 Filo was re-established as a weapons, tactics and STANEVAL squadron, using Phantoms of 131 Filo (as well as NF/F-5s from 133 Filo).

At the beginning of the 1990s another batch of 40 F-4E Phantoms was received from the USAF. The aircraft were delivered as a reward for the use of Incirlik Air Base and the co-operation of Turkey during the Gulf War. The first four aircraft were handed over at Eskişehir on 25 March 1991, the last ones delivered by 29 April 1992. The F-4Es were drawn from various sources, most of them from the 35th TFW /USAF 'GA'/'WW' (George AFB), 457th TFS /AFRES 'TH' (Carswell AFB) and 110th TFS /ANG 'SL' (St Louis IAP). More than one third of this batch came from the 110th TFS Missouri ANG, whose 'Hill Gray II'-camouflaged Phantoms had a sharkmouth painted on their noses. As it was thought to be more effective for the interception role, the 'Hill Gray II' camouflage scheme was maintained on all 40 Phantoms. The sharkmouth remained on the ex-Missouri ANG Phantoms, and was applied to the remaining (25) aircraft. The aircraft were divided between 112 Filo and 173 Filo, which transferred most of their earlier machines to 111, 171 and 172 Filos. The few remaining aircraft in 112 were slowly repainted in the 'Hill Gray II' camouflage scheme, including sharkmouth.

In the autumn of 1991 the Luftwaffe (German Air Force) decided to withdraw its entire RF-4E fleet over a period of two years. Under Project KAAN, an agreement was made between the Turkish and German Governments to transfer a part of the RF-4E fleet to Turkey, which planned to acquire 47 RF-4Es (34 aircraft for operational use and 13 aircraft as spares sources). The operational RF-4Es received a basic life-extension upgrade by MBB, which also changed the APQ-99 radar for the APQ-172. In 1992, the German Government postponed the deal due to the conflict between the Turkish Government and the PKK (Kurdish Labour Party), which deteriorated at that time.

The Luftwaffe's AG 51 was disbanded on 17 September 1992, after which deliveries of the first 12 aircraft for spares recovery started in December. Delivery of the first updated operational Phantoms to Turkey did not occur until May 1993. The initial batch of 17 aircraft was delivered to Eskişehir and assigned to 113 Filo. Soon afterwards, the Bora Kita was formed at Eskişehir, a conversion 'flight' for new recce pilots to be assigned to 173 Filo at Erhaç. However, 173 Filo did not receive any of the former Luftwaffe RF-4Es before the second half of 1994. The reason was the concern of the German Government about the use of former German military equipment by the Turkish military against the PKK. With the introduction of these 'Recce Rhinos', the air defence F-4s transferred to 172 Filo.

After the training of 173 Filo pilots had been terminated, the Bora Kita was disbanded and some of the Phantoms were flown to Erhaç to strengthen 173 Filo. The two reconnaissance squadrons now fly around 20 RF-4Es each. All former Luftwaffe examples maintained their lizard colour scheme for a while, but after overhaul at 1st HIBM the aircraft were painted in the SEA camouflage. The RF-4Es which were acquired as spares source all went to the 1st HIBM at Eskişehir.

Another important step was placed in 1995 when the Turkish Government signed a contract with Israeli Aircraft Industries (IAI) for a large-scale modernisation programme of 54 F-4Es. All initial deliveries were destined for 111 Filo, and by the 1st week of December 2001 the squadron transferred around half of its inventory to the Erhaç-based 171 Filo, these being the first 2020s for this squadron. The final 2020s were delivered in 2003, bringing the final number to 27 aircraft per Filo.

Over the years the THK received 182 F-4Es and 55 RF-4Es. Approximately 125 F-4Es and 35 RF-4Es are currently operational.

Phantom training

Located in central Anatolia, just two hours south of the Turkish capital Ankara, is the 3rd main jet base of the THK, Konya (3 AJÜ). This huge training facility houses over 80 aircraft comprising Northrop N/F-5A/B Freedom Fighters and McDonnell Douglas F-4E Phantoms. These aircraft are divided over four squadrons. The 'Rhinos' are assigned to both 131 and 132 Filo.

The 131 Filo 'Edjer' is solely tasked with the training of graduates from the primary training units at Izmir-Çigli or NATO training bases in the US. The first of a total of six months will be spent on ground training. All students start with a theoretical phase of three weeks and a week of simulator training, which covers around 15 hours. This is followed by the flying phase in which student pilots fly approximately 70 hours in the Phantom. The squadron is comprised of 50 instructor pilots who normally train around 25 students (including some WSOs) per training syllabus.

According to Filo Commander Besir, the unit has around 20 operational aircraft at its disposal. These THK workhorses have an average of over 7,000 hours flying, with some even

1 AJÜ's all-weather interception unit is 112 Filo – the 'devil' squadron. Initially flying aircraft from the first Phantom batch, the squadron is now in its 30th year of F-4 operations. This aircraft is one of those delivered from USAF stocks in 1991 as 'thanks' for Turkish assistance during the Gulf War.

reaching 9,000 hours. However, in spite of extensive maintenance on the airframes at the 1st HIBM, their age is now beginning to show. In co-operation with Boeing, the THK developed a formula to extend the operational lifetime per aircraft by stripping the airframe and assessing individual parts for metal fatigue. By this concept the amount of hours available has been extended, enabling training operations with these aircraft to continue for several more years.

Opposite the 131 Filo ramp is the other Phantom unit, 132 Filo. Named 'Hançer', it is the only fighter squadron in the Turkish Air Force to operate two types of aircraft, flying a mix of 10 F-4Es and 10 (N)F-5A/Bs. 132 Filo focuses on STANEVAL (STANdards and EVALuation) flights, academic research and development, and various advanced training courses. STANEVAL flights have the main objective of observing and evaluating procedures during the planning, briefing, leading and debriefing of various tactical missions. This enables the monitoring of tactics and techniques of other squadrons, as well as instruction in new tactics and developments.

The second main task of 132 Filo involves academic research and development, which includes the examination and study of new weapons and tactics, and International/NATO and Turkish publications. Furthermore, academic courses (like TLP) and exercises are followed so that improvements to tactical procedures can be incorporated into THK operations. 132 Filo also conducts various training courses such as weapons and tactics courses, F-4E Instructor Pilot training courses and air combat courses. The principal aims of these courses are to increase leadership skills and tactical flying capabilities, and to improve mission planning and tasking capabilities.

In the future the squadron may receive more advanced equipment in the form of F-4E 2020s and F-16C/Ds, but there are no firm plans at present. Aircraft from both 3 AJÜ units would operate in the fighter-bomber role during a wartime scenario.

Air defence Phantoms

Although the backbone of the Turkish air defence is now the F-16C/D, the THK is still able to count on a force of two large squadrons of Phantoms tasked with the same role, divided between Eskişehir (1 AJÜ, 112 Filo) in the west and Erhaç (7 AJÜ, 172 Filo) in the east. Each unit possesses around 24 'Hill Gray II'-camouflaged F-4Es. Both units have a NATO and national QRA task, and therefore two Phantoms are on 24-hour duty armed with four AIM-7H Sparrows and four AIM-9M/P Sidewinders. The Phantoms hold a 'B' type scramble alert with a launch time of 15 minutes, whereas the F-16s fulfil the 'A' type scramble and are able to be airborne in five minutes.

Besides the ADF role, both Filos are secondarily tasked with air-to-ground duties, which are practised on a regular basis, including participation in national bombing competitions. The last few years has also seen some international participation by both 112 and 172 Filos in exercises organised by NATO HQ AIRNORTH – Clean Hunter 2001 (112 Filo) and the NATO Air Meet 2000 (172 Filo). During the 1991 Gulf War 172 Filo was part of a Turkey-based Combat Task Force (CTF) with Italian RF-104G Starfighters and German Alpha Jets. The Phantoms were also used to escort E-3 AWACS aircraft over the Iraqi border, as well as fly CAP missions. During Provide Comfort and Northern Watch some F-4s were transferred to Incirlik AB to operate in packages as well as on CAP sorties.

Turkey's operational Phantom fleet is divided between two bases: Eskişehir and Erhaç. Each base has three squadrons, one tasked with air defence, one with fighter-bomber duties and one for reconnaissance. These F-4Es are from 172 Filo 'Şahin', the air defence squadron at Erhaç-Malatya.

An F-4 from 172 Filo sits on alert at Erhaç, with AIM-9P Sidewinders and AIM-7 Sparrows. The two air defence squadrons maintain a continuous alert, with a requirement to be airborne within 15 minutes of receiving a scramble order.

There are no immediate planned changes for either air defence squadron. Under current planning the ADF Phantoms are scheduled to be replaced by the F-35 Joint Strike Fighter from 2015 at the earliest. As it is likely to prove hard for these FY66/67/68 aircraft to operate effectively for another dozen or so years, it is likely that these 'Rhinos' will need to undergo a modernisation programme, although no plans have been made public.

Reconnaissance Phantoms

Currently, the THK has two recce squadrons with RF-4Es at its disposal. The Eskişehir-based 113 Filo is tasked with operations within the 1st Tactical Air Force Command (headquarters in Eskişehir) while the 173 Filo based at Erhaç has the responsibility for the 2nd Tactical Air Force Command (headquarters in Diyarbakir). The latter operates only former Luftwaffe Phantoms, while 113 Filo operates a mix of the original RF-4Es plus former Luftwaffe examples.

All RF-4Es have three different kinds of cameras, an AAD-5 Infra-Red (IR) line scanner, a KA-56 vertical low-level high-speed panoramic camera and three KS-87 optical cameras (forward- and oblique-facing). In addition, the aircraft are also capable of carrying the Goodyear UPD-4 SLAR (Side Looking Airborne Radar) which is mounted in a part of the low-*g* centreline fuel tank (remaining space can still be used for fuel storage).

As the RF-4 reconnaissance system still uses wet-film technology, both units have their own ground stations which are capable of developing two films at a time. In less then 30 minutes, the films are removed from the Phantoms, developed and analysed. Developing IR film is more time-consuming and is performed using a Turkish system which differs from NATO equipment. According to the operator, this Turkish process is much faster. The IR gives the Phantom limited night capability. Turkish recce Phantoms have a limited weapons capability with free-fall Mk 82/84 bombs.

Reconnaissance crew have gained significant combat experience over recent years, especially in 173 Filo. During Northern Watch the Phantoms flew many recce missions along the Iraqi border. All missions were flown from Erhaç as no Forward Operation Location (FOL) was used. In the conflict in eastern Turkey, 173 Filo was mainly tasked with the search for locations and movements of PKK terrorists. ern A lot of experience was gained during these operational missions including the development of visual reconnaissance tactics.

This experience is passed on to new crew of both reconnaissance squadrons. New pilots will first fly in the front seat and after becoming formation leader they will move to the back seat to concentrate on reconnaissance and WSO tasks, making all pilots multi-capable. Actually, this measure was taken because of the shortage of WSOs in the (R)F-4 fleet.

Currently, the capability of the RF-4E is deemed suitable to meet requirements, but the THK is studying various possibilities. The future will see the introduction of real-time imaging, un-manned recce flights and, in a much later phase, satellite imaging.

For the moment, the THK is modernising the RF-4Es, enabling them to operate with real-time capabilities. These modernisations, which are developed by Elbit/El-Op, include nose-mounted video camera, rear cockpit updates including a single Multi Function Display (MFD), and integration of a new EO/IR LOROP (Electro-Optical/IR Long Range Oblique Photography) system. This new system is embedded in the original G-139 centreline-mounted pod which was developed for the RF-4C. The state-of-the-art EO/IR sensor system features real-time high-resolution reconnaissance and high-capacity ground coverage in both visible and IR spectral bands for aircraft flying at ranges up to 100 km (62 miles) from the target.

This RF-4E of 113 Filo carries the El-Op EO-IR/LOROP pod on the centreline. Only seven reconnaissance Phantoms have been modified to carry this equipment, which gives the RF-4E the ability to gather imagery from long distance. In the near future a datalink will be added to allow the real-time transmission of imagery to a ground station.

***Above:* 113 Filo is the recce Phantom squadron at Eskişehir. This aircraft wears a special badge on the fin to commemorate 113,000 hours of Phantom flying by the squadron, in addition to the Filo's 'spook' badge (left) on the nose.**

***Below:* 173 Filo is at Erhaç and operates only ex-Luftwaffe aircraft, now repainted in standard Turkish 'jungle' colours. This aircraft is in full squadron markings, carrying the Filo badge (right) on the tail and another marking on the nose.**

At a later stage the THK will receive the datalink system which will enable the Phantoms to transfer real-time images to a (mobile) Image Exploitation Station. The THK is also studying the introduction of the EO/IR LOROP system in its F-16 fleet. So far only two updated pods have been purchased and are undergoing tests with 113 Filo. Under current plans only the seven remaining FY77 RF-4E Phantoms of 113 Filo will undergo this modernisation. There is no publicly available information about plans to upgrade the ex-Luftwaffe aircraft, which would allow 173 Filo to also gain EO/LOROP capability.

F-4E 2020

During the 1990s the Turkish Air Force initiated a study for the modernisation of its F-4Es which were flying in the ground attack role. The National Defence Ministry approached prospective companies to bid for the upgrade. Principal competitors were Rockwell International Projects of the United States, Daimler Benz Aerospace of Germany and IAI Industries of Israel. The programme was finally assigned to IAI Industries, an obvious choice because IAI already had much experience in the modernisation of the IDF/AF Phantom fleet in the ground attack role to Phantom 2000 (Kurnass) standard. The extensive friendship built up between between Turkey and Israel might have also played a part in the decision.

In August 1995 a preliminary agreement was signed between the Turkish Government and the Lahav Division of IAI to modernise 54 F-4Es to Phantom 2020 standard. A final contract was not signed before December 1996 due to lack of funding to finance the entire programme. These funding problems were partly solved by a greater involvement of Turkey in the modernisation programme. It was decided to upgrade 26 Phantoms in Israel while the remaining 28 aircraft were to be upgraded in Turkey. IAI was authorised to offer US$340 million state-backed guarantees for private-sector loans.

On 6 December 1996, IAI received its biggest ever foreign contract from the National Defence Ministry in Ankara. The entire deal, worth US$ 632.5 million, included not only the primary modernisation of 54 Phantoms but also support and test equipment, spares, technical documentation and training of pilots and technicians. An additional contract of US$80 million was signed for the acquirement of 64 Rafael Popeye 1 Anti-Surface Missiles. Other armament include AGM-65G Maverick, GBU-10/12

Features which readily distinguish the 2020 from an unmodified F-4E are the large blade antenna on the spine and the bullet fairing for the Elta RWR on the fin-tip. This pair of 2020s is from 171 Filo at Erhaç.

and AIM-9L/M Sidewinder (the aircraft no longer uses the AIM-7 Sparrow).

The modernisation is based on the IDF/AF Phantom 2000 project and is divided into two parts: a structural upgrade and an avionics upgrade. The structural upgrade is carried out to extend the structural life of the airframe by 6,000 hours. All Phantoms which were to be modernised had around 5,000 hours on average. Every aircraft first undergoes a non-destructive test programme to check the condition of the airframe. All parts are checked by means of X-ray and other methods (real-time and magnetic penetrating ISP). When needed, new parts are installed. All renewed parts are under the guarantee of IAI, including parts manufactured in Turkey. These new parts are rechecked before installation. Some structural improvements are made to all aircraft including:

- replacement of the wing fold ribs by new and stronger ones
- replacement of attachment fittings and engine mountings
- replacement of canopy sill bar
- replacement of most hydraulic and pneumatic lines and tubes
- approximately 20 km (12.4 miles) of wires are replaced, saving 750 kg (1,653 lb)
- Fuel tank upgrade, including some reinforcements

Many out-dated and antiquated (and unreliable) avionics were to be replaced with more modern systems, such as:

- the front cockpit is re-equipped with one multi-functional display (MFD) while the WSO position receives two
- installation of an El-Op 976 wide-angle HUD and a HOTAS system
- high-performance Elta Electronics EL/M-2032. The EL/M-2032 radar is a multi-mode fire control radar and was originally developed for the IAI Lavi
- IAIC (Improved Avionic Interface Computer) mission computer. The mission computer controls all communication, navigation and armament
- new navigation equipment, including advanced GPS/INS, connected to the mapping mode
- avionics package is managed by a dual Mil Std 1553B databus which integrates all the systems
- Central Air Data Computer manufactured by Astronautics
- UHF (large new black antenna on rear of fuselage) and IFF packages
- Airborne Video Tape Recorder (AVTR)
- An integrated Elisra SPS-1000 and Elta passive/active EW sub-systems

Elbit Systems carries responsibility for complete integration of all avionics systems.

The first two Phantoms to be upgraded were flown to the IAI Lahav facility at Ben Gurion in February 1997. Two years later, on 11 February 1999, the Phantom 2020 prototype (73-1025) took off for a 55-minute test flight. Testing of the first pre-production aircraft was carried out by a team of IAI and THK pilots. The first two Phantom 2020s were officially delivered to 111 Filo at Eskişehir on 27 January 2000. Fifteen Turkish instructor pilots went to Israel for a one-year training programme. On their return, these pilots formed a conversion unit within 111 Filo, which was known as the 'Terminator' Kita, operating both 'old' and 'new' Phantoms. The 'Terminators' (also referred to reverentially as the 'Father Squadron') were tasked with the development of handbooks, checklists and a

The 'Panthers' of 111 Filo were the first THK unit to receive the Phantom, acquiring F-4Es to replace F-100s at Eskişehir in late 1974. They have remained at the base ever since, and are now tasked as 1 AJÜ's fighter-bomber squadron. The first F-4E 2020s arrived in 2000, and 111 Filo parented the 2020 conversion flight, known as the 'Terminators'. This F-4E 2020 carries an SUU-20 dispenser with Mk 76 'blue bombs', which simulate the ballistics of the Mk 82.

At Erhaç the fighter-bomber unit is 171 Filo 'Korsar', which converted to the 2020 from December 2001 after training had been undertaken with 111 Filo. The 2020 introduced a significant upgrade in precision attack capability and much improved bombing accuracy when compared with the unmodified F-4E. Among its weapon options is the AGM-65 Maverick, a training round of which is carried by this aircraft.

training programme for the future 2020 crew. The course to convert pilots to the new Phantom 2020 lasted three months, with new courses starting every month. As soon as this complex task had been completed and sufficient instructor pilots and WSOs were available, the 'Terminator' Kita was dissolved. It must be stressed that this name was only involved with the training unit, and was not the official THK name of the F-4E 2020.

Also located at Eskişehir, the 1st HIBM began work on its Phantom 2020 programme. All update packages were delivered by IAI, and the whole modernisation process of a single aircraft took about a year. A few 1st HIBM technicians were trained by IAI in Israel while others received their training in Turkey. A Turkish technical assistant remained in Israel to smooth communications between IAI and 1st HIBM. A few Israeli personnel were assigned to the Eskişehir programme, mainly for logistics and in case of any technical problems or malfunctions. At the end of April 2001 111 Filo had to send to 1st HIBM 17 Phantom 2020s upgraded by IAI, due to problems with fuel lines and fuel valves. These problems caused the destruction of several engines, but was identified and solved by IAI, and compensation of over US$600 million was paid.

The first Phantom 2020 upgraded by 1st HIBM, 77-0306, made its first test flight on 18 June 2001 before being handed over to the THK (111 Filo) on 3 July. At any given time around 13 aircraft were in the 1st HIBM workshops, which delivered around one per month. Deliveries initially went to 111 Filo 'Panter', but from 2002 171 Filo 'Korsar' also received 2020s from 1st HIBM.

In May 2001 171 Filo sent its first pilots to Eskişehir for training. In the first week of December 171 Filo received its first batch of 13 aircraft, drawn from the 111 Filo inventory. After a period of little activity, 'Korsar' was re-established on 21 December during an official ceremony at Erhaç.

The upgrade programme was completed in July 2003. Currently, each unit has an F-4E 2020 strength of 27 aircraft. Operational readiness is judged by THK headquarters staff, and by March 2003 111 Filo received NATO officials who evaluated the unit's operational status. Unfortunately, 111 Filo lost its first F-4E 2020 on 11 November 2002, although thankfully the crew ejected safely. What caused the 'Rhino' to crash remains undetermined.

Both units have been approved by the headquarters in Ankara to apply squadron markings. 111 Filo applied a large panther's head marking on the nose section, while 171 Filo opted for a skull and cross-bones applied on the tail, and a black and white checkered fin-stripe.

The Turkish Air Force plans to continue operations with the F-4E 2020 until at least the year 2020, the stated replacement being the Lockheed Martin F-35 Joint Strike Fighter. With the F-4E 2020, the THK possesses a formidable ground attack aircraft which will form the backbone of the attack force until the end of the next decade.

Cees-Jan van der Ende and Marco Muntz

Bottom: The 2020 programme has breathed new life into the Phantom in the fighter-bomber role, and it is possible that the air defence force may also receive an update in the future. It is expected that some form of airframe overhaul will be required if they are to meet their planned out-of-service date of around 2015.

These badges are worn by F-4E 2020 aircrew from 111 Filo ('Panter' – panthers, left) and 171 Filo ('Korsar' – pirates, below).

Russian Federation

Part 3: 4th, 5th, 6th, 11th and 14th Armies

4th Krasnoznamionnaya Armiya VVS i PVO

4th Army of the VVS and PVO, awarded with Red Banner HQ: Rostov-on-Don

The army was established in 1998 as the result of the merger of the former 4th Vozdushnaya Armiya (VA) of the VVS and the 12th Otdelnyi Korpus (OK), Independent Corps of the PVO. The 4th VA VVS was formed on 22 May 1942 and during World War II undertook 340,000 sorties, dropped 2 million bombs and destroyed or damaged more than 5,000 enemy aircraft. After the war, the 4th VA remained in Poland, where it was stationed until the collapse of the USSR and of the Warsaw Pact. In 1992 the 4th VA headquarters was moved to Rostov-on-Don and the army absorbed the units withdrawn from Germany, Poland and Belarus. Since 1994, the 4th VA has borne the main burden of fighting in Chechnya.

The 4th Army's zone of responsibility equates to the territory of the North Caucasus Military District. Apart from its air units, it has three regiments of surface-to-air missiles – the 1536th Zenitnyi Raketnyi Polk (ZRP) at Rostov-on-Don, the 1537th ZRP at Novorossiysk, and an unknown ZRP at Volgograd – as well as radio technical units. The commanding officer of the 4th Army is Major General Vladimir Gorbas' (appointed 7 August 2002).

1st Gvardeyskaya Stalingradskaya ordena Lenina dvazhdy Krasnoznamionnaya ordenov Suvorova i Kutuzova Smeshannaya Aviatsionnaya Diviziya

1st Stalingrad Composite Air Division of Guards awarded with Lenin order, Red Banner (twice), Suvorov and Kutuzov orders – Krasnodar

The former 1st Attack Air Division was equipped only with Su-25 aircraft, but in 2002 two regiments of Su-24 tactical bombers were attached to the division, which was renamed as the 1st Composite Air Division.

368th Shturmovoy Aviatsionnyi Polk

386th Attack Air Regiment – Budyonovsk

The regiment was formed on 12 July 1984 at Zhovtnevoye, Ukraine, and then moved to Kalinov airfield, also in Ukraine. During 1986/87, one squadron operated in Afghanistan as part of the 378th ShAP at Bagram air base. On 5 September 1988, the 386th ShAP was transferred to Demmin-Tutow air base in East Germany. On 8 July 1993 the regiment was withdrawn from Germany to Budyonovsk. Since 1994, the 368th ShAP has taken part in Chechnya operations.

461st Shturmovoy Aviatsionnyi Polk

461st Attack Air Regiment – Krasnodar

Formed in 1992/93, the regiment has been involved in Chechnya fighting since 1994.

960th Shturmovoy Aviatsionnyi Polk

960th Attack Air Regiment – Primorsko-Akhtarsk

Until 1993 this unit was named the 960th Uchebnyi Aviatsionnyi Polk (UAP) or Training Air Regiment. It belonged to the pilots' school at Krasnodar and was equipped with 120 L-39C jet trainers (as of 1991). In 1993, the regiment was transformed into the 960th IAP with MiG-29 fighters. Around 2000 the unit was re-equipped again, this time with Su-25 attack aircraft, and gained its present name of the 960th ShAP.

559th Bombardirovochnyi Aviatsionnyi Polk

559th Bomber Air Regiment – Morozovsk

In the 1960s and 1970s, the regiment was stationed at Finsterwalde in East Germany as the 559th APIB, with MiG-17Fs and Su-7Bs. In 1976 it re-equipped with MiG-27 fighter-bombers. On 22 March 1993 it withdrew from Germany to Morozovsk, where it received Su-24M bombers and was renamed as the 559th BAP. Since 1994 it has been active in Chechnya operations.

959th Bombardirovochnyi Aviatsionnyi Polk

959th Bomber Air Regiment – Yeysk

For the past several decades this unit was the 959th UAP training regiment of the pilots' school in Yeysk, and until 1993 it operated L-39C trainers. Around 1993/94, the 959th UAP was re-equipped with Su-24M bombers transferred to Yeysk from Belarus after the disintegration of the USSR and was renamed the 959th BAP.

3rd Gvardeyskiy Rostov-Donskiy Istrebitelnyi Aviatsionnyi Polk

3rd Rostov-Don Fighter Air Regiment of Guards – Krymskaya near Krymsk

Until recently this unit was known as the 562nd IAP. The regiment flew Yak-28P interceptors from the 1960s to the late 1980s, when it converted to Su-27s. Around 2000, the regiment took over the number and tradition of the 3rd IAP of Guards.

The 3rd IAP was raised to Guards status on 6 December 1941. During World War II, the regiment flew LaGG-3, La-5 and Yak-3 fighters. After the war, the 3rd IAP was stationed in the Russian far east, at Nikolayevka and then at Khabarovsk. During 1952/53 it took part in the Korean War with MiG-15bis fighters and for two years was stationed at Sanshilipu in China. In May 1955 the regiment moved to Poland (first to Szprotawa and in 1961 to Krzywa). In September 1960, the unit was transformed into the 3rd APIB with MiG-17s, followed by Su-7B and MiG-27 fighter-bombers. In December 1988 the unit became the 3rd BAP with Su-24 bombers. In June 1992 it withdrew from Poland to Lebiazhye air base in Russia, and disbanded (the aircraft were given to the 1st BAP).

19th Gvardeyskiy Istrebitelnyi Aviatsionnyi Polk

19th Fighter Air Regiment of Guards – Millerovo

The unit came into being in 1939 as the 145th IAP, and on 4 April 1942 it received its present number and was ranged among the Guards. During World War II, the 19th IAP used P-40 Tomahawk, P-39 Airacobra and La-5FN fighters. After the war, the regiment was stationed in Germany (Mirow-Larz) with MiG-17Ps. In the 1960s, the unit was transformed into the 19th APIB with Su-7B and then MiG-27 fighter-bombers. In March 1993 the regiment withdrew from Germany to Millerovo, where it was rearmed with MiG-29 fighters and transformed again into the 19th IAP.

As the 4th Army shoulders the brunt of operations in Chechnya, it is an 'attack-heavy' organisation, with a large former army aviation contingent with helicopters like this armed and armoured Mi-8MTV (above), and three regiments with Su-25s (left).

31st Gvardeyskiy Nikopolskiy Krasnoznamionnyi ordena Suvorova Istrebitelnyi Aviatsionnyi Polk

31st Nikopol Fighter Air Regiment of Guards awarded with Red Banner and Suvorov order – Zernograd

The unit was founded in spring 1941 as the 273rd IAP, and on 22 November 1942 it was ranged among the Guards and received the new number of 31. During World War II, the regiment used Yak fighters. After the war, it was stationed in Germany at Falkenberg (Alt Lonnewitz) and since 1989 has been equipped with MiG-29 fighters. In June 1993 the 31st IAP withdrew from Germany to Zernograd, Russia.

11th Krasnoznamionnyi ordenov Suvorova i Kutuzova Vitebskiy Otdelnyi Razvedyvatelnyi Aviatsionnyi Polk

11th Vitebsk Independent Reconnaissance Air Regiment awarded with Red Banner, Suvorov and Kutuzov orders – Marinovka

The regiment was formed on 19 July 1942 at Goroshino near Moscow with Pe-2R reconnaissance aircraft. After the war, it was stationed at Jekabpils in Latvia and operated Il-28R reconnaissance aircraft. In 1954 the 11th RAP moved to Neu-Welzow air base in East Germany. In 1966 it re-equipped with Yak-27Rs and two years later with Yak-28s (two reconnaissance squadrons with Yak-28Rs and one electronic warfare squadron with Yak-28PPs). In 1986 the regiment received reconnaissance Su-24MRs and some time later several EW Su-24MPs. During 1991 to 1993, the regiment gradually withdrew from Germany, the EW squadron going to Chortkov in Ukraine and the other squadrons to Marinovka in Russia. Since 1994 the 11th ORAP has participated in the fighting in Chechnya.

535th Otdelnyi Smeshannyi Aviatsionnyi Polk

535th Independent Composite Air Regiment – Rostov-on-Don

The unit serves the headquarters of the 4th Army and operates transport An-12s, An-26s, and Mi-8s. Recently, the 535th OSAP absorbed the EW helicopter squadron (the 286th OVE REB) at Zernograd, which had been an independent unit until around 2000.

55th Otdelnyi Sevastopolskiy Vertoliotnyi Polk

55th Independent Sevastopol Helicopter Regiment – Korenovsk

The regiment originates from the 807th ShAP (later the 807th APIB) stationed at Brody, Belarus, and equipped with MiG-15bis aircraft. On 20 June 1961 the unit was transformed into a helicopter regiment known as the 55th OVP. It initially used Mi-4s, then Mi-6s and, in the 1970s, it equipped with Mi-8s and Mi-24s. From 1981 the 55th OVP was stationed in Poland at Legnica and Bagicz air bases. In May 1992 it withdrew to Korenovsk in Russia. The regiment operated for three years with Russian peacekeeping forces in North Ossetia and Abkhazia, and since 1994 has taken part in fighting in Chechnya. The regiment is composed of two squadrons with Mi-24 and one squadron with Mi-8 helicopters.

4 KA VVS i PVO HQ: Rostov-on-Don

UNIT (1ST LEVEL)	UNIT (2ND LEVEL)	BASE	INVENTORY/NOTES
1st SAD		Krasnodar	
	368th ShAP	Budyonovsk	Su-25
	461st ShAP	Krasnodar	Su-25, Su-22
	960th ShAP	Primorsko-Akhtarsk	Su-25
	559th BAP	Morozovsk	Su-24
	959th BAP	Yeysk	Su-24, L-39C
51st KPVO		Rostov on Don	
	3rd IAP	Krymskaya	Su-27
	19th IAP	Millerovo	MiG-29
	31st IAP	Zernograd	MiG-29
11th ORAP		Marinovka	Su-24MR
535th OSAP		Rostov on Don	Mi-8, An-12, An-26
Army Aviation component			
325th OTBVP		Yegorlykskaya	Mi-8, Mi-26
55th OVP		Korenovsk	Mi-24, Mi-8
487th OVP BU		Budyonovsk	Mi-24, Mi-8
326th OVE		Bataysk	Mi-8

Two of the 4th KA's three air defence fighter regiments have MiG-29s (this is a 'Fulcrum-C'). Both were formerly based in East Germany with the 16th Air Army, although the 19th was then a fighter-bomber unit with MiG-27s.

325th Otdelnyi Transportno-Boyevoi Vertoliotnyi Polk

325th Independent Transport-Combat Helicopter Regiment – Yegorlykskaya

This regiment undertakes most of the transport missions in Chechnya and is equipped with Mi-8 and Mi-26 helicopters. Until the early 1990s it was stationed at Culukidze, Georgia. The regiment's Mi-6s were replaced by Mi-26s in 2001.

487th Otdelnyi Vertoliotnyi Polk Boyevogo Upravleniya

487th Independent Helicopter Regiment for Battle Control – Budyonovsk

The regiment was stationed at Prenzlau, Germany, until August 1992, when it withdrew to Budyonovsk. Since 1994 the 487th OVP BU has been active in combat operations in Chechnya. The unit consists of two squadrons of Mi-24 and one squadron of Mi-8 helicopters.

326th Otdelnaya Vertoliotnaya Eskadrilya

326th Independent Helicopter Squadron – Bataysk

Until the early 1990s, the squadron was stationed at Kirovabad (now Ganca) in Azerbaijan with Mi-8, Mi-24 and Mi-6 helicopters, as well as transport aircraft. In 2003 the squadron was reduced to six Mi-8 helicopters. Due to the incorporation of army aviation into the VVS, the squadron likely will soon be disbanded or incorporated into the 535th OSAP.

In Chechnya the Mi-26 is valued for its heavylift capability, especially for airlifting large numbers of troops and for medevac. It was the crash of an Mi-26 in Chechnya on 19 August 2002, with the loss of 119 on board, that spurred the move to place army aviation under air force control.

5th Armiya VVS i PVO

5th Army of the VVS and PVO HQ: Yekaterinburg

The 5th Army of VVS and PVO was formed in 2001 from the 5th Independent Corps of VVS and PVO, which came into being in 1998 as the successor to the 5th Corps of PVO. The Army has no combat air units – only three regiments of anti-aircraft missiles (in Yekaterinburg, Engels and Samara), as well as radio technical units. The 764th IAP fighter regiment with MiG-31 interceptors stationed at Bolshoye Savino falls within the zone of responsibility of the 5th Army but is not subordinated to it; rather, it reports to the High Command of the VVS. In June 2001 the Commander-in-Chief of the VVS declared that the 5th Army would be reinforced by a bomber air regiment, a fighter air regiment, an attack air regiment and a reconnaissance air squadron, although there has been no information about the implementation of this intention.

The zone of responsibility of the 5th Army includes the Volga-Ural Military District, which is inside Russia at the border between Europe and Asia (the southern part borders Kazakhstan). Kant air base in Kyrgyzstan, which was created in October 2003, is also subordinated to the 5th Army. The commanding officer of the 5th Army of VVS and PVO is Lieutenant General Yevgeniy Yuryev (2003).

Note: Some reference sources on the composition of the 5th Army mention the 239th OSAP at Karabash near Chelyabinsk, equipped with An-12, An-26, Mi-6 and Mi-8 transports, but this unit was probably disbanded several years ago.

128th Otdelnaya Smeshannaya Aviatsionnaya Eskadrilya

128th Independent Composite Air Squadron – Koltsovo near Yekaterinburg

The squadron serves the 5th Army headquarters, providing transport.

320th Otdelnaya Transportnaya Aviatsionnaya Eskadrilya Poiskovo-Spasatelnoy Sluzhby

320th Independent Transport Air Squadron of Search-Rescue Service – Uprun (Troitsk) near Chelyabinsk

An independent transport air regiment (number unknown) was founded in January 1963 at Kedainiai (Lithuania) with Il-14 aircraft. In December 1965 that regiment moved to Yuzhno-Uralsk (now Troitsk), and in 1970 it received An-24s, followed in 1974 by An-12s. In October 1980 the regiment moved to Kustanay in northern Kazakhstan. The unit was employed in searching for landing spaceships and cosmonauts (Russian spaceships usually land in the southern Urals or on territory belonging to Kazakhstan). Among its achievements, in 1975 the regiment provided security for the launch, travel and landing of a joint US-Soviet Apollo-Soyuz mission. In January 1998 the regiment was reduced to squadron size and renamed the 320th OTAE. Three months later, the squadron moved back to Russia, to Troitsk air base.

793rd Otdelnyi Vertoliotnyi Polk

793rd Independent Helicopter Regiment – Kinel'-Cherkasy

In Soviet times this regiment was stationed at T'elavi, Georgia, but after the fall of the USSR it moved to Russia. Presently, the 793rd OVP consists of two squadrons with Mi-8 and one with Mi-26 helicopters.

237th Otdelnaya Vertoliotnaya Eskadrilya

237th Independent Helicopter Squadron – Bobrovka

This unit dates to the 1950s, when it was formed as the 473rd OVP helicopter regiment. In 1962 the regiment with 33 Mi-4 helicopters was deployed to Cuba as part of a large Soviet contingent, which resulted in the so-called Cuban Missile Crisis. Until 1998/99 the regiment was stationed at Ozinki in the Saratov region, with about 35 Mi-24s and 15 Mi-8s. Then, the regiment was reduced to squadron size, renamed the 237th OVE and moved to Bobrovka near Samara (about 200 km/ 125 miles from Ozinki). Until 1996/97 Bobrovka air base was occupied by the MiG-23-equipped 683rd IAP of the PVO and the 4022nd AB storage facility with MiG-21, MiG-23, MiG-25 and Su-15.

5 A VVS i PVO HQ: Yekaterinburg

UNIT	BASE	INVENTORY/NOTES
128th OSAE	Koltsovo near Yekaterinburg	An-26
320th OTAE PSS	Uprun (Troitsk) near Chelyabinsk	Mi-8, An-24, An-12
Army Aviation component		
793rd OVP	Kinel'-Cherkasy	Mi-8, Mi-26
237th OVE	Bobrovka	Mi-24, Mi-8

6th Krasnoznamionnaya Armiya VVS i PVO

6th Army of the VVS and PVO, awarded with Red Banner HQ: St Petersburg

The army was established on 1 June 1998 following the merger of the 76th VA of the VVS and the 6th OA of the PVO, both with headquarters in St Petersburg. The 6th Army's zone of responsibility equates to the territory of the Leningrad Military District. In the past, this was a very powerful army, particularly its defensive component – regiments of interceptors and anti-aircraft missiles. It had two tasks: the protection of the most industrialised, European part of the USSR against possible American attack from the north via the North Pole, and the coverage of deployment of nuclear submarines stationed on the Kola Peninsula.

These two tasks are no longer as important and, therefore, the 6th Army has been considerably reduced. Many units were disbanded during the VVS/PVO merger in 1998, and cutbacks continue. In 2001, the 470th IAP at Afrikanda with Su-27 fighters and the 174th IAP at Monchegorsk with MiG-31 interceptors were disbanded, followed in 2002 by the 180th IAP at Gromovo with MiG-31s. Apart from the aircraft units, the army has at least five SAM regiments. The commanding officer of the 6th Army of VVS and PVO is Lieutenant General Gennadiy Torbov (since April 2000).

9th Gvardeyskiy Odesskiy Krasnoznamionnyi ordena Suvorova Istrebitelnyi Aviatsionnyi Polk

9th Odessa Fighter Air Regiment of Guards awarded with Red Banner and Suvorov order – Kilp-Yavr

Until 2001, this fighter regiment at Kilp-Yavr was numbered 941. The 941st regiment was formed in 1952 as a part of Northern Fleet aviation and was equipped with MiG-17 fighters. In 1957 the unit was allocated to the PVO. From 1959 it was equipped with Su-9 interceptors, and in 1970s it received MiG-23Ms. In autumn 1985, it was one of the first regiments to convert to Su-27 fighters, which it still operates. One of this unit's Su-27s collided in mid-air (without serious damage) with a Norwegian P-3B Orion in September 1987.

In 2002, the 941st IAP inherited the number and tradition of the 9th IAP of Guards which, after the collapse of the USSR, remained at Andijon air base in Uzbekistan. This unit was formed in 1938 in Ukraine as the 69th IAP, with I-16 fighters. On 7 March 1942, the regiment was ranged among the Guards and received the new number of 9. When World War II ended the unit was on German territory, and after the war it was stationed in Uzbekistan. It flew MiG-19 fighters during the 1950s and 1960s, and in 1980s received Su-27s. In October 1992 the regiment was taken over by Uzbekistan.

458th Gvardeyskiy Polotskiy ordena Suvorova Istrebitelnyi Aviatsionnyi Polk imeni Leninskogo Komsomola

458th Polotsk Fighter Air Regiment of Guards awarded with Suvorov order and named after Lenin's Komsomol – Savvatiya near Kotlas

The regiment derives part of its tradition from the 485th IAP formed in July 1941; in March 1943, the regiment was ranged among the Guards with the new number of 72. During World War II, the regiment used MiG-3, Hurricane, Yak-1/-7 and P-39Q Airacobra fighters. Later, the 72nd IAP took part in the Korean War and then moved to Anderma air base in the north of Russia. In December 1986, the 72nd IAP received MiG-31 interceptors.

Another unit involved in this regiment's

From its base to the south of Murmansk, the 98th ORAP flies the MiG-25RB (like this aircraft) and the Su-24MR on a mix of maritime and overland reconnaissance duties.

history is the 445th IAP named after Lenin's Komsomol, which was stationed at Savvatiya (Kotlas) air base and equipped with Tu-128 long-range interceptors (1960s and 1970s) and then with MiG-25Ps (from 1980) and MiG-25PDs.

In 1993, the 72nd IAP from Anderma was transferred to Kotlas and disbanded; the 445th IAP was disbanded also. At the same time, the new 458th IAP was established and took over the colours, awards and traditions of both IAPs, and the MiG-31 fighters of the 72nd IAP. In 1998, the 458th IAP received more MiG-31s from the disbanded 518th IAP at Talagi. The commanding officer of the 458th IAP is Colonel Vladimir Apraksin (2004).

159th Gvardeyskiy Novorossiyskiy Krasnoznamionnyi ordena Suvorova Istrebitelnyi Aviatsionnyi Polk

159th Novorossiysk Fighter Air Regiment of Guards awarded with Red Banner and with Suvorov order – Besovets near Petrozavodsk

The regiment was awarded Guards honour on 14 April 1944, and during World War II it operated La-5 and La-7 fighters. After the war, the 159th IAP remained in Poland, first at Malbork, then at Brzeg and, from February 1952, at Kluczewo. The regiment operated Yak-15, MiG-15, MiG-17, Yak-25M and MiG-21 jet fighters, until in June 1987 it re-equipped with the Su-27. In July 1992, the 159th IAP was moved from Kluczewo to Besovets in Russia.

Many air units were stationed at Besovets air base in the past, all part of the PVO. Until 1972 it housed the 641st IAP with Yak-28P interceptors (later moved to Rogachevo). Afterwards, Besovets was the base for the 991st IAP and then the 57th IAP with Su-15 interceptors. In 1987 the 57th IAP moved to Norilsk, and Besovets air base was handed over to the VVS.

177th Moskovskiy Istrebitelnyi Aviatsionnyi Polk

177th Moscow Fighter Air Regiment – Lodeynoye Pole

The regiment was formed in June 1941 near Moscow. One of its pilots was Hero of the Soviet Union Viktor Talalikhin, who on the night of 6/7 August 1941 rammed a German He 111 (he himself was slightly injured). From 1948 the regiment was stationed at Yaroslavl and was equipped with Yak-17 and MiG-9 fighters. In 1950/51 the 177th IAP took part in the Korean War using MiG-15 and MiG-15bis fighters. After that war, the regiment moved to Lodeynoye Pole air base, where it remains. During the 1950s and 1960s, the regiment was equipped with MiG-19s (a MiG-19P acts as the base's gate guard now), then with Su-9s and MiG-23s (in 1990 the unit had 38 MiG-23 fighters). At the beginning of the 1990s, the 177th IAP converted to the Su-27. In 2002 it was awarded with the honorary title 'Moskovskiy'. The commanding officer of the regiment is Colonel V. Kravchenko (2003).

Note: The 177th IAP of Guards, which existed in the past, was a different unit.

67th Bombardirovochnyi Aviatsionnyi Polk

67th Bomber Air Regiment – Siverskiy-2 (Ordynka)

The regiment was formed on 5 December 1968 and for the next decade it flew Su-17 fighter-bombers as the 67th APIB. After rearming with Su-24 bombers, the regiment was renamed the 67th BAP. In July 1992 another unit, the 89th BAP, moved from Szprotawa in Poland to Siverskiy air base, where it disbanded and its Su-24Ms were adopted by the 67th BAP. The regiment day is 5 December. Since October 2003 the commanding officer of the 67th BAP is Colonel Nikolay Ledenev.

722nd Bombardirovochnyi Aviatsionnyi Polk

722nd Bomber Air Regiment – Smuravyevo near Gdov

Smuravyevo airfield was built in 1953 and in the 1980s it housed the 722nd APIB with MiG-27 fighter-bombers. In 1990, this regiment re-equipped with Su-24M bombers and was renamed the 722nd BAP.

98th Otdelnyi Gvardeyskiy Vislenskiy Krasnoznamionnyi ordena Kutuzova Razvedyvatelnyi Aviatsionnyi Polk

98th Independent Visla Reconnaissance Air Regiment of Guards awarded with Red Banner and with Kutuzov order – Monchegorsk

The regiment was formed in 1942 as the 4th reconnaissance air regiment with Pe-2 and

The 6th KA's area of responsibility covers the strategically important St Petersburg area and the western portion of the northern coastline. It is not surprising that long-range interceptors proliferate, with three former PVO regiments operating the Su-27 (above) and one flying the MiG-31 (right). The latter are based at Kotlas, from where they can cover the northern coast. The Su-27s cover the Kola Peninsula and the St Petersburg district.

6 KA VVS i PVO HQ: St Petersburg

UNIT (1ST LEVEL)	UNIT (2ND LEVEL)	BASE	INVENTORY/NOTES
21st KPVO		Severomorsk	
	9th IAP	Kilp Yavr (Poliarnyi)	Su-27
	458th IAP	Savvatiya (Kotlas)	MiG-31, MiG-25U
54th KPVO		Taytsy	
	159th IAP	Besovets (Petrozavodsk)	Su-27
	177th IAP	Lodeynoe Pole	Su-27
67th BAP		Siverskiy-2	Su-24
722nd BAP		Smuravyevo (Gdov)	Su-24
98th ORAP		Monchegorsk	MiG-25RB/U, Su-24MR, MiG-31
138th OSAP		Levashevo	Mi-8, An-12, An-26, Tu-134
147th OVE REB		Pushkin	Mi-8PPA
Army Aviation component			
332nd OVP BU		Pribylovo	Mi-24, Mi-8
85th OVE		Alakurtti	Mi-24, Mi-8
714th BRV		Kasimovo	Mi-24, Mi-8

Pe-3bis aircraft. On 17 June 1943, the regiment was ranged among the Guards and given the new number of 98. In the 1950s and 1960s it flew reconnaissance Il-28Rs, followed by Yak-27Rs and Yak-28Rs. In 1990 the regiment operated Su-17R and MiG-25RB aircraft, and now has one squadron of MiG-25RBs and one squadron of Su-24MRs. The regiment inherited several MiG-31 interceptors from the disbanded 174th IAP, which was stationed at Monchegorsk until 2001.

138th Otdelnyi Smeshannyi Aviatsionnyi Polk
138th Independent Composite Air Regiment – Levashevo

The regiment serves the 6th Army headquarters and is equipped with a dozen or so Mi-8 helicopters and various transport aircraft (An-12, An-26, Tu-134).

147th Otdelnaya Vertoliotnaya Eskadrilya Radio-Elektronnoy Bor'by
147th Independent Helicopter Squadron for Electronic Warfare – Pushkin

Independent electronic-warfare helicopter squadrons have recently been included in composite regiments serving individual air armies. Only the 6th Army of VVS and PVO still has such a squadron, with several Mi-8PPA helicopters.

332nd Otdelnyi Gvardeyskiy Vertoliotnyi Polk Boyevogo Upravleniya
332nd Independent Helicopter Regiment of Guards for Battle Control – Pribylovo

This unit's origins are in the 158th IAP fighter air regiment formed in autumn 1940. On 7 July 1943, the regiment was ranged among the Guards and received the new number of 103. In the 1950s, the 103rd IAP was deprived of its fighter aircraft and received Mi-4 helicopters (later also Mi-6s), and was given the new number of 332. It was known as the 332nd OTBVP transport-combat helicopter regiment and was equipped with Mi-8s and Mi-26s until around 2000, when its transport Mi-26s were replaced by combat Mi-24s and the regiment was renamed the 332nd OVP BU.

85th Otdelnaya Vertoliotnaya Eskadrilya
85th Independent Helicopter Squadron – Alakurtti

This unit is the former 485th OVP regiment stationed at Brandis, Germany, and is equipped with Mi-24V, Mi-24R/K and Mi-8MT helicopters. In May 1992 it moved to Alakurtti in Russia. Around 2000, the regiment was reduced to squadron size and became the 85th OVE.

714th Baza Rezerva Vertoliotov
714th Helicopter Reserve Base – Kasimovo

In 1990 the 93rd OVE squadron was based at Kasimovo with Mi-8 and Mi-24 helicopters. In 1992 the 172nd OVP helicopter regiment with Mi-8s and Mi-24s withdrew from Parchim air base in Germany and moved to Russia, initially to Kaliningrad and then to Kasimovo. This regiment absorbed the 93rd OVE squadron. The 172nd OVP was extant at Kasimovo until around 2000, when it was disbanded and its helicopters handed over to the new 714th BRV reserve base.

The Su-24 'Fencer' (this is an Su-24MR) is a mainstay of 6th KA operations, with two fighter-bomber regiments based near St Petersburg and close to the Estonian border, plus Su-24MR reconnaissance aircraft further north on the Kola peninsula.

11th Armiya VVS i PVO

11th Army of the VVS and PVO HQ: Khabarovsk

The 11th Army of VVS and PVO was created in 1998 from the units of the 11th Army of the PVO (with the exception of a regiment of MiG-31s at Kamchatka, which was taken by the Navy) and the 1st Air Army of the VVS. In addition to aircraft units, the 11th Army includes three regiments of air defence missiles (in Khabarovsk, Komsomolsk and Vladivostok), as well as radio technical units. The commanding officer of the 11th Army is Lieutenant General Igor Sadofyev (born in 1956), who was formerly the second in command of the 4th Army of VVS and PVO in Rostov.

Like the other armies, the 11th is still subject to a reduction in air units. In recent years, two combat regiments have been disbanded: the 404th IAP at Orlovka and the 523rd BAP at Vozzhayevka. Orlovka (or Vernoye) air base was one of the newest in Russia, built in the early 1970s. The 404th Tallinn Fighter Air Regiment awarded with Kutuzov order was equipped with MiG-23 fighters, MiG-29s (from 1987), and Su-27s that it received in 1996 from the disbanded 871st IAP at Smolensk. In August 2000 the 404th IAP itself was disbanded and the airfield is now empty. The aircraft and crews of the 404th IAP were incorporated into the 60th IAP at Dzemgi, creating the new 23rd IAP in the process.

The 523rd Orsha Red Banner Bomber Air Regiment of Guards awarded with Suvorov, Kutuzov and Aleksandr Nevskiy orders was formerly stationed at Vozdvizhenka. In May 1959 the 523rd regiment was the first unit in Soviet aviation to receive Sukhoi's new Su-7 fighters, and in 1965 it picked up Su-7BKL fighter-bombers. Until the early 1990s, the 523rd APIB (as the regiment was by then known) used successive versions of this aircraft, up to the Su-17M3. The regiment then adopted the Su-24M and changed its name again, to the 523rd BAP. It moved from Vozdvizhenka to Vozzhayevka, probably in the 1990s, and was disbanded in 2002.

Until the 1990s, Vozzhayevka air base also housed a fighter air regiment with MiG-29s and a reconnaissance regiment with Su-24MRs; the base is no longer used.

11 A VVS i PVO HQ: Khabarovsk

UNIT (1ST LEVEL)	UNIT (2ND LEVEL)	BASE	INVENTORY/NOTES
23rd KPVO		Vladivostok	
	22nd IAP	Centralnaya Uglovaya (Artem)	Su-27
	530th IAP	Sokolovka (Chuguyevka, Sandagou)	MiG-31, MiG-25PU
25th DPVO		Komsomolsk na Amure	
	23rd IAP	Dzemgi	Su-27
83rd BAD		Komsomolsk na Amure	
	277th BAP	Khurba	Su-24
	302nd BAP	Verino (Pereyaslavka)	Su-24
303rd SAD		Ussuriysk	
	18th ShAP	Galionki	Su-25
	187th ShAP	Chernigovka	Su-25
	799th RAP	Varfolomeyevka	Su-24MR, MiG-25RB (?)
257th OSAP		Khabarovsk Bolshoy	An-12, An-26, Mi-8
Army Aviation component			
unknown OVP		Sokol (Dolinsk)	Mi-8
319th OVP BU		Chernigovka	Mi-24
364th OVP		Srednebelaya	Mi-8, Mi-24, Mi-26
825th OVP		Garovka-2 (Birofeld ?)	Mi-8

22nd Gvardeyskiy Krasnoznamionnyi Istrebitelnyi Aviatsionnyi Polk
22nd Fighter Air Regiment of Guards awarded with Red Banner – Centralnaya Uglovaya (Uglovoye Northwest, Artem) near Vladivostok

The regiment's tradition derives from the 6th IAP created on 6 December 1918. From 1934 it was stationed at the town of Artem. On 26 September 1945, the regiment was given Guards status and received the number 22. After World War II, the 22nd IAP flew interceptors: Yak-25s, Su-9s and then, until the mid-1990s, MiG-23MLDs. In 1994, one of the Su-27 squadrons of the disbanded 47th IAP at Zolotaya Dolina was assigned to the 22nd IAP, and more Su-27 fighters came later. The regiment day is 6 December.

Russians call this airfield Centralnaya Uglovaya, but on American aviation maps it is named Uglovoye Northwest. The town where the airfield is located is Artem. There is another airfield in Artem, called Knevichi by Russians and Artem North by Americans; a naval air regiment is stationed there and it is also the civil airport for Vladivostok.

530th Istrebitelnyi Aviatsionnyi Polk
530th Fighter Air Regiment – Sokolovka (Chuguyevka, Sandagou)

This is the unit to which Lieutenant Viktor Belenko belonged, who on 6 September 1976 defected to Japan with his MiG-25P. This air base is called Sokolovka or Chuguyevka (the distance between these two places is 5-7 km/3-4.5 miles) by Russians, and Sandagou (a settlement with this name is 15 km/8.7 miles away) on American maps. Since September 1988 the 530th IAP has operated MiG-31 interceptors.

23rd Istrebitelnyi Aviatsionnyi Polk
23rd Fighter Air Regiment – Dzemgi (Komsomolsk North)

The 23rd IAP was formed in 2000 as an amalgamation of the 60th IAP at Dzemgi and aircraft and crews which were relocated there after the disbandment of the 404th IAP at Orlovka. The 60th IAP was established in 1938 and was based at Dzemgi until the unit merger. Dzemgi is also the factory airfield of the KnAAPO aircraft plant at Komsomolsk na Amure. As such, in 1985 the 60th IAP became the first unit to be equipped with the new Su-27 (formerly, the regiment used Su-15s). 23rd IAP Su-27s are currently undergoing modernisation to Su-27SM standard, the first in the Russian air force.

277th Mlavskiy Bombardirovochnyi Aviatsionnyi Polk
277th Mlava Bomber Air Regiment – Khurba (Komsomolsk South)

This regiment's early history is unknown, but from 1956 the 277th BAP was stationed in East Germany at Eberswalde-Finow air base and was equipped with Il-28 bombers. In 1965 it received Yak-28I supersonic bombers. In 1969, as the result of Soviet-Chinese border conflicts, the regiment moved to Khurba near Komsomolsk, 300 km (186 miles) from the Chinese border. Since 1976 it has operated Su-24 bombers.

302nd Bombardirovochnyi Aviatsionnyi Polk
302nd Bomber Air Regiment – Verino (Pereyaslavka)

This unit was formed in May 1941 at Khabarovsk as the 302nd IAP fighter regiment. In the past, several units were stationed at Verino air base, including the 300th APIB with MiG-27D fighter-bombers (in the 1980s) and the 302nd APIB with Su-17M4 aircraft; a squadron from the latter regiment fought in Afghanistan in 1988/89. At the beginning of the 1990s, only one regiment, the 302nd, remained at the base. In 1992, it received Su-24M bombers from the disbanded 42nd BAP (which came back from Poland) and was renamed the 302nd BAP.

18th Gvardeyskiy Vitebskiy dvazhdy Krasnoznamionnyi ordena Suvorova Shturmovoy Aviatsionnyi Polk Normandie-Nieman
18th Vitebsk Normandie-Nieman Attack Air Regiment of Guards awarded with Red Banner (twice) and Suvorov order – Galionki

This regiment came into being in 1938 in Garovka near Khabarovsk as the 6th IAP. After

In terms of area, the 11th Army has the biggest responsibility, covering the Russian Far East with the exception of the Kamchatka peninsula, which is administered by the navy. Apart from two Su-25 regiments located in the Vladivostok area, close to Korea and China, the assets are all long-range interceptors and fighter-bombers, such as the Su-24M (above), MiG-31 (left) and Su-27 (below).

Saab 37 Viggen

Sweden's Thunder

For over 30 years the Viggen – which, according to Norse mythology, is the thunderclap resulting from the strike of Thor's hammer – has rung out over Sweden, a resolute deterrent against any potential aggression. Today, thanks to three major upgrade programmes (AJS 37, JA 37D and SK 37E), the fighter Viggen is in the prime of its life and the older versions remain highly valued assets in their more specialised roles. The Viggen has also become an invaluable bridge to the next generation of combat aircraft. Yet the Viggen era is drawing swiftly to a close, the aircraft having become a victim not of obsolescence, but of rapidly shrinking defence budgets.

Above: Two JA 37D(I)s cruise over a typical northern Swedish landscape in 2004. Although it is due to be retired by the end of the year, the latest Jakt-Viggen version remains one of the most capable interceptors in the world.

Left: All Flygplan 37 aircrew are presented with this stylised Viggen badge. The white version is worn by the pilots of first-generation aircraft, whereas JA 37 pilots have the yellow version (with RWR antennas added).

Below left: A 'family portrait' captures the four remaining Viggen versions in service, all flying with 1 Div./F21 (although the JA 37D nearest the camera retains its 1 Div./F17 eagle markings).

Below: From any angle, the Viggen is one of the world's most instantly recognisable aircraft.

Sweden's defender – the Viggen was borne when the Cold War between East and West was raging, and was very much a product of its time. It was designed solely with the Swedish national interest in mind, but could have achieved success in the export market had government policy been less restrictive. Thirty-seven years after the prototype (above right) first flew, the aircraft is still an integral part of Flygvapen operations, although it is to be the main casualty as the air force is slashed in size to a tiny fraction of its Cold War force levels.

According to the current plans of Svenska Flygvapnet (the Swedish Air Force), the Saab 37 Viggen will be retired by the end of 2005, bringing to a premature end a highly effective service career which began when the Försökcentralen (test centre) at Malmen initiated the training of an instructor cadre in early 1972. In the intervening years the aircraft became the backbone of the Swedish air force, serving in all the major combat roles, and a source of fierce national pride. In its later years the Viggen has contributed enormously to the development of its successor, the JAS 39 Gripen, a process which is ongoing today. Indeed, many of the later improvements applied to the Flygplan (Aircraft) 37 were developed with application to the Gripen in mind. This has, of course, also been a two-way street, with Gripen items finding their way into the Viggen to maintain the older aircraft's combat viability while the JAS 39 was being brought into service.

Viggen - the first 30 years

To set the recent upgrade programmes in some context, it is necessary to review the Viggen's initial development and history. What eventually crystallised as Flygplan 37 can be traced back to the early 1950s, when Saab engineers began studying a follow-on design to the double-delta J 35 Draken, itself still some years away from first flight. While the Draken was primarily a fighter, its successor was initially seen as a replacement for the A 32 Lansen in the attack role. As the years passed, the new aircraft metamorphosed into a multi-role aircraft. Not only would it replace the Lansen, but it would also supplant the Draken in the fighter and reconnaissance roles. It would be designed to excel in all its given tasks, rather than have secondary roles thrust upon it for which it was less than ideal. Today, the development of true multi-role aircraft is taken for granted, but in the late 1950s this was a brave and innovative step in a world dominated by single-role types. Subsequent high-level programmes, notably the Panavia Tornado (with which the Viggen is often compared), validated the lead taken by the Swedish team.

Designers at the Svenska Aeroplan Aktiebolaget (Swedish Aircraft Company – Saab) plant at Linköping undertook a range of studies, numbered in the 1500 series. A variety of powerplants was examined, including lift fans for vertical take-off versions, but in December 1961 the Försvarets Materielverk (FMV – Defence Material

Administration) specified that a derivative of the Pratt & Whitney JT8D turbofan would be the new aircraft's engine, to be built under licence by Svenska Flygmotor (later Volvo Flygmotor, then Volvo Aero) at Trollhättan. This engine was essentially an airliner engine, normally powering the Boeing 727 and Douglas DC-9, and considerable work would be needed to develop it into a fighter engine, including the provision of afterburning. However, with the engine specified, Saab could finalise its proposal, which was submitted in the form of Project 1534. On 28 September 1962 it was formally accepted for development.

In designing the Flygplan 37, Saab engineers not only had to produce a multi-role aircraft, but also one which conformed to the unique requirements of Flygvapen operational doctrine. At the heart of this was the ability to operate from very short, dispersed strips and to be maintained by conscripts and reservists in the open – often in sub-zero conditions. Never a company to be scared of taking unconventional steps, Saab schemed an unorthodox aircraft with an angular delta wing and large foreplanes, the latter playing a large part in keeping landing speeds low. The wings/foreplanes combined with a powerful afterburner and a unique thrust-reverser (the 37 was the first aircraft to have both features) to fully satisfy the short take-off and landing requirements, while the airframe was liberally covered with access panels to ease maintenance. The developed JT8D powerplant, fitted with an afterburner section bigger than the engine itself, was designated RM8A.

Just as the 37 looked unconventional and futuristic on the outside, so the internal systems also represented an ambitious and futuristic approach. To cater for true multi-role capability and single-pilot operation, the aircraft's systems were built around a central computer, the CK 37, which connected and controlled many of the aircraft's functions, including the Ericsson PS-37/A radar and weapon system. For the pilot, this meant that all nav/attack data could be programmed into the computer and the aircraft and its displays were configured accordingly. Although the capabilities of the CK 37 are distinctly pedestrian by today's standards, the Flygplan 37 can be regarded as the first multi-role aircraft to be designed around a 'weapon system', and laid solid foundations for future developments.

Named Viggen, Flygplan 37 first took to the air on 8 February 1967, with Erik Dahlström at the controls. Following the completion of eight prototypes, including one two-seater, production began with the baseline AJ 37 (Attack – Jakt, attack – fighter) version, which was intended for a primary anti-surface role with Rb 04 and Rb 05 missiles, and various bombs and rockets. Rb 75 Maverick capability was added in 1975. A total of 108 AJ 37s was produced, the first being rolled out in November 1970 and first flown on 23 February 1971. Delivery of the first production aircraft to Försökcentralen occurred on 21 June 1971. In January 1973 conversion began of the first operational squadron: 2 Division of F7 at Såtenäs, at a time when Svenska Flygvapnet still carried the 'Kungliga' ('Royal') prefix – dropped in 1974. A total of six squadrons converted to the type from the Lansen.

As well as the AJ 37s, 17 production SK 37 (Skol, school) two-seat trainers were built, in addition to the prototype, these being completed without radar but with some attack capability and delivered from June 1972. The SH 37 (Spanings Havsövervakning, reconnaissance coastal surveillance) had a PS-371/A radar 'tweaked' for sea surveillance and greater range, reconnaissance capability through podded systems, and increased computing power. Twenty-seven were built, delivered from June 1975. The SF 37 (Spanings Foto, reconnaissance photo), delivered from April 1977, was built with a camera nose replacing the radar, and some cockpit changes. It accounted for 28 aircraft, raising 'first-generation' Viggen production to 188, the last of which (an SF 37) was delivered on 1 February 1980.

Viggen operators - Svenska Flygvapnet

Unit	Callsign	Variant	from	to
F4 Jämtlands Flygflottilj, Östersund-Frösön				
1 Div	David röd	JA 37, JA 37D	1983	2004
2 Div	David blå	JA 37, JA 37D	1985	2004
TIS/TK		SK 37, SK 37E	1997	2003
F6 Västgöta Flygflottilj, Karlsborg				
1 Div	Filip röd	AJ 37	1977	1993
2 Div	Filip blå	AJ 37	1978	1993
F7 Skaraborgs Flygflottilj, Såtenäs				
1 Div	Gustav röd	AJ 37, AJS 37	1973	1998
2 Div	Gustav blå	AJ 37, AJS 37	1973	1997
3 Div	Gustav gul	AJ 37	1974	1977
TIS/GSFU		SK 37	1972	1974
F10 Skånska Flygflottiljen, Ängelholm-Barkåkra				
1 Div	Johan röd	AJS 37, AJSF 37, AJSH 37	1993	2001
F13 Bråvalla Flygflottilj, Norrköping-Bråvalla				
1 Div	Martin röd	SF 37, SH 37	1976	1994
2 Div	Martin blå	JA 37	1980	1994
F15 Hälsinge Flygflottilj, Söderhamn				
1 Div	Olle röd	AJ 37, AJS 37	1975	1997
2 Div/TIS	Olle blå	SK 37	1974	1997
F16 Upplands Flygflottilj, Uppsala				
2 Div	Petter blå	JA 37, JA 37D	1986	2003
3 Div	Petter gul	JA 37, JA 37D	1987	2003
F17 Blekinge Flygflottilj, Ronneby-Kallinge				
1 Div	Quintus röd	JA 37, JA 37D	1981	2003
2 Div	Quintus blå	SF 37, SH 37, AJSF 37, AJSH 37	1978	2002
F21 Norrbottens Flygflottilj, Luleå-Kallax				
1 Div	Urban röd	SF 37, SH 37, AJSF 37, AJSH 37, JA 37D, SK 37E	1979	2005
2 Div	Urban blå	JA 37	1985	2002
3 Div	Urban gul	JA 37	1982	1998

Jakt-Viggen: the second-generation

During the early development of the Viggen one role was conspicuously absent: that of fighter. This was partially due to the greater complexities of producing a fighter version, but more importantly due to the lower priority afforded to replacing the Draken, which was performing adequately in the air defence role. A similar path was also later followed by Europe's other 'multi-role combat aircraft' team, which was building the Tornado. Nevertheless, low-key work on the JA 37 (Jakt – Attack) Jakt-Viggen continued behind the scenes, and by late 1972 development was placed in full swing.

***Above left:** Not the result of a landing accident, but the staging of one so that emergency crews can practise their job. A number of retired Viggens, like this AJ 37 at F16, are used for ground instruction and emergency training.*

Two Viggen 'specials' – in 2001 F16's 2 Div. (Petter Blå) painted up a JA 37 in patriotic colours (opposite page, top) for the 75th anniversary of the Swedish air force. The well-known 'red ghost' AJS 37 (below, with its Saab stablemates) was painted by 1 Div./F10 (Johan röd) and was flown until mid-2001, by which time the squadron was converting to the Gripen. The red Viggen is now at the Flygvapen museum at Malmen.

Above: F13 at Norrköping was the first wing to equip with the Jakt-Viggen. This example is seen carrying the original standard missile armament of BAe Dynamics Rb 71 Sky Flash and Rb 24J (AIM-9J) Sidewinder.

Below: The last JA 37 Mod C Viggens, such as this F16 aircraft, were withdrawn in 2003. This modification state entered service in 1992, introducing automatic gun aiming which slaved the autopilot to the radar's air combat mode.

Below: In 2003, one of 2 Div./F16's JA 37 Mod Cs was painted with the 1st Maintenance Squadron badge on a red tail, to join the squadron's other display aircraft. The aircraft flew for the last time on 8 October before being preserved at Uppsala.

Although it looked like a Viggen, the JA 37 was a very different beast under the skin, partly reflecting the different nature of its role but also the advances made possible by its delayed development. These centred on a new central computer, Ericsson PS-46/A radar and a revised cockpit with new displays. There were also some external differences: for instance, the extended fin of the SK 37 two-seater was fitted, and the aircraft was a few inches longer than the AJ 37 (the stretch accommodated between the canard trailing edge and wing leading edge), a result of fitting the improved RM8B engine.

Development of the RM8B began in late 1971 as a collaboration between Volvo Flygmotor and Pratt & Whitney. The two principal aims were to increase thrust and to improve the engine's surge tolerance in manoeuvres and at high altitude. The latter aim was achieved by replacing the first stage of the low-pressure compressor with a third fan stage, while increased thrust was achieved by higher turbine inlet temperatures, in turn requiring the design of a new high-pressure turbine and a four-nozzle combustion system in each of the nine flame tubes. The RM8B was 8.2 cm (3.2 in) longer than the RM8A and weighed 150 kg (330 lb) more, but its afterburning take-off thrust rose to 125 kN (28,089 lb) from 115.6 kN (25,977 lb). Dry thrust of the RM8B is 72.1 kN (16,202 lb).

JA 37 avionics

Avionics were built around the Saab CD 107 central computer, a licence-built version of the Singer-Kearfott SKC-2037 offering five times the processing capability of the CK 37 in the first-generation Viggen. Singer Kearfott also supplied a rapid-alignment inertial navigation system. The flight control system was the Honeywell/Saab SA 07, and was the first digital FCS to enter production. The cockpit display system – collectively known as the Ericsson EP-12 – comprised a 28° head-up display, a head-down radar display, and a tactical situation display. Ericsson's PS-46/A radar was Europe's first pulse-Doppler unit, and offered a wide variety of air-to-air modes (plus air-to-ground ranging). Its look-down range against small low-flying targets was in the region of 50 km (31 miles).

JA 37 weaponry comprised the BAe Dynamics Rb 71 Sky Flash semi-active radar-homing missile and Rb 24/74 Sidewinders. It also introduced a single Oerlikon KCA 30-mm cannon housed in a bulged fairing under the fuselage. The gun was offset to port, while its 150 rounds of ammunition were housed to starboard. Firing a large round at a muzzle velocity of 1050 m (3,445 ft) per second, the KCA was much harder-hitting than contemporary 30-mm guns such as the ADEN. Rounding out the JA 37's armament options were six-round M70 pods for 135-mm rockets, of which four could be carried.

Elements of the JA 37 were tested by five of the original AJ 37 prototypes after modification. The seventh Viggen prototype, redesignated 37-71, was the first to fly with an RM8B engine and lengthened fuselage, on 27 September 1974 with Per Pellebergs at the controls. The only new-build Jakt-Viggen prototype first flew with Ulf Frieburg at the helm on 15 December 1975. Gösta Sjöström flew the first production aircraft on 4 November 1977 and deliveries began (to F13) in June 1980. From June to October 1981 the JA 37 was grounded and deliveries suspended while an engine problem was fixed. The last of 149 production JA 37s – and the 338th Viggen – was delivered to Flygvapnet on 29 June 1990, bringing to an end the production phase of the aircraft's history.

Mod 11/12 - looking forward to Gripen

Although at the time of its debut the Viggen was heralded as a multi-role aircraft, it could be more accurately described as a multi-role design, with role-specific versions being tailored from a baseline aircraft, albeit sharing significant commonality. On the other hand, its successor – the JAS 39 Gripen – is a true multi-role aircraft. This single type has been created to perform any of its assigned missions with minimal changes, which are restricted to little more than different armament/stores loadouts. For the IG JAS team which was established to develop the Gripen, with Saab at its core, this represented a new series of technical challenges, but continued development of the Viggen could not be ignored. The avenues being explored for the JAS 39 would also provide many of the answers to the question of maintaining the Viggen's viability in a world of ever more sophisticated threats and of greater political confusion.

By the time Viggen production ended in 1990, the JAS 39 Gripen had already been in the air for 18 months and development was proceeding. Nevertheless, over the decades the pace of new aircraft development had become increasingly tardy due to growing complexity and shrinking budgets, and it was obvious that the Gripen would not realistically enter service for several years. Furthermore, the rate at which aircraft are built has also slowed down dramatically (in the early 1950s the Linköping factory was churning out a J 29 Tunnan every day!), and it would take many more years for the Gripen to be available in sufficient numbers to fully replace the Viggen (and few remaining Drakens). By the time the JAS 39 was in full service, the Viggen – as built – would have fallen way behind the 'threat curve'.

The answer, of course, lay in upgrading the aircraft. Relatively inexpensive by comparison with procurement of new machines, upgrading the Viggen would not only restore the aircraft's combat effectiveness, it also had the added attraction of rapidly fielding in an operational aircraft technologies destined for the JAS 39. As well as allowing technical appraisals of these systems in an operational scenario, it also allowed pilots and maintainers to gain experience on them prior to fielding on the Gripen. This became one of the key drivers behind the Mod 11/12 programme, which resulted in the AJS 37 Viggen.

Anti-ship missile

The spur to development of the AJS 37 was the appearance of the Saab RBS 15F missile, developed primarily with the Gripen in mind. This weapon was created by Saab Dynamics as a smart, turbojet-powered, subsonic anti-ship missile. It flies very low, even measuring the sea state to set its altitude so that it will not fly into waves, and has a straight-line range of around 200 km (125 miles). The onboard computer can store a large number of waypoints, allowing it to be programmed to fly sophisticated attack profiles, including deliberate 'misses' and approaches over or around islands, while its radar is highly resistant to jamming. The Ku-band radar is very sensitive, and its algorithms are effective at discriminating its target in a cluttered environment – a key feature in its main role of attacking vessels operating in and around the many archipelagos which fringe Sweden's Baltic coastline.

Initially the RBS 15F was to be integrated with the Viggen by means of changes in the individual pylons, this being covered by Mod 11. However, a more ambitious upgrade to the weapons system (Mod 12) was also being studied, and the two were rolled into what became the Mod 11/12 programme. At the heart of Mod 11/12 lay increased computer memory, the installation of a Mil-Std 1553 databus and Mil-Std 1760 weapons interface, which allowed all the pylons to be capable of taking a variety of weapons without reconfiguration. In operational terms, any Viggen with the new system was theoretically capable of flying the attack, fighter or reconnaissance missions – hence the new AJS 37 (Attack – Jakt – Spanings) designation (the equally applicable 'JAS' designation was ruled out to avoid confusion with the Gripen).

The Mod 11/12 programme expanded the self-defence capability of the recce Viggen. The AJSF 37 routinely operates with the KB (BOZ-100) chaff/flare pod under one wing, usually counterbalanced by a U22/A (Erijammer 200) deception jamming pod under the opposite wing. This AJSF 37 is seen after further modification for the SWAFRAP role, which added GPS and a second radio. Note that the original four-tone splinter camouflage is beginning to show through the new grey paint around the thrust reverser.

Thor's hammer

As well as allowing the modified Viggen to fire and launch the RBS 15F, the aircraft was also capable of launching the BK 90 (DWS 24/39) stand-off glide dispenser for sub-munitions, another weapon developed for the Gripen. Named Mjölner (Thor's hammer) and developed with MBB (subsequently DASA and EADS-Deutschland), the BK 90 has 24 launch tubes which fire the sub-munitions laterally. The MJ 1 sub-munition is a 4-kg (8.8-lb) anti-personnel/soft target weapon, while the MJ 2 is an 18-kg (39.7-kg) anti-armour weapon. The sub-munitions fall by parachute, and are ejected so that they produce an even coverage.

Carrying a U22/A jamming pod, an AJS 37 from F7 leads an AJSH 37 from F21 during an exercise. Both variants had been updated to carry the BK 90 and RBS 15F weapons, but the AJS 37 could additionally fire the Rb 75 Maverick. F7 retired its AJS 37s in 1997/98 to become the first JAS 39A Gripen wing.

Above: An AJS 37 from F7 is seen with a maximum load of four 600-kg (1,323-lb) BK 90 (Bombkapsel m/90) Mjölner weapons. Each of the dispenser's 24 12-cm (4.7-in) tubes can hold three sub-munitions. Range of the BK 90 when launched at Mach 0.9 and low level is about 10 km (6.2 miles) in straight flight, or around 5 km (3.1 miles) with a 90° turn. Guidance is by inertial system and radar altimeter. The sub-munition pattern covers an area approximately 250 m (820 ft) wide and 400 m (1,312-ft) long.

Right: A mixed formation of Viggens from 1 Div./F21 breaks away. The AJSH 37 carries two BK 90s.

Introducing these weapons on the Viggen allowed experience to build in their use long before they could be fielded on the JAS 39. In the AJS 37 the ability to fire the Maverick was retained, along with its repertoire of unguided bombs and rockets. In the fighter role the aircraft could carry up to six Rb 24J (AIM-9J) or Rb 74 (AIM-9L) Sidewinders, as opposed to the previous two. Another improvement was an enhanced radar warning receiver, which not only provided warning, but also a recording function so that threat data could be fed back into the intelligence apparatus. The AJS 37 could also be used with the PLA planning tool, of which more later.

Original plans for the AJS 37 programme envisaged 115 first-generation aircraft being modified to a near-common standard, but in the event budgetary restraints resulted in a less ambitious programme covering 98 aircraft with reduced commonality. Forty-eight AJ 37s became AJS 37s (with additional radar search capability), 25 SH 37s became AJSH 37s (retaining sea search functions but with enhanced attack options), and 25 SF 37s were raised to AJSF 37 standard. The latter type kept its camera nose, but added some fighter capability in the form of Rb 24/74 Sidewinders.

Running from 1993 to 1997, the AJS modernisation programme was instrumental in shifting Flygvapen thinking away from role-specific communities and into a single organisation. This experience was vital in preparing the service for the Gripen, while providing – along with the upgrades also being applied to the JA 37 Jakt-Viggen – a common weapons inventory with the new aircraft.

Fighter developments

When the JA 37 entered service in 1980 it was a first-rate fighter, but developments move swiftly in this arena, and the aircraft was to be the subject of an ongoing series of improvements. The first, known internally as Mod A, involved the aircraft being prepared to accept a fighter-to-fighter link which allowed aircraft within a flight to share targeting and weapon information in the air without voice communications. This link was first tested in 1982, and became operational in 1985.

In the following year a ground collision warning system was added, and in 1987 flare dispensers became operational. The same year the JA 37 received a boost in its weaponry in the shape of the Rb 74 (AIM-9L) all-aspect Sidewinder, which superseded the Rb 24J and was chosen for the Viggen after development was stopped of the indigenous Saab Rb 72. The datalink was expanded to allow the uplink of secondary information, with air targets prioritised, from ground control stations.

Addition of Rb 74 and the datalink resulted in the JA 37 Mod B, which also had its airbrakes locked shut. This step was driven primarily by the need to free up the stick (where the airbrake control was situated) for additional switches. As one experienced test pilot quipped: "We never

This JA 37C of 2 Div./F16 was painted up for air display work. As well as the fuselage/overwing stripes, it had an Indian on horseback badge underneath.

This gaggle of JA 37 Mod Cs was photographed during the last flight by 2 Division of F16 at Uppsala, in August 2003. The formation is led by the striped display aircraft. As is amply demonstrated, not all Jakt-Viggens were painted in the air superiority grey scheme, although all JA 37D(I)s are painted in the later two-tone scheme.

needed the airbrake anyway … we always had this," gesturing towards the control column in a reference to the Viggen's acknowledged rapid energy loss when under *g* loading.

Improvements continued into the 1990s, with a multi-target track capability becoming operational in 1990, and automatic gun-aiming in 1992. The latter feature slaved the aircraft's control system to the radar so that highly accurate shots could be taken, especially at long distances where the auto-gun system was more accurate than the pilot could be. In 1995 chaff dispensers were added. With the auto-gun modification, the Jakt-Viggen became the JA 37 Mod C. Finally, in 2001 the JA 37 Mod D became operational, an update programme of such magnitude that the service designation was changed to JA 37D.

Before discussing the JA 37D, it is worth examining the process by which modifications were introduced to the fleet in the later years of the Viggen's career. Proposals were first discussed by a System Development Board, which included representatives from the manufacturers (Saab, Ericsson etc), the operational test and evaluation unit (which consists of experienced personnel with recent front-line experience), the FMV and its flight test centre (FMV:Prov). A Project Board then convened to decide upon what course to take and a directive was then issued to industry to begin development. The new system was then passed to FMV:Prov for testing and refining, in conjunction with industry, and when it had been cleared for use it was produced and fielded by Flygvapnet.

While the above might appear to be a bureaucratic and laborious process, the reality of the Viggen programme in its later years is that the whole process involved just a handful of individuals who worked closely together, all with intimate knowledge of the aircraft and its systems. It helped that FMV:Prov and Saab are located on opposite sides of Sweden's aviation capital of Linköping, with just a few minutes' drive between the two facilities. This very tight development/test group resulted in very short loops between the parties, and very short development times. Industry was paid on a time basis, rather than by complex procurement contracts, resulting in a cheap product, and the group had the benefit of working with a solid, mature platform. By contrast, the Gripen programme remains a major production programme of enormous scope and huge national significance (just as the Viggen had been decades before). Many more agencies have their say, and decisions have to be taken at very high level: as a consequence it takes much longer to field modifications.

Old foes meet in friendship – four Jakt-Viggens from F16 escort two Su-27s, two Su-30s and their Il-76 support during a visit to F16's base at Uppsala in 2000. The Sukhois were from the 4th TsBPiPLS at Lipetsk. During the Cold War Viggens regularly intercepted Su-27s over the Baltic: indeed, it was a Viggen pilot who was almost certainly the first person in the West to see a 'Flanker' up close.

Mod D – the ultimate Viggen

Despite the constant series of rolling improvements made to the JA 37, a number of factors arose in the 1990s which dictated a more radical overhaul of the aircraft. Among these was the end of the Cold War, which demanded more interoperability with other forces as Sweden began to play a more prominent role in peace-

Armed with four Rb 99s, two Rb 74s and a U95 pod, this Viggen (37347) was the first of two 'prototype' JA 37D conversions which were performed by the FMV at Malmen. Note the BOY401 flare dispensers under the inner wing. The aircraft was further modified to become an 'interoperable' JA 37D(I).

JA 37D 37347 fires an Rb 99 during the AMRAAM integration trials at RFN Vidsel, Sweden's large overland weapons trials range in the Norrbottens region. The aircraft wears markings for accurate photo-calibration of the test launches, of which four were undertaken.

keeping operations, and another was the need to bridge a perceived gap in the national defence structure until the Gripen was fully functional in the air defence role. In preparation for Gripen, Flygvapnet was devising and implementing its FV2000 operational concept of infrastructure and communications, and it was felt desirable to adapt the JA 37 for FV2000. In operational terms, the slow but sure enhancement of Russian fighter forces, exemplified by the development of the Vympel R-77 active radar missile, had eroded the JA 37's ability to defend its home nation. However, it should be noted that many believed that the JA 37 continued to work well in the 'Russian' scenario.

In answer to these new challenges, the JA 37D programme was launched. As with the AJS 37 modification before it, Mod D would draw heavily upon technology used in or destined for the JAS 39, and in turn would feed back technology and experience into the Gripen programme. Development got under way in the mid-1990s, handled by the same tight group which had rapidly fielded the earlier improvements.

To cater for the many improvements planned for the Mod D, a new avionics architecture was developed. The 'old' JA 37's system was based around a CD 107 central computer, which had neither the power nor memory to support all the planned modifications. In the Mod D it was replaced by a new unit designated CD 207. Developed by Saab and based on three Intel 486 processors, the new computer is programmed using the Assembly language, as opposed to the Focap language used by the CD 107. The central computer works in conjunction with the new ANP 37 stores management system, both being connected to other avionics elements by two Mil Std 1553 databuses. One bus handles the nav/comms/display functions, connecting the GPS, communications and colour display to the central core of the system, while the other is the weapons bus, linking the pylons and radar. The pylons have Mil Std 1760 weapons interfaces, primarily to allow the carriage of the Rb 99 (AIM-120) AMRAAM.

Although it was studied, TARAS – an L-band communications system based on NATO's Link 16 and used in the Gripen – was not installed in the JA 37D. This was most likely a result of the short remaining service life of the aircraft, and possibly because it was more applicable to the national solution rather than multi-national operations. Nevertheless, the JA 37D retains excellent datalink communications, allowing it to exchange data with other fighters, ground control centres and the S 100B Argus AEW aircraft.

Colour display

For the pilot, the most notable difference in the 'full-spec' JA 37D is the TI 237 multi-function colour screen, also known as the Colour Tactical Indicator, which was introduced in 2001 and provides a large tactical situation display on the right-hand side of the instrument panel. The screen measures 6 x 8 in (15.2 x 20.3 cm), has a resolution of 120 dpi and can display colours in up to seven layers. The head-down radar display screen in the centre is retained, allowing the pilot to easily work the radar manually if required. Adding the CTI required that some engine instruments had to be moved to free up space. All pilots agree that the CTI provides a huge step up in situational awareness, and gives the Mod D's cockpit the feel and functionality of the very latest fighters. "You don't realise how much you miss it till it's taken away," enthused one vastly experienced test pilot. Another reported occasional confusion when demonstrating the JA 37D's cockpit to visiting foreign pilots who were unfamiliar with Sweden's fighters – "This

F4 Jämtlands Flygflottilj

Like most Flygvapen wings, F4 is named for the Landskap (region) in which it is located. The airfield lies on the island of Frösön, in the Störsjon lake, close to the town of Östersund which is the regional centre. A civil terminal serves the city.

When Kungliga Svenska Flygvapnet (the Royal Swedish Air Force) was formed from various army and navy aircraft-operating units by a Riksdag (parliament) bill on 2 July 1925, the new service was to consist of four operational air corps (Flygkåren). The fourth of these was to be established at Frösön with reconnaissance for the army as its primary purpose, and with air defence as a secondary role. It was formed the following year, its early equipment including the CVM S 1 (S 21/25), Junkers Trp 2A (W 34) and Fokker S 6 (C.V), some of which operated on floats in the summer and skis in the winter. A 1936 reorganisation of the air force introduced the wing system, and the 4:e Flygkåren was reborn as F4, a light bomber unit. During World War II it flew the ASJA/CVM/Götaverken B 4 (Hawker Hart), Saab B 5 (Northrop 8A-1) and the Saab B 17.

In 1947 F4 was re-roled as a fighter unit and received J 26s (NA Mustangs). It also operated an ambulance flight equipped with the Noorduyn Tp 78 (Norseman). In 1952 jets arrived in the shape of the J 28 (DH Vampire), followed in 1956 by the indigenous swept-wing J 29E/F. The 'Barrel' served until replaced by the J 32B Lansen in 1966. F4's 1 Division converted to the J 35D Draken in 1969, but it was not until 1974 that 2 Division also adopted the Mach 2 fighter.

The Viggen arrived at Frösön in August 1983 when F4 became the fourth wing to convert to the JA 37. 1 Division was the first to convert, being declared operational in July 1984. In April 1985 2 Division also achieved operational status, many of its Drakens eventually going to Austria. The Jakt-Viggens were administered by the Luftförsvarssektor Nedre Norrlands (Lower Norrland Air Defence Sector), whose HQ was also at Frösön. In 2004 the wing was undergoing conversion to the Gripen, with Viggen operations due to cease in December.

These are the badges of F4's two squadrons: 1 Div./David röd (left) and 2 Div./David blå (right). The F4 wing crest consists of an elk, a wolf and a bird.

is the Viggen, right? Are you sure it's not the Gripen?" was a common response. Experience with the cockpit was fed directly into the JAS 39 development programme. In fact, the colour screen was 'exported' into the Gripen, being installed in Batch 3 aircraft.

The Viggen's cockpit also gained a new look through the adoption of a light grey background to the instrument panel, giving a much brighter 'feel'. Pilots report that the JA 37D Viggen has a "very good cockpit, very ease to use. The buttons are big and easy to see. Everything is in the right place."

To cater for inertial system failure, the JA 37D employs a function called Synthetic Attitude Heading Reference System. SAHRS is based mainly on the normal flight control system sensors (gyros and accelerometers) and works by implementing a strap-down inertial algorithm in the FCS. The main driver for the SAHRS function in the JA 37 was to have a better flight data back-up system than possible with 'iron' instruments for when the inertial system fails. The generation of inertial systems based on gimbals are not very reliable, and before SAHRS was implemented the instrumentation would 'fall' all the way down to standby instruments at INS failure. A version of SAHRS was incorporated into the Gripen which, with limited space available in the three-screen cockpit, had to dispense with back-up instruments altogether.

A key feature of the JA 37D is its ability to carry the Ericsson Saab U95 ECM pod, as also used by the SK 37E and described more fully later. In the JA 37D installation the pod occupies pylon V7H (starboard centre wing pylon). Thought was given to housing the jamming equipment internally, but there was no room for it without decreasing fuel capacity, and the modification would have been expensive. As it is, the pod is aerodynamically clean and affects performance only slightly, but it does take up one weapons pylon.

In tactical terms, the U95 decreases the effective range of enemy fighter radars. Flygvapen planners acknowledged that the 'old' JA 37/Rb 71 Sky Flash combination was 'outgunned' by some newer fighter systems, which could lock on and launch missiles before they came within the lethal range of the Viggen. Using jamming to degrade the performance of the opposing radars brought the opposition well within the Viggen's lethal zone, restoring the JA 37's effectiveness in the air defence role. U95 remains a very valuable tool, but it was especially so when the Jakt-Viggen force was waiting for the AMRAAM, which would provide it with a long-range capability of its own.

Introducing AMRAAM

In January 1998 Sweden became the first country outside NATO to receive the AIM-120 AMRAAM, and its acquisition followed the abandonment of the Active Sky Flash programme. This had begun as the Rb 71A, and as its name suggested, it combined the Sky Flash body with an active radar seeker. After abandonment of the Rb 71A, Saab collaborated with BAe Dynamics (today part of MBDA) on the Rb 73. This was subsequently overtaken by the Rb 91 programme. Although this did not bear fruit, the experience gained was put to good use in the current Meteor programme for the Gripen, in which Saab and MBDA are major partners.

Designated Rb 99 in Swedish service, the Hughes (now Raytheon) AIM-120B AMRAAM introduced a quantum leap in air defence capability and tactical options. Its active

The highly distinctive canard delta shape of the Viggen was driven by the need to satisfy the conflicting requirements of high performance and short-field capability, the foreplanes contributing a significant amount of lift at low speed. Although the JA 37 has a digital flight control system, it is aerodynamically stable. It thus shares the same drawback as other conventional delta-winged fighters: a high energy bleed-off when manoeuvring hard. This underside view highlights the aircraft's six missile-carrying pylons, the position of the underside slit for the thrust-reverser, and the small airbrakes. Since the JA 37 Mod B the airbrakes have been inoperative.

Left: Two JA 37D(I)s cruise over the Norrbottens region in the north of Sweden. For much of the year the snow coverage and frozen lakes add a new hazard to low-level flying, as it is more difficult to discern terrain contours. Flygvapen pilots are acknowledged as among the world's best at flying at very low level over such difficult terrain.

War-base operations have long been the cornerstone of Swedish defence and are still practised, despite the ending of the Cold War which took away much of the reasoning behind the dispersal concept.

radar seeker allows the launch aircraft to turn away once the missile has acquired the target itself. The modes in which it can be used are varied, and can be exploited to the full by the 'full-spec' Mod D Viggen. AMRAAM employment is particularly suitable to the datalink-driven beyond-visual-range air defence operations for which Flygvapnet has become renowned.

While the Jakt-Viggen's air defence repertoire expanded, the secondary attack function – provided by rocket pods – was quietly dropped, although ground strafing with the KCA cannon continued to be practised.

Mod D development

As with previous Viggen modifications, the development of the JA 37D followed an iterative process, with functionality enhancements being matched by new software editions. The first JA 37D (serial 37347), with Edition 34:01 software, flew from the Försökcentralen on 4 June 1996 with Mj. Jerry Pousette at the controls. Following the end of production in 1990, the focus of Viggen testing was shifted away from Saab's plant at Linköping's civil airfield 'across town' to FMV:Prov at Malmen. The first development task was to recreate the old functions of the JA 37 Mod C – albeit with a fictitious AMRAAM capability – with the new avionics system, a task achieved by mid-1998. By mid-1999 a real AMRAAM capability had been incorporated, although initially this was a basic integration. Subsequent software editions added more functions, allowing the missile to be fired in varying modes.

Flying the JA 37

Berndt Weimer was involved with the JA 37 from the earliest days of testing, and went on to command the first operational JA 37 squadron – 2 Div./F13 – during its conversion period. He is also a Viggen and Gripen test pilot, and is better qualified than most to talk about the aircraft from the pilot's point of view.

On QRA duty we had a 60-second alert requirement. We had to be airborne in less than a minute from receiving orders to launch. Obviously that meant a cockpit alert, but even on the ground we were connected into the air defence network through the datalink as a passive member of the net. With power on and the link established, we could see the air picture, relayed to us by the air operations centre. When the call came, we already had a very good idea of what the situation was. This was very different to the Draken days when we would launch before knowing what was really going on. We usually used ground support equipment to provide power, but the aircraft could sit with the engine running at idle for extended periods. It used very little fuel.

Engine start normally takes about 25 seconds – it's a big engine – but we could then taxi straight to the runway and take off. In a standard configuration with two or four missiles, plus a drop tank, lift-off was normally achieved after about 1000 m [3,280 ft], the aircraft rotating at 290 km/h [180 mph]. The normal take-off weight was around 17 tonnes, with the maximum set at 19.2 tonnes. The length of the take-off run was largely governed by the temperature – on a really cold winter's day up north we could get airborne in as little as 300 m [985 ft]. The radar would already be in standby mode and it was switched on straight after take-off. Almost as soon as we were airborne we were ready to fight.

Against a typical Soviet attack package, we would aim to fly out supersonically at low level – around 300 m – and attack from head-on in a four-ship. The approach would be 'silent', using the RWR and cues from the ground radars before we attacked. We had separate datalinks for GCI control and between the four-ship. After flying through the enemy formation, we would aim to fly a 'lazy eight' and re-attack any survivors from behind. At this point we would be in Zone 3 afterburner and travelling so fast that, as we flew back through the formation, we would have sufficient overtake to be safe from any missiles.

Although we expected our serious 'trade' to come from the East, for a time on F13 we had a choice practice target twice a week in the shape of the SR-71 Blackbird – and it was free of charge! It would approach at Mach 3.2, but would have to slow to around Mach 2.3 to make a tight turn near our usual exercise area around Gotland. Using GCI warning, we were regularly in the air and climbing to meet it. We got very good at getting near it, with about 10 seconds available to get lock on and fire. The missile simulations showed it was quite possible to shoot it down. The Sky Flash did not work too well at those altitudes, but the Sidewinder would have been excellent since the SR-71 was so hot.

Our friends on the recce squadron [1 Div./F13] went one better, or so we thought. They were probably a bit fed up with the 'fighter boys' and their SR-71 tales. One day they sent us some photos taken by a Viggen – looking down on the top sides of the SR-71. That set us back a bit, but it was soon revealed that the 'Blackbird' had suffered a technical problem, and had been forced to descend.

When manoeuvring the Viggen's initial turn rate is excellent, but it loses energy fast and the sustained turning rate is poor. In a 180° turn at maximum *g* the aircraft goes from around Mach 0.9 to 300 km/h [186 mph]. However, in my mind the initial rate is much more important in air combat – the ability to

Right and below: Jakt-Viggens from F4 take off from a road during war base exercises. In time of tension Flygvapen units leave their peacetime training bases and deploy to a number of war bases. These usually consist of a central bare-base airfield (often a civil field or former military base) with a number of outlying strips – either further airfields or stretches of highway. Many of Sweden's roads have strengthened sections and hard-standings for flying operations

Right: A JA 37D(I) from F4 lands on a road which runs along a narrow strip between the trees. This type of operation demands short-field performance and stable approach characteristics, which were key criteria of the Viggen design from the outset. The thrust-reverser is critical to the short-field requirement, in turn requiring a special landing technique. In the early days of the Viggen the thrust-reverser and nosewheel steering could cause instability on the runway, resulting in at least two accidents, including one in which the no. 4 prototype skidded and turned over on the side of Linköping's runway. Changes in the systems cured the problem. Viggens are equipped to use the mobile Tactical Instrument Landing System (TILS), akin to the US Navy's carrier approach system. This allows precise landings even at dispersed bases.

A pair of fighter Viggens launches from Luleå. During the mainwheel leg retraction sequence the strut is shortened to reduce the amount of space needed to stow the landing gear. For quick reaction in the interceptor role the Jakt-Viggen has a rapid-aligning INS which cuts the time needed to get all systems 'on line' to less than a minute.

During the course of Mod D development four Rb 99s were fired at FMV:Prov's RFN (Robotförsöksplats Norrland) facility at Vidsel, the largest overland test range in Europe. Of these only one was live, the others having telemetry equipment installed. One Rb 71 Sky Flash was also fired to validate the weapon on the aircraft, although it is rarely used on the JA 37D. As well as launches, the test team also employed representative missile rounds which cannot be fired, but which replicate the functions of the weapon right up to the moment of launch.

A total of 34 JA 37Ds was produced, two converted by Saab/FMV at Malmen, and 32 by Flygvapen servicing units. The aircraft entered service in 2001 and served with F4, F16, F17 and F21. Today the last active aircraft are concentrated in 2 Div./F4 at Östersund-Frösön, from where they will operate until December 2004.

point the nose rapidly. Handling is generally good, but it took a lot of work to get the transonic handling right. Early aircraft suffered from quite pronounced pitch changes going through the transonic regime, but the engineers worked with the flight control system over time and made the handling much better.

Compared to the 'first-generation' Viggens, the JA 37 has much better Alpha performance. The older aircraft, with the RM8A engine and two fan stages, are limited to 18° Alpha, whereas the JA 37 is service limited to 23°. It is actually capable of 31 or 32°, and I have flown it there in tests. This higher Alpha capability is largely a function of the RM8B engine, which has three fan stages and better surge tolerance. The Viggen is difficult to superstall [a problem which greatly afflicted the Draken], and there is lots of warning. The elevons have much more authority and recovery from a potential superstall situation is much easier. Unlike with the Draken, there is no need to conduct dedicated superstall training on the Viggen.

From the systems point of view, the JA 37 has an excellent man-machine interface, and in some respects it is easier to operate in the air defence role than the Gripen. The radar is very easy to work, especially in a jamming environment. Retaining the radar display screen allows an experienced operator to rapidly control the radar manually, rather than automatically via the tactical situation display which does not present first-generation radar data. The Viggen's RCS [radar cross-section] and visual signature are large, but its IR [heat] signature is relatively low, a by-product of fitting the thrust-reverser assembly.

The HUD [head-up display] was a big step forward at the time. Originally the engineers proposed a single screen which moved forward and aft to cater for different flight regimes, but they settled on a double glass unit. The HUD is very good in the fight, but it is small.

Bringing the Viggen home

There are two ways to land a Viggen. The standard method for a full-size runway employs an approach at about 12° Alpha (depending on weight) and a flared touchdown followed by aerodynamic braking, with the nosewheel held high off the runway so that the wing forms a large airbrake. The short-field landing, an integral part of the Viggen's ability to operate from dispersed locations, is far more demanding and needs good piloting skills. Approach is flown at a steeper angle – 15.5° – usually flown manually but less experienced pilots can use the automatic speed control system. The thrust-reverser is pre-armed so that it is deployed when the nosewheel touches.

As soon as the mainwheels touch the runway the stick is pushed forward so that the thrust-reverser is activated on nosewheel contact. However, with the stick forward and reverse thrust applied, the effect is to tip the aircraft's weight too much on to the nosewheel, lifting the mainwheels up. To counter this, as soon as the reverse thrust is applied, the pilot must pull back the stick to push the weight back on to the mainwheels – but not too much, as you don't want to lift the nosewheel. This balancing act requires good technique, especially in a crosswind or on ice when the aircraft is even more directionally unstable. All Viggen pilots are well versed in the use of this technique, which has proved safe even on icy or wet runways.

Another thing to watch for is over-braking – if the wheelbrakes are used too hard they can catch fire, unlike the brakes fitted to more modern aircraft like the Gripen. Using the short-field technique, the Viggen typically pulls up in about 400-450 m [1,310-1,475 ft], again depending on weight and temperature.

By contrast, the Gripen is a much easier aircraft to land (particularly on ice), a function of its lighter weight and, most importantly, its advanced flight control system which automatically flares the aircraft and corrects for crosswind. One of the beauties of the JAS 39 is that it makes all pilots good pilots.

The rear mainwheels have just touched the runway as a JA 37D(I) lands at Luleå. The centreline 1400-litre (308-Imp gal) tank is nearly always carried. Note the deflection of the flaps on the canard foreplanes.

Stepping into the new world

Until the early 1990s the cornerstone of Sweden's defence posture was a strong armed neutrality. While relations with their Scandinavian neighbours and the rest of western Europe were good, the Swedes kept a respectful distance from them in military matters and remained outside any formal defence pacts. Their borders and airspace were regularly policed against all intruders, as they remain to this day. However, from the end of World War II Sweden's defences generally looked east – to the Soviet Union, and south – to the Warsaw Pact Baltic nations. This led to a general alignment towards the West, and some high-level tacit agreements, such as that which allowed Britain's V-bombers to begin their low-level attack runs on the Soviet Union while overflying southern Sweden. The nation also routinely looked to the UK, US and France for the military equipment it could not produce itself.

The ramifications of the break-up of the Soviet Union, and the new world which emerged thereafter, were felt as keenly in Sweden as they were in other parts of the world. The nation's semi-isolationist stance was no longer valid – it was time for Sweden to 'come in from the cold' and formalise to some degree the unspoken alignment that had existed in the days of the Cold War. A major step was joining NATO's Partnership for Peace (PfP) programme, and by the end of the 1990s Flygvapnet was keen to participate in peacekeeping operations, just as it had done more than three decades previously when J 29 fighters and S 29 reconnaissance aircraft had operated in the Congo under UN control. Another important factor in the softening of Sweden's 'neutrality' was the need to maintain its defence industry through increased exports as domestic contracts were cut as a result of vastly reduced threat levels.

For Flygvapnet, the 1990s ushered in a period of change and challenge. "In the old days," opined a Viggen pilot, "the operational doctrine was simple. Everything was black and white – if it wasn't Swedish, shoot it down. Now things are much more grey." As the Russian threat all but vanished overnight, new operational procedures were put in place to reflect the reduced tension, especially when intercepting foreign aircraft. But, as the air force stood at the beginning of the decade, it would have found any international peace support operations (PSO) involving tactical aircraft difficult to undertake.

Decades in the isolation of neutrality had meant that, although no less capable, Swedish forces had evolved in different ways to those of NATO. Operational procedures, even for such relatively simple functions as joining a pattern and landing, were different. Swedish aircraft only carried one radio as they and ground control were always on the same frequency, whereas NATO aircraft had two. In many instances the navigation equipment could not handle co-ordinates outside the area of Sweden and its immediate surrounds – this did not reflect any reduction in performance, but in the days of a purely national defence there

Above: This AJSF 37 carries the standard defensive equipment of a KB dispenser and U22/A jamming pod. As well as their international role, the SWAFRAP AJSF 37s maintained a national commitment and took part in national exercises. For 1 Div./F21 the main war base exercise location is the civil airfield at Jokkmokk.

The rapid conversion required to get the AJSF 37s ready for the SWAFRAP commitment included a smart new paint scheme. Externally, the main physical difference was the addition of a swept-back blade aerial on the spine for the second radio, which was necessary for NATO operations in which communications with air and ground are conducted on separate frequencies.

was simply no need to cater for operations far from home. To undertake multi-national PSO missions, some modifications were clearly needed, to both the aircraft and the standard operating procedures. At the forefront of the 'interoperable' push was the Viggen.

SWAFRAP

On 21 June 2000 the Swedish government reached a decision that it would form a rapid reaction force known as SWAFRAP (SWedish Air Force RAPid reaction). The force was to be ready to deploy anywhere in Europe or its proximities within 30 days' notice, under the PfP umbrella. As far as Flygvapnet was concerned, this required the establishment of two units, one equipped with the Tp 84 (C-130) Hercules, and another with AJSF 37 'recce' Viggens. The SWAFRAP AJS 37 mission was entrusted to F21 at Luleå, whose wing commander also wore the 'hat' of SWAFRAP AJS 37 commander.

For the reasons given above, the AJSF 37s in service with F21's 1 Division were not equipped for multi-national operations, and required a hasty modification to get them ready. The main elements were the installation of GPS and a second radio, together with changes to the navigation system to allow operation in 'Bullseye' mode (explained later). The small Saab engineering team was given a matter of weeks to prepare the mod and to get it into an aircraft. Trials were successful, and around 10 AJSF 37s were modified in time to undergo an evaluation exercise in Sweden in the autumn of 2000. This validated their capabilities, and SWAFRAP AJS 37 was declared ready for deployment on the scheduled date of 1 January 2001.

With Sweden having not undertaken any multi-national tactical operations since the early 1960s, SWAFRAP AJS 37 called on some help, notably from the RAF's No. 41 Squadron, which flies Jaguars in the reconnaissance role from Coningsby. The Jaguar pilots had extensive experience of multi-national operations, and were of great assistance to the Swedes from an operational point of view. Drawing on this experience allowed the new Viggen unit

***Left and below:** SWAFRAP AJS 37's first major deployment was to the NATO PfP Exercise Co-operative Key, held in Bulgaria in 2001. Thanks to their rapid build-up of knowledge in combined air operations, AJSF 37 personnel have been instrumental in the establishment in Sweden of a new training/simulation facility which provides instruction in the command and control of large multi-national packages.*

Aircraft 37347 was one of the first two JA 37D(I) conversions, and was given this scheme while undergoing tests with the FMV at Malmen. On the tail it carries the legend 'Ed 36' to signify the software edition that covered the addition of interoperable features to the 'baseline' JA 37D (edition 34).

The majority of the JA 37D(I) conversions were performed in Flygvapnet's own workshops (Flygverkstad). Unlike on the SWAFRAP AJSF 37s, which adopted English-language warnings, the external markings of the JA 37D(I) remained in Swedish ('Fara' = danger).

to adapt its capabilities to the NATO style of conducting operations.

Resplendent in a freshly applied two-tone grey low-visibility scheme, the AJSF 37s participated in a number of PfP exercises, including Co-operative Key in Bulgaria, but were not called upon for any operational deployments. On 30 September 2003 SWAFRAP AJS 37 stood down, leaving a gap in Sweden's commitments to NATO PfP until 1 January 2004, when a Gripen-equipped squadron of F17 at Ronneby took up the SWAFRAP role, deploying for its first overseas multi-national exercise (to Bardufoss in Norway) in April 2004.

JA 37D(I) – interoperable fighter

Before the SWAFRAP Viggen commitment ended, it was felt desirable to have an organic fighter capability to provide protection for the 'recce' Viggens. Accordingly, in 2002 a number of JA 37Ds were added to 1 Div./F21 (F21 had previously operated two divisions of JA 37s, but one of these units was disbanded and the other converted to the Gripen). The requirement for the fighter Viggen to be interoperable with other air forces had been one of the original drivers behind the Mod D programme, but the realisation of this came in the second step of Mod D development, resulting in the JA 37D(I) – where 'I' stood for interoperable (also sometimes written JA 37Di).

Again, the basis of this was the addition of a second radio (FR31 analogue speech) and GPS, but there were other important changes. The navigation system was modified so that it could handle position co-ordinates on a global basis, and also to work in a number of modes that were commonly in use by other air forces, such as 'Bullseye' operations. These use a pre-specified point on the map (usually changed daily) as a 'Bullseye' point from which other positions are expressed as a distance/bearing. Similarly, the system was modified to handle 'Rolex' times (using an arbitrarily set time as a benchmark from which to measure other times). Earlier improvements also had application in the JA 37D(I), such as the ability to measure flight data in feet, knots and nautical miles (used by most other air forces) as opposed to the metric system used by Flygvapnet. The JA 37D(I) also introduced a change to the Data Transfer Unit used in the Gripen, which increased the recording time of tactical operations to around 90 minutes.

Software Edition 36:01 covered the first JA 37D(I) configuration, a 'prototype' of which (aircraft 37415) first flew at Malmen on 30 August 1999 with the FMV's Mj. Jerry Pousette at the controls. A further software upgrade was introduced by Edition 36:02, first flown by Pousette in aircraft 37347 on 25 June 2001. The final software edition to enter service was 36:03, which was first flown by Mj. Magnus Johansson on 16 December 2003 and which covered some small fixes to 36:02. To complete the Viggen modification story, Edition 36:04 – at the time of writing almost certainly the last software iteration – was first flown by Mj. Nils Lundqvist on 10 March 2004 in aircraft 37326. This remains a development programme only, and is not scheduled for deployment with the operational fleet.

Viggen jammers

In the mid-1990s, with upgrade programmes for all the combat versions of the Viggen – AJ, JA, SF and SH – in full

Although better known for their electronic warfare duties, the 10 SK 37E upgraded two-seaters also had an important training role until Viggen pilot conversion ended in 2003. While Saab's avionics team worked on bringing the SK 37's systems up to the same standard as those of the Mod 11/12 AJS 37, the EW suite was developed independently, drawing on a mixture of equipment used by the JA 37D, AJS 37 and J 32E ECM Lansen.

swing, attention also turned to the upgrade of the small number of SK 37 trainers. The SK 37E aircraft came about as the result of two seemingly exclusive requirements: firstly, the provision of a trainer which was more closely related to the upgraded mission aircraft and, secondly, a replacement for the Saab J 32E Lansen 'electronic aggressor' machines, which were retired from service with F16M at Malmen on 1 July 1997.

To allow the SK 37E to perform both of its intended roles Saab came up with an ingenious scheme in which the rear cockpit could be reconfigured to either trainer or EW configuration, simply by lifting out one instrument panel and replacing it with another. The 'conversion' could be accomplished at unit level in about two days. The 'trainer' instrument panel is essentially a mirror of that of the front cockpit, although some EW equipment is permanently situated in the right-hand side console. The EW panel has controls for the jammers and rudimentary flight instruments. A key addition was an FR31 radio, to allow the electronic warfare operators to converse on their own frequency. A large swept blade aerial on the aircraft's spine denoted its presence.

Saab forged ahead with the engineering programme and began work on a 'prototype' conversion to aircraft 37814. The first task was to provide an updated trainer with avionics similar to those of the AJS 37, including GPS and Mil Std 1553 databus. These avionics needed testing before any electronic warfare equipment could be added, so development of the EW gear proceeded in parallel with clearance of the basic aircraft. The SK 37E's first flight was undertaken by Mj. Sievert Ohlander at Malmen on 25 June 1998. Clearance of the EW systems was mainly entrusted to the specialists in that role and to the first operational unit. Saab undertook the first conversion, but the remaining nine aircraft were converted by the air force in F21's workshops at Luleå between 1998 and 2000, leaving four unmodified SK 37s to continue in a pure training role until they were retired.

From a hardware point of view, the most difficult aspect was creating an electronic 'window' in the nosecone for the nose-mounted jammer. SK 37Es feature a prominent line on the nose to indicate the position of the dielectric material (SK 37 trainers were not fitted with radar). Software integration was less problematic, since the EW equipment runs on a discrete system, and does not interface with the aircraft's own avionics. The same basic aircraft software is used for either mission. While the aircraft were in for conversion, the opportunity was taken to replace the worn-out intakes.

EW systems

Mission equipment for the electronic warfare role comprises a series of jammers, mechanical countermeasures and an RWR system, together with associated controls and displays. The radar warning receiver is the APP 27, taken from retired AJ 37s with antennas installed in the tail-cone and leading-edge bullet fairings. This provides a threat direction display, together with an aural warning. In the nose is a G24 jammer, which gives frequency jamming and noise jamming in the C-band (5.1 to 5.9 GHz), for use against surveillance radars, and S-band (2.7 to 3.2 GHz). The G24 was taken from the J 32E and provides coverage in the frontal sector in a cone of roughly 90°.

On one of the wing pylons the SK 37E can carry a U22/A deception jamming pod, also taken from the AJS 37 Viggen. It responds to signals according to a threat library, and jams across a frontal sector of about 120°. The main podded system, however, is the U95 (as used in the JA 37). Taken together with the GMA rear cockpit control and display unit, the U95 is known as the AXEL system when installed in the SK 37E. The highly advanced pod has three antennas to provide jamming across 120° sectors to the front, left and right, the three combining to give 360° coverage. There is also a high-gain antenna facing forward which can give high-power jamming across a 20° sector. The system works in the X-band (6.8 to 10.5 GHz) and is primarily used

An SK 37E leads a pair of JAS 39As. The talents of the Stör-Viggen form an important element of the EW training for all three arms of the Swedish forces – army, navy and air force. Consequently, the SK 37E is in high demand and features prominently in exercises.

The SK 37E programme was based on the systems architecture of the Mod 11/12 to enable the updated two-seater to be more representative of service aircraft for the training role. The EW package was grafted on in a separate sub-system, making the integration relatively simple. The front cockpit is essentially unchanged from the days before modification, the pilot having no control over the EW suite. This aircraft is one of the six left in service with F21 at Luleå.

Apart from the yellow line painted on the radome to signify the dielectric 'window' for the nose-mounted G24 jammer, the main external difference of the SK 37E compared with an unmodified trainer is the spine-mounted blade aerial for the second radio. This was provided to allow the EW operators to converse on their own frequency while the pilot retains the standard radio for regular flight-related communications.

against fighter radars. It can work in three jamming modes: noise, signal delay and Doppler. The GMA provides the operator with a display showing threats, sectors and friendlies, with ID information. A smaller display shows the jamming activity as commanded by the operator. Finally, the underwing KB pod (BOZ-100) can carry 35 kg (77 lb) of chaff and 50 flares simultaneously.

Known colloquially as the Stör- (jammer, literally interrupting) Viggen, the SK 37E was allocated to the TIS/TK-Grupp (Typingskolan/Telekrig Grupp, type conversion and electronic warfare group), part of F4 at Östersund-Frösön. The first pair was handed over on 1 September 1999 and all 10 had been received by mid-2000. By the summer of 2003, when the requirement for Viggen type conversion had ended, the SK 37Es had transferred to 1 Div./F21, where they remain today operating solely in the EW role.

Inside the Stör-Viggen

The rear cockpit of the SK 37E can be configured with either standard instruments for the training role, or the RMA electronic warfare control and display set. Today all the SK 37Es are configured for the EW role as there is no longer a training requirement. As well as controls for the jammers and mechanical countermeasures, there is a large screen showing the tactical situation fed by the RWR, and a smaller screen displaying the SK 37E's own jamming activity. Additional EW controls are located in the right-hand console which is a permanent fixture. The SK 37E retains the two hooded windows for the direct-view periscopes, which look forward through the two 'horn' fairings mounted on either side of the divide between the two cockpits.

Compared to the old Lansen, the SK 37E offers many gains and a few losses in the EW role. The Viggen is much faster, allowing it to keep up with strike aircraft when operating in the escort role, and is also much more comfortable for the operator (although offering a poorer view). The anti-fighter jamming capability is hugely increased thanks to the AXEL system, and the Viggen can use flares (the J 32E could only dispense chaff). On the minus side, the Lansen had a MERA communications jammer but this equipment was not fitted in the SK 37E, mainly due to a lack of space.

Viggen in its twilight years

At the height of its service the Viggen equipped 18 *divisioner* (squadrons) in eight *flottiljer* (wings). By the start of 2004 only three squadrons were left: two with F4 at Östersund flying JA 37s (one converted to the Gripen in the spring of 2004) and one with F21 at Luleå. This last unit, 1 Division, is destined to be the last Viggen unit and, until very recently, was unique in operating all four current versions. In a friendly 'swipe' at their Gripen-operating colleagues of 2 Div./F21, just down the taxiway at Luleå, they boast of being the first – and so far only – true 'JAS' squadron in the air force, albeit using different Viggen variants to achieve the 'JAS' effect. Nonetheless, an examination of 1 Div./F21's operations provides an excellent insight into the operations of the Viggen community as a whole during its later years, and the opportunity to review each of the roles in detail.

Based at Kallax airfield, a few kilometres south of the city of Luleå, F21 Norrbottens Flygflottilj has played host to the Viggen since 1 Spaningsflygdivisionen officially began operating the SF 37 and SH 37 on 1 July 1979. Two fighter squadrons were added in the 1980s (2 Division in 1982 and 3 Division in 1985). The 1st squadron remained solely tasked with the reconnaissance role until recently, when the EW and fighter roles were added as the Viggen force contracted.

Commanded by Captain Dick Henriksson, in the spring of 2004 1 Division had 17 pilots on the unit, including four undergoing training and a small number of reservists flying the JA 37D (which was withdrawn from the unit in June 2004). All of the squadron's pilots fly the recce/attack/EW roles, but around half are additionally trained in the fighter role. The squadron's principal focus is reconnaissance, accounting for approximately 60 per cent of the sorties generated. Around 10 per cent are attack sorties and 30 per cent are fighter missions. The EW role is performed on an as-required basis. To meet its requirements the squadron has the following aircraft available for operations: eight/nine AJSF 37s, five AJSH 37s and six SK 37Es. Until June 2004 six JA 37D(I)s, were also on strength, kept and maintained in a separate hangar from the 'first-generation' machines, with a dedicated maintenance team.

Unlike in some air forces, maintenance is the responsibility of a separate unit – 1 Kompanie/F21 – with around 50 personnel. Although it reports through a different chain of command, in reality the Kompanie and squadron work closely together. The Kompanie provides flightline support crews, service crews (specialised into different branches such as electronics, engines etc), weapons handling/storage, reconnaissance equipment support and daily checks of flight safety equipment. An important consideration in the drawdown of the Viggen fleet is the planning of maintenance and flying so that the remaining hours on the aircraft are kept level across the fleet, while staying capable of meeting the remaining demands.

Assigned to the squadron is an intelligence officer. In peacetime the officer's primary task is to support training and provide education on foreign systems. The intel officer also provides a conduit for new information between the air force tactical command's intelligence structure and the unit. Another task is setting threat scenarios for exercises and training sorties. Under operational circumstances, the intelligence officer briefs pilots on the latest information, and reports back to higher command with new findings after missions have been flown.

Tasking

In the 'Swedish' way of doing things, most of the tasking is generated by the squadron itself, with the exception of exercises. Major missions are handed down from the air force command at Uppsala, but it is left to the unit to devise its own training syllabus and procedures to meet those demands. There are no flying instructors on the unit, *per se*, with all combat-ready pilots being qualified to instruct others. This approach allows all the experienced pilots to fly with the junior pilots, and provide their input accordingly, rather than the development of a close instructor/pupil relationship. The unit is also self-evaluating, with the commander assessing whether the squadron as a whole is capable of meeting its objectives. This is slowly changing to an assessment of individual pilots, based on a common air force requirement, as practised in other air arms.

By adopting this means of tasking and training, the squadron develops an individual approach to operations, heightened by the fact that, unlike other air arms, Flygvapnet does not use a 'tour' posting system, and pilots will often fly all their careers with one squadron. The approach has enormous benefits when operating within the unit, as it is very flexible – especially in the air – and allows the squadron to concentrate on its tactical tasks. However, the approach also means that the unit may have considerably different SOPs from even other Swedish units, and it can be difficult for new pilots coming into the squadron to adapt.

The SK 37Es were first issued to F4 at Östersund, assigned to a joint training/electronic warfare group within the wing. A few unmodified SK 37s were also on the strength of this TIS/TK-Grupp. As the Viggen force contracted, the SK 37Es were transferred to F21 and the SK 37 trainers were retired. As well as jamming pods (U95 shown), the SK 37E can also carry a Saab BT-53 laser reflector pod on the outer pylon which is used in conjunction with a ground-based training system for anti-aircraft artillery to record simulated 'hits'.

F21 Norrbottens Flygflottilj

Military aviation in northern Sweden was not commonplace until 1941, when an air base corps was established at an airfield near the village of Kallax, across the Luleälven (Luleå river) a few miles to the south of Luleå town. A few transport aircraft and air ambulances were flown to serve the needs of the army garrison and local population, although the base's primary purpose was to allow the reinforcement of the Lappland area. F21 was established as a sub-wing, the last of Flygvapnet's permanent numbered wings (F22 was the unit established to fly J 29s and S 29s in the Congo). In 1949, 1 Division of F21 was formed as a reconnaissance unit, initially equipped with the Saab S 18A but subsequently operating the S 26 Mustang, J 28 Vampire and, from 1954, the Saab S 29C Tunnan. On 1 October 1961 F21 gained its first fighter equipment when 2 Division was formed by the expedient measure of moving in a squadron of J 32 Lansens from F12. In 1963 F21 was raised to full wing status.

Both squadrons traded in their equipment for Drakens in the 1960s, 1 Division acquiring S 35Es for reconnaissance in 1966 while 2 Division received J 35Ds in 1969. In 1973 a third squadron was added, 3 Division standing up to fly Saab SK 60s in the light attack role. The Viggen era began on 1 July 1979 when the reconnaissance unit officially started operations with the SF 37 and SH 37. 3 Div. received JA 37s in October 1982, declaring itself operationally ready in July 1983, while 2 Div. converted to the Jakt-Viggen from May 1985, becoming combat-ready in April 1986. One of the Jakt-Viggen squadrons, 3 Div., was disbanded in 1998 while the other converted to the Gripen in 2002, leaving 1 Div. to continue as the sole Viggen squadron at F21.

As well as its Viggen and Gripen squadrons, F21 has a Sambandsflygrupp (liaison flight) equipped with a handful of SK 60s in both two- and four-seat configurations, and a single Tp 100C (Saab 340B) for transport duties. The latter arrived in April 2004 to replace a Tp 101 (Beech King Air 200) which had been in use for some time. A military flying club flies civil-registered, ex-air force Saab SK 50 Safirs.

F21 occupies a sizeable complex on the southern side of Kallax's runway, which is the longest in Sweden. It was extended in the early 1990s to cater for an expected boom of refuelling stop-overs for aircraft flying from the US to the Far East across northern Russia. Such services failed to materialise, and the runway extension remains a matter of some controversy in the region to this day.

This is the view from the approach to Kallax's single runway – at 3350 m (10,991 ft) it is Sweden's longest (orientated 14/32). F21's sizeable complex is to the right (south) of the runway, while to the left is the small but busy civil terminal. Beyond the far threshold is the Luleå archipelago and the Gulf of Bothnia which, at a latitude of 65° 35′ N, remains frozen for several months of the year.

Nonetheless, the commercial side of Kallax is very busy with regional traffic, a function of Luleå's position as the biggest town in the north and an important industrial/mining centre.

F21's heritage is reflected by an S 29C mounted on a pole on the road into the base (left), plus an SH 37 and S 35E Draken at the main gate (above). Additional aircraft are preserved on the base.

Inside the Viggen's final lair – three SK 37Es and a pair of AJSF 37s shelter from the northern Swedish climate in the Luleå-Kallax hangar of 1 Kompanie/F21, which maintains the last squadron of first-generation Viggens. While the Kompanie 'owns' and services the aircraft, 1 Division operates them and provides the aircrew from its facilities upstairs.

With the growing accent on interoperability, the system is changing slowly to a more standardised approach. There is a desire for growing contacts with other air arms which have, hitherto, been extremely limited due to Sweden's foreign and defence policies, despite the proximity of capable, 'friendly' neighbours. Recent developments have added the Arctic Fighter Meet (exercise with Finland) and the Nordic Air Meet (Denmark/Finland/Sweden/Switzerland/Norway) to the squadron's calendar as a means of expanding its international expertise.

Training

In early 2004 the last four pilots to convert on to the Viggen were nearing combat-ready status with the squadron. Following basic/advanced training (around 250 hours in the Saab SK 60), they were the last students to go through the TIS/TK-Grupp at F4 (Östersund) before the transfer of the SK 37Es to Luleå. Type conversion comprised around eight hours in the SK 37 with an instructor, followed by solo flights in the SK 37, after which the students began flying the JA 37, learning basic tactical skills, aerobatics and two-ship formations. For final check rides it was back to the SK 37 two-seater. The TIS course was typically 50 hours. Once with the operational unit, the students began learning the recce/attack trade in depth, initially concentrating on low-level flying and navigation before a tactical dimension, including simulated threats, is added. An advantage of the base at Luleå is that the low-flying areas and ranges are virtually on the doorstep, with no precious flying hours wasted in transit.

New pilots build up to full combat-ready status after about 18 months. During this time the new pilots do take part in exercises, based on the level of their experience. As this grows, so the pilots are cleared down to lower heights and faster speeds. New pilots on the squadron generally start at around 200 m (656 ft) and Mach 0.7, and then step down in increments to 30 m (98 ft) at above Mach 1. The Flygvapen 'hard deck' is 20 m (66 ft). (An interesting feature of the Viggen is that the HUD 'cleans up' below 100 m/328 ft altitude, removing all but essential symbology to improve the pilot's visibility). Following the achievement of combat-readiness, pilots then continue to train to become qualified as pairs or four-ship leaders. In addition to training flights, there is a Viggen simulator at F7 (Såtenäs), which is occasionally used for emergency procedures training and checks.

Of course, the squadron is also fully capable of flying from austere 'war bases' – comprising a mix of other airfields and highway strips – with the ability to move out by road everything needed for continued operations in around 30-40 minutes. Although off-base operations have been a cornerstone of the Swedish defence policy for decades, the receding threat and tightened budgets mean that the chance to train in off-base operations has diminished dramatically. The squadron has not undertaken any road ops for some while, but has used the airport at Jokkmokk as an austere location.

Daily ops – briefing, planning and debrief

On flying days the squadron generally has two 'pushes'. The morning's operations begin with a mass briefing for all the pilots at 0730, covering the weather, runway state, status of diversions etc, and the nature of the morning's flying. The afternoon operations begin with a similar brief

Below right: For two years between 2002 and 2004 1 Division achieved unique status by operating all four surviving Viggen variants (left to right: SK 37E, AJSF 37, AJSH 37 and JA 37D).

1 Div./F21 styles itself as the 'Akktu Stakki' (lone wolf) squadron, and its crew room boasts a stuffed wolf. Combat-ready aircrew wear the full squadron badge (above, based on the unit's callsign Urban röd – Uniform red) and red scarves, while those still undergoing training wear an alternative badge (below) and pink scarves.

at 1230. If flying night operations, the squadron usually briefs at 1600. A Flygvapen requirement for peacetime training is a met brief at least every four hours, and the provision of military SAR cover. In Luleå's case the latter is provided by HKP 10 Super Pumas on standby at the army's helicopter base at nearby Boden (before Sweden's military helicopters were brought under a single command the air force-manned SAR detachment was at Luleå).

Mission planning

For detailed planning of a recce/attack mission, the task begins with traditional planning on a map. The main route is plotted on a 1:500,000 scale map, taking into account the targets and simulated ground-based air defence (GBAD) zones. An entry point into the low-level portion is established, with holding points, so that deconfliction with other aircraft can be achieved by time separation. Target approach details can be plotted on a 1:50,000 map if required.

With the mission planned, the pilot then moves to the PLA planning and evaluation system. This is a computer with a pre-loaded map. By either clicking on the map or typing in co-ordinates, the pilot enters the mission plan into the computer. GBAD zones can be programmed in, as well as the normal 'avoid' areas. Speeds and altitudes are specified for each leg of the route, and the aircraft version, fuel load and weapons are entered. The computer then calculates fuel burn, giving the pilot figures to verify as each of the waypoints is reached. The calculated data is loaded on to a cartridge, which is then loaded into the aircraft's own navigation system. In the JA 37D the PLA also has an added

Above: The 'interoperable' JA 37D(I) has a second radio for international operations, as denoted by the large blade antenna on the spine, and a GPS antenna.

Above: 1 Div./F21's Jakt-Viggens operated from their own hangar, with a team of support staff separate from that for the first-generation aircraft. Most of the aircraft retained the markings of their former units, either F16 or F17. Note the U95 jamming pod.

Left: Two JA 37D(I)s lift off at the start of a training sortie. During the winter months a urea-based compound is sprayed on the runway to prevent ice formation. When the Viggen rotates the afterburner plume burns off this compound, producing a signature trail of smoke and steam.

Despite the harsh climate, flying continues unabated all year, the Viggen proving easily adaptable to operations from snow- and ice-covered surfaces and in the numbing cold. In the final months of Viggen service the AJSF 37 is arguably the most important of the versions as it provides Flygvapnet with all of its tactical reconnaissance capability. That is due to change in early 2006 when the first of nine SPK 39 multi-sensor pods are to be fielded on the Gripen with the SWAFRAP unit at F17 Ronneby. The Gripen's pod uses some of the wet-film cameras taken from the Viggen, to which is added a Recon/Optical CA270 medium-altitude camera. Further sensors planned for the pod include a long-range oblique camera and infra-red linescan, while it is intended to make the imagery digital.

function of being able to display appropriate checklists on the cockpit screen at pre-specified points along the route. Using the pre-loaded data, the aircraft's central computer generates navigation symbology in the HUD. Waypoints are usually recognisable geographic features, such as a promontory on a lake, so that the pilot can visually check the navigation system and adjust accordingly if there is any drift. As with all the squadron's operational equipment, the PLA is always ready to be moved into the field if required.

On return from the mission, the aircraft's data is downloaded on to a cartridge and then reviewed using the UTB training and analysis system. This computer-based aid replays the aircraft's movements throughout the sortie, and shows radar (if fitted) displays and HUD imagery. The three-screen system allows a detailed playback of the mission. By downloading data from more than one aircraft and running it simultaneously, the UTB allows air combat to be reconstructed, viewed on a three-dimensional display. The system evaluates kill probability of simulated missile launches, and also allows the operator to run hypothetical manoeuvres to show what would have happened if the pilot had taken different options. UTB provides the squadron with a very powerful training tool, and most sorties are reviewed and debriefed in this way.

JA 37 – Jakt-Viggen

In 2002 1 Div./F21 added six JA 37D(I)s to its roster, primarily to add fighter 'teeth' to the reconnaissance unit but also to contribute to the national air defence network. All the fighter squadrons share the task of standing a QRA (Quick Reaction Alert), usually at a base in southern Sweden, closer to where airspace incursions are more likely to occur. This usually means Ronneby, in the south-east corner of the country, or Visby on the island of Gotland. QRAs are occasionally staged at Uppsala, close to the capital. There are also QRAs established for special occasions, such as NATO exercises in neighbouring Norway. NATO pilots who are unfamiliar with the area are quite likely to stray into Swedish airspace, so it is not uncommon for some fighters to deploy to Kiruna, beyond the Arctic Circle, to be on hand to guide any wanderers back to the right side of the border.

AJSF 37 camera installation

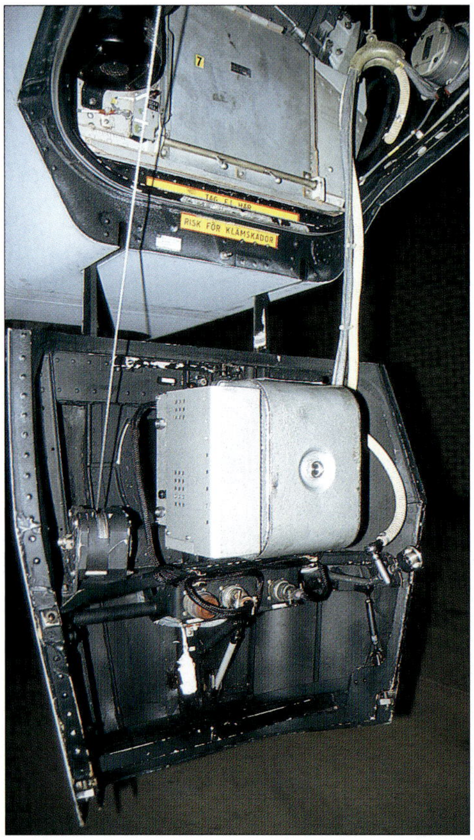

Right: The undernose fairing has a window for the forward-facing oblique 'forward-fire' SKa 24C, part of the low-altitude set.

Above: As part of the low-altitude sensor set, an SKa 24C camera with 120-mm lens is mounted on either side, peering outwards at 90° to the direction of flight and canted downwards.

Left: The lower hatch door is used as the mount for the 58-mm SKa 24 vertical wide-angle camera, which chiefly provides cueing information for the 600-mm high-altitude cameras. In the fuselage behind the door is the VKa-702 IRLS, above which are the two side-facing cameras.

Above: There are three access hatches to the recce system. The main drop-down hatch allows access to the low-altitude set and IRLS, while two upper nose hatches provide access to the film magazines of the high-altitude cameras. The latter are mounted in a split-vertical arrangement, peering through twin windows in the flat-bottomed nose section (below).

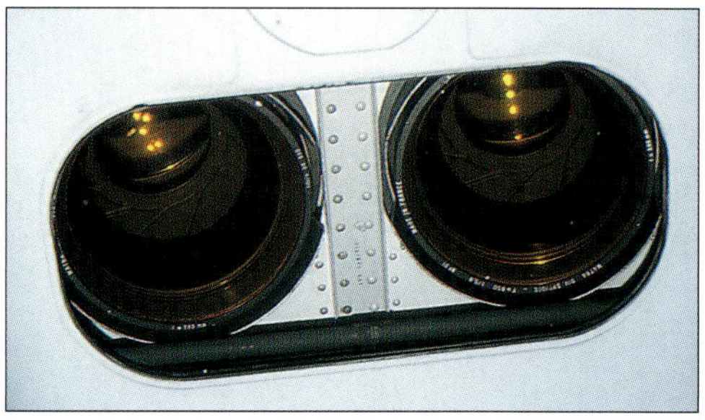

Viggen cockpit

Despite having reached the eve of its retirement, the JA 37D(I) has one of the most modern cockpits to be found in a service fighter. The massive colour tactical indicator (CTI) on the right-hand side presents the Jakt-Viggen pilot with a comprehensive display of the tactical situation, based on data fused together from the aircraft's own sensors (radar, RWR), system-generated map and pre-loaded information (waypoints, fuel estimates, latest intelligence information etc), combined with data from other assets (ground stations, other 'friendly' fighters and Argus AEW platform) which is received via the datalinks. As well as a tactical situation display, the large screen can be used to display checklists and aircraft systems data. A useful feature is the automatic presentation of checklists at pre-planned points along the aircraft's routes, the display being commanded by the aircraft's central computer based on data loaded before the flight from the PLA mission planning tool. The central head-down display (target indicator, or MI) is used for displaying first-generation data from the radar, rather than having it 'fused' with other data, as in the Gripen. Fused data is, of course, presented on the CTI.

***Above:** To accommodate the new colour screen two engine instruments were relocated to the lower left-hand side of the main panel, below the attitude indicator and replacing a weapon pylons control/display, while a third was moved to an angled position between the main and starboard panels. This view highlights the double screen of the head-up display.*

***Below:** Provided for comparison with the JA 37D, this photo shows the cockpit of the JA 37 Mod C, with small right-hand monochrome tactical display.*

***Left:** This is the cockpit of the now-retired AJS 37. The 'attack' Viggen had a central display for the nav/attack radar and a small scope for Maverick imagery. The AJSH 37's cockpit is similar, although it does not have Maverick capability. The pilot sits on a Saab RS-37 ejection seat.*

***Above:** In the AJSF 37 the radar screen of the 'attack' Viggen is replaced by the direct-view optics for the downward-looking periscope, used for sighting the high-altitude cameras. The SWAFRAP-modified aircraft have a GPS control panel in the starboard side console.*

PS-46/A development chronology

Development of the JA 37 is inextricably linked with that of the Ericsson PS-46/A radar, which has been undertaken in a series of incremental improvements throughout the Jakt-Viggen's career. The radar itself is an X-band pulse-Doppler unit with a variety of modes. As well as various air-to-air modes (range-while-search/**RWS**, priority target tracking/**PTT**, single-target tracking/**STT**, and air combat/**ACM**) which use a medium-PRF waveform with Doppler filtering, the radar can also be used to detect maritime targets (**SEA**) using a low-PRF waveform and Doppler, and for air-to-ground ranging (**AGR**). The following is a chronology of the major developments since the basic radar was first produced in 1978:

1978 (software edition D) – basic functions

1980 (software edition G) – introduced RWS thresholds, a wide-scan air combat mode and air-to-ground ranging (AGR)

1982 (software edition H) – introduced 120° two-scan bars for RWS and improved AGR

1985 (software edition J) – an improved radome increased RWS performance, while in the STT mode clustered target detection was introduced. The Rb 71 SkyFlash link was improved.

1990 (software edition K) – this significant radar iteration introduced a number of enhancements, chief among which was the ability to track two targets simultaneously in the PTT mode. For air combat the antenna scan rate was increased to 120° per second, while new Doppler filtering improved STT performance. The radar could also track the rounds from the JA 37's cannon

1992 (software edition L) – enhancements in the signal processor allowed the provision for automatic gun aiming (**AGA**), in which the radar provided commands to the aircraft's autopilot for accurate gun aiming. The ECCM (electronic counter-countermeasures) were significantly improved, and the radar could give missile warning if a target being tracked in the PTT mode fired at the JA 37

1996 (software edition M) – a new tracking and radar control computer provided a sea target mode and a look-up mode in the RWS function, the latter using low-PRF waveform without Doppler filtering. A search-while-track (**SWT**) mode was added, in which the radar can continue searching for targets while it is tracking a priority target. In SWT, track update intervals and dwell times are adaptively selected to meet the requirements of the situation, and tracked targets can be outside the search volume.

1999 (software edition N) – driven by the need to integrate the Rb 99 AMRAAM missile, the PS-46/A underwent a major transformation. The first step was to modify the radio frequency generator to provide an AMRAAM link. This modification also allowed the development of a **Passive** mode in both RWS and STT modes, in which the radar can detect and identify other radars using their sidelobes, without the PS-46/A transmitting any signal. The range of the emitting target can be estimated in two ways: in autonomous operation the JA 37's radar can calculate the range by triangulating the returns in relation to deviations in its own course. Alternatively, using two JA 37s a range and position of the emitting radar can be estimated by using the data from both aircraft and triangulating accordingly via the datalink

2001 (software edition O) – the second part of the AMRAAM integration was achieved in hardware terms by modifying the receiver, expanding the PS-46/A's capability to match that of the new missile. The primary improvement was an increase in RWS range to 120 km (74.6 miles). Passive detection was extended to the PTT mode, while in STT a new function was introduced, known as Active-Passive Mode (**APM**). This automatically combines the best features of both to enhance tracking

2003 (software edition P) – in the final PS-46/A development, APM was extended to the PTT mode

The bulky profile of the Viggen's nose allows a large antenna dish to be fitted. The radar's antenna is approximately 81 cm (32 in) in diameter.

Oerlikon KCA 30-mm cannon

The KCA cannon, or 30-mm Automatkanon (Akan) m/75 to give its official Flygvapen title, fires 360-g (12.7-oz) projectiles at a rate of 1,350 per minute. Combined with the high muzzle velocity of 1050 m (3,445 ft) per second, the projectile has a kinetic energy around eight times that of the ADEN cannon. The cannon is self-contained in a gondola (left), part of which hinges down for access, with the gun offset to starboard. The gondola incorporates a hardpoint on the centreline for the carriage of a drop tank.

Countermeasures

Above: RWR antennas are fitted to the wing leading edges and at the rear of the 'nib' fairing above the jetpipe. The standard system for first-generation Viggens is Apparat 27, although the JA 37 has a refined version.

Below: For mechanical countermeasures the KB (BOZ 100) dispenser pod is available on first-generation Viggens, also employed by the SK 37E in its EW training role. The JA 37 has BOY401 (CelsiusTech BOP/A) flare dispensers on unused weapons pylons under the wings aft of the mainwheel bays, and BOL451 chaff dispensers (BOL 300) in the rear of the missile launchers.

Above: The U95 is a highly capable jamming pod with 360° wide-sector coverage and a high-gain narrow-sector beam. It is used by both the JA 37D and SK 37E, employed primarily as an anti-fighter defence. This Stör-Viggen also carries a laser reflector pod.

Two AJSF 37s cruise at medium altitude. Both carry acquisition rounds for the Rb 74 Sidewinder, of which two are usually carried for self defence. For many years the Swedish reconnaissance mission was performed almost exclusively at very low level, which was considered to be the most survivable method when operating without the benefit of 'blanket' fighter cover. Now that Flygvapnet also trains for multi-national operations, in which such cover is likely to be provided, the recce Viggens also practise the medium-level mission. The addition of GPS in the SWAFRAP-modified aircraft allows them to fly these missions very accurately.

A QRA detachment usually involves four aircraft being on alert for two weeks. A Tp 84 Hercules usually supports the deployments. During the two weeks the Viggens are kept on constant alert with the cannon armed. If there are reasons to warrant a higher state of alert, missiles are brought to readiness near the aircraft, ready for immediate loading, although this practice is rare. Of course, in times of real tension – as happened routinely throughout the Cold War – aircraft can be held on alert fully armed.

Viggen interceptor

In its JA 37D form the Jakt-Viggen is a superb interceptor. Not only does it react, climb and fly fast, it has a powerful radar, superb datalink and excellent cockpit display which combine to give the pilot unrivalled situational awareness and the ability to present the tactical situation in a number of ways (aircraft-based, 'Bullseye'-based etc), together with information on the target aircraft and other 'friendlies'. For a standard intercept of an unknown aircraft, two Viggens would launch and position themselves to approach the unknown aircraft from behind. The pair would be spread so that the target was bracketed laterally, with the lead aircraft 1-2 km in front of the wingman. The lead aircraft would identify the unknown aircraft while being watched by the wingman behind, who is in a position to react if the 'unknown' turns 'hostile'.

In a 'hot' intercept the Viggen would use the Rb 99 as its primary armament, attacking from head-on. The AMRAAM is an active radar weapon, receiving support from the fighter's radar during the initial part of the fly-out before assuming control of the attack itself at a certain range. (At short ranges it can be fired in active mode from the start). At the moment of target 'hand-over', the missile receives the latest target position information from the aircraft's radar and proceeds to use its own radar to look at the correct position, acquire the target and fly itself to the intercept. This means of operation gives the AMRAAM shooter the opportunity to break away while the missile is still in flight, something not possible with the earlier generation of missiles such as the Rb 71 Sky Flash.

Most versatile of the remaining Viggen variants is the AJSH 37. As well as its primary sea search/ maritime attack role it can operate as an emergency fighter (with up to six Sidewinders) and as an attack platform. It also has some photo-reconnaissance capability through a forward-facing podded system, as carried here. This SKa 24 camera and 600-mm lens are mainly used for photographing ships ahead of the aircraft, after they have been located by the PS-371/A radar.

Recce pair – an AJSH 37 (left) flies with an AJSF 37. This pairing is used for overwater reconnaissance by day, the AJSH using its maritime-optimised radar to guide the camera-equipped AJSF towards the target for a low-level photo-run.

Pooling the capabilities of the Rb 99, radar and datalink also gives the JA 37D (and JAS 39) the ability to undertake 'bistatic' attacks, in which the target position data and missile support is provided by another radar. This could be a GCI radar, or the S 100B Argus AEW aircraft, or more likely another JA 37D or JAS 39. A typical scenario might involve a Viggen (or Gripen) using its radar from great distance to supply target information via the datalink to a 'shooter' aircraft operating 'silently' much nearer the target.

While this capability is indeed impressive, pilots downplay the capability as it has some limitations. The most obvious of these is the need for really good positional correlation between the 'players', in turn requiring everyone to have their navigation systems working perfectly. Any small inaccuracies at missile launch will be magnified during the fly-out so that when the target is 'handed over' to the missile, it might be outside the limits of the weapon's radar, which has a very small look angle ("like looking through a straw"). Nonetheless, the capability presents a whole new tactical dimension to air defence, made possible through the extraordinary advances made by the Swedes in datalink technology and employment.

Spanings-Foto – AJSF 37

Thanks to their SWAFRAP role, 1 Div./F21's best-known aircraft are its AJSF 37 photo-recce Viggens. The SWAFRAP commitment ended on 30 September 2003, with the recce Viggens reverting to the national tasks which they performed before and during the SWAFRAP period. Part of that commitment is the provision of an aircraft on a two-hour reconnaissance QRA.

Although its central avionics system is relatively modern, thanks to the Mod 11/12 upgrade programme of the mid-1990s, the AJSF 37 nevertheless comes from an earlier generation in terms of its operational equipment. The camera installation comprises seven sensors in three groups. The 'low-altitude' set comprises three SKa 24C cameras, each with a 120-mm (4.72-in) focal length lens. Two of these cameras peer obliquely sideways through windows on either side of the nose, while the third looks obliquely through a forward-facing window in the tip of the nose. Further back, behind the side windows, are the main sensors of the high-altitude system. These are two SKa 31 cameras in a split-vertical arrangement (to give stereoscopic interpretation capability), each with a 600-mm (23.6-in) focal length lens. They look downwards through a large rectangular window. Supporting them is a fourth SKa 24 with a 58-mm (2.28-in) lens, which looks vertically downwards to give a wide-angle image. This is employed mainly to provide orientation reference for the images from the 600-mm system. Between the two vertical camera installations is a VKa 702 infra-red linescan, which records on to film.

There is also a night photography capability using pods carried under the intakes. Three SKa 34 cameras, each with a 75-mm (2.95-in) lens, are mounted in the port-side pod (Mörkerspaningskapsel – literally, darkness reconnaissance capsule) providing a fan coverage of approximately 120° beneath the aircraft. In the rear of the pod are two windows for the flash unit, which is powered by capacitors mounted in the starboard pod (Blixtcapsel – flash capsule). Operational stores are rounded out by chaff/flare pods and jammers on the inner wing pylons, and Sidewinders for self-defence on the outer or intake stations.

The pilot has a central screen in the cockpit linked to a periscope, which looks vertically downwards through a

Above and right: 1 Div./F21 retains a handful of AJSH 37 aircraft. They are important as they represent the last attack capability in the Viggen force, and have proved useful in the tactical development and build-up of experience in employment of the RBS 15 anti-ship missile and BK 90 stand-off munitions dispenser – both important weapons for the Gripen. As well as these guided weapons, the AJSH 37 can carry a variety of unguided bombs, rockets and 30-mm ADEN gun pods.

Left: In May 2004 1 Div./F21 participated with its SK 37Es in Exercise Elite. Carrying KB and U95 pods, the Stör-Viggens operated from Lechfeld in Germany as part of a multinational exercise which involved full COMAO (combined air operations).

window covered by an excrescence on the lower starboard side. This is used for alignment of vertical photography. The cockpit has controls to switch on the various sensor sets (low-altitude, high-altitude, IRLS), and the main camera on/off button is located on the control column.

Recce Viggen pilots train for both the traditional 'fast and low' reconnaissance mission and the medium-level tactics that have flourished in recent conflicts. The medium-level mission places the Viggen outside most of the GBAD threats, but assumes almost complete air superiority. In its own national operations, Flygvapnet does not have the scale of resources available to, say, the USAF, and cannot guarantee the level of air superiority which the medium-level mission demands. Accordingly, the fast and low approach is still widely used, using extreme low altitude to avoid the attention of fighters, and speed to defeat ground-based defences. In previous times the recce mission was usually flown as a singleton, but today the aircraft usually fly in pairs.

In the low-level mission there are two main tactics employed. For a lightly defended target the low-altitude camera set is used, the 'forward-fire' camera photographing targets lying immediately in the aircraft's path while the side oblique cameras photograph targets to the side. However, this implies a direct or near overflight of the target. Against better defended targets, the AJSF 37 pilots employ the high-altitude camera system with the aircraft in a turn so that the cameras peer obliquely sideways. The cameras have an effective slant range of about 10 km (6.2 miles), providing a significant stand-off distance from the target. For better-quality photos the Viggen can get in closer to the target but stay lower. A general rule of thumb for stand-off distance/altitude is 1:10 – at 10 km out from the target the Viggen flies at 1000 m (3,280 ft), at 3 km (1.86 miles) out the aircraft only climbs to 300 m (984 ft).

An SK 37E poses at Lechfeld alongside a Tornado ECR from the resident JBG 32. Greek F-16s also flew from the base during Exercise Elite. As Europe's only fast-jet electronic warfare aircraft, the SK 37E has been highly prized during these exercises, bringing with it an unmatched ability to provide jamming support for strike packages. The Stör-Viggen can perform the co-operative SEAD role, working in conjunction with dedicated lethal SEAD assets such as the ECR.

Using the high-altitude system in a bank is an elegant solution to the problem of providing good photos while keeping the Viggen as far away from point-defence threats as possible. The impressive slant range of the cameras provides the pilot with an infinite range of altitude and stand-off distance options to match the mission's tactical considerations. These can include threat dispositions, nature of the target and its surrounds, and the type of imagery required.

On return from the mission the film cartridges are hastily removed from the aircraft, easily accessible through a drop-down hatch under the fuselage, and hatches above the two high-altitude cameras. The film can be rapidly processed and analysed in field-deployable intelligence centres. Despite the age and old technology of the system, it is very effective. NATO commanders were particularly impressed by the quality of the photos produced during PfP exercises.

AJSH 37 – Spanings-Havsövervakning

As well as its overland requirements, the recce Viggen is also used for maritime reconnaissance, for which it is part-

Most of the SK 37Es retain the traditional 'fields and meadows' four-tone splinter camouflage, but two wear a two-tone grey scheme. The value of the SK 37E in the EW training role – for land, sea and air forces – is such that the six-aircraft fleet is in constant demand. The flight hours of the jammer Viggens are now carefully husbanded to ensure that they can meet their remaining commitments.

The EW operators form their own separate group attached to 1 Div./F21, and have their own patch based on the squadron's 'Urban röd' callsign and the EWOs' internationally recognised crow symbol.

This page: In June 2004 1 Div./F21 flew its last sorties in the JA 37D(I), leaving 2 Div./F4 as the last operator of the Jakt-Viggen. Shortly before the type's withdrawal from service at Luleå, this aircraft was given special 'midnight sun' markings, with the squadron badge aft of the cockpit. The fuselage-side marking is a stylised 'JA', in the shape of a Viggen, superimposed on a 'U' (for F21). Several of 1 Div./F21's JA 37 pilots were reservists.

nered by the AJSH 37. Developed for the sea surveillance/maritime attack mission, the AJSH 37 has a radar 'tweaked' for maritime search, and in daylight hours is often used in partnership with the AJSF. Flying as part of a pair, the AJSH uses its radar to plot ship positions, and then leads the AJSF toward the desired target for a photo-run.

At night the AJSH 37 flies alone, as pairs flying at night raises serious flight safety considerations. The AJSH can carry the same night reconnaissance pods as the AJSF to provide a photo capability. For daytime photography it can carry a forward-facing SKa 24 camera with 600-mm lens in a pod under the starboard intake.

As well as its reconnaissance tasks, the AJSH 37 is important to 1 Div./F21 as it provides the squadron with its only attack capability (apart from the JA 37's cannon). Thanks to the Mod 11/12 upgrade, the maritime Viggen can carry the RBS 15F anti-ship missile and the BK 90 munitions dispenser. The latter could be used against area targets, a typical example being an amphibious landing force. Training in RBS 15F and BK 90 attacks is of great value to Flygvapnet, as these are two of the Gripen's main weapons. In fact, the only primary Gripen weapon not used by the Viggen is the Rb 75 Maverick, this capability having been lost with the retirement of the AJS 37.

SK 37E – Telekrig

1 Div./F21's fourth Viggen version is the SK 37E electronic warfare platform, with six available. These aircraft are a training asset, with no formal operational tasking. However, they are extremely capable jammers and could, in theory, be used operationally if the situation demanded it. They are available to all branches of the armed services – army, navy and air force – to provide an EW environment for anti-aircraft assets, aircraft and fighter controllers. They feature prominently in large-scale exercises, and are also used occasionally by the FMV in support of trials.

The aircraft are flown by regular squadron pilots, but the back-seat operators form their own distinct group attached to the squadron. In fact, there is only one full-time operator in the squadron, the other seven operators being drawn in from their 'day jobs' with other units when required. EW knowledge at this level is highly specialised, and new operators being brought into the group will already have extensive experience in the subject area. Training of a new operator takes around eight months.

In training exercises the SK 37Es are used in a number of ways. They usually fly in the escort jamming role, operating alongside strike aircraft against fighters or GBAD

units, or both. The Viggen is an excellent platform for this role as it is easily fast enough to keep up with its charges. They can also be employed in the deception jamming role, and in co-operative SEAD (suppression of enemy air defences) missions. In the latter they will use their jammers to degrade or 'flush out' GBAD radars so that they can be attacked by other aircraft. Depending on the nature of the exercise and the area to be covered, the SK 37E jammers work as singletons, pairs or four-ships.

Following participation in a NATO exercise in Germany, the SK 37Es have drawn strong international praise for their capabilities, as they represent Europe's only fighter-based EW aircraft. In May 2004 a detachment returned to Germany for Exercise Elite, which involved full Comao (combined air operations) with several nations.

On the back of the universal acclaim for the SK 37E, from both home and overseas, studies were undertaken into keeping the Stör-Viggens flying for several years yet, but the costs involved in maintaining such a small fleet are huge, and it appears most likely that they will bow out with the rest of the fleet. Slated for retirement in December 2005, the airborne EW capability is not due to be reinstated until around 2008, when an EW version of the JAS 39D theoretically becomes available. Over the years – through its operations with the J 32E and SK 37E – Flygvapnet has built up an impressive EW knowledge base. Rather than let this expertise 'die' with the Viggen's retirement, the intention is to maintain an EW cadre on the ground until the EW Gripen enters service.

Retirement

Today's JA 37D(I) is one of the best interceptor/BVR fighters flying. There exists a general consensus among operational and test pilots that the Jakt-Viggen is in the prime of its life, and in certain air defence scenarios is currently more capable than the JAS 39. Of course, the Gripen is a programme still in the formative years of its

Above: Beyond 2004 this scene of F4 JA 37D(I)s being serviced in the dispersal area of a war base will be a thing of the past. The JA 37 was designed to be 'turned' between missions in no more than 10 minutes by a team of one officer and six conscripts.

Above: A 2 Div./F21 JAS 39A waits at the holding point while a JA 37D(I) from 1 Div. lands at Luleå-Kallax. F21's first squadron is scheduled to be the the last of eight Gripen units to form, in January 2006, but this conversion may not take place due to the effects of the 2004 defence review.

Left: Gripen is replacing the Viggen in all its major roles. The JAS 39 has already assumed the attack role from the AJS 37, and by 2006 will have completely assumed fighter, maritime and recce duties, as represented by this mixed formation from F21.

TECHNICAL BRIEFING

HH/MH/SH-60 Seahawk

Pitted against the Boeing Vertol BV-179, Sikorsky Aircraft's model S-70B was selected as the winner of the US Navy Light Airborne Multipurpose System (LAMPS) Mk III programme in September 1977. Developed from the US Army's UH-60A utility helicopter, the Seahawk was assigned the designation SH-60B. LAMPS Mk III was an integrated system comprising both airborne and ship-based equipment, and was very different from previous projects.

Uniquely, IBM Systems Integration Division had been named as the prime contractor for the LAMPS III programme in 1974, and was responsible for overall systems integration. (IBM Systems Integration Division – then known as IBM Federal Systems – was sold to Loral Corporation in March 1994. Loral was subsequently acquired by Lockheed Martin in April 1996 and became Lockheed Martin Federal Systems. It is currently known as Lockheed Martin Systems Integration). Sikorsky Aircraft was assigned responsibility for the air vehicle, and served as the principal sub-contractor to IBM. Five YSH-60Bs were ordered and the initial example made its first flight from Sikorsky's Stratford, Connecticut, facility on 12 December 1979. The third aircraft, which was the first equipped with complete mission avionics, arrived at the Naval Air Test Center at NAS Patuxent River, Maryland, for testing in June 1980.

Rather than operating from aircraft-carriers, the SH-60B was designed as a force multiplier operating from smaller aviation-capable ships, including guided-missile frigates, destroyers, guided-missile destroyers and guided-missile cruisers. LAMPS Mk III was intended to provide the ships with the ability to locate and engage targets beyond the horizon, complementing the earlier LAMPS Mk I system and its Kaman SH-2F Seasprite helicopters. Flown by a three-man crew, comprising a pilot, co-pilot/airborne tactical officer (ATO) and a sensor operator (SO), the aircraft was equipped for a variety of missions, including anti-submarine warfare (ASW) and anti-ship surveillance and targeting (ASST). Today these missions are referred to as Undersea Warfare (USW) and Surface Warfare (SUW). For search and rescue missions the aircraft was equipped with a rescue hoist, and a 6,000-lb (2722-kg) capacity cargo hook enabled it to carry out vertical replenishment (VERTREP) duties.

The first fleet readiness squadron (FRS) was established at NAS North Island, California, on 21 January 1983 and the Navy accepted the initial production aircraft the following month. Helicopter Antisubmarine Light Squadron Four One (HSL-41) received its first Seahawk on 28 September 1983 and the SH-60B's maiden deployment began in February 1985.

***Left:* A tiger-striped SH-60B assigned to the 'Battle Cats' of HSL-43 flies over the Hotel Del Coronado near its home station at NAS North Island.**

***Right:* An HSL-42 SH-60B lifts off from the flight deck of USS Stephen W. Groves (FFG 29) while conducting counter narcotic operations in the Caribbean Sea in December 2003. The Seahawk is equipped with FLIR Contingency Kit (FCK) modifications, an AAS-44 FLIR being mounted on the starboard weapon pylon.**

Left and right: From 1993 the SH-60B gained an impressive anti-ship capability for the SUW role in the form of the Kongsberg AGM-119B Penguin missile. The weapon is seen here being fired by an HSL-51 aircraft (left) and carried by a Seahawk from HSL-43 (right). Both SH-60Bs have received the AHP upgrade which added FLIR in the nose. Note how the missile's wings fold for carriage.

IBM/Sikorsky delivered 186 SH-60Bs to the Navy by the time production ended in 1996, and 148 remain in operational service. SH-60Bs currently serve with 10 operational and two fleet readiness squadrons (FRS), as well as a single naval reserve squadron. Depending upon operational commitments, each 'fleet' squadron operates 9-15 aircraft that are normally divided into two-aircraft detachments. With fewer ships to support, the reserve squadron operates just five SH-60Bs. The detachments generally deploy aboard aviation-capable ships assigned to carrier strike (CVSG) or surface strike groups (SSG), or are tasked individually.

While retaining much commonality with the UH-60A, the SH-60B differed significantly from the utility model. The UH-60A's General Electric T700-GE-700s engines were replaced by T700-GE-401 engines, each rated at 1,690-shp (1260-kW) for an additional 136-shp (101-kW). They were tailored for marine operations. Additionally, the SH-60B carried 592-US gal/3,848-lb (2241 litres/1745 kg) of fuel that increased the radius of action over the Blackhawk, and the UH-60's armoured seats were deleted. A redesigned main rotor system featured a rotor brake and an electrically operated folding system. The tail wheel was moved forward, and a folding tail was incorporated to reduce the Seahawk's footprint on smaller flight decks. The landing gear was also simplified due to a reduction in the landing sink rate from 42 ft (12.8 m) per second to just 12 ft (3.66 m) per second. A recovery, assist, secure and traverse (RAST) system allowed the helicopters to be recovered in rough seas. Although later removed, an emergency flotation system was also incorporated.

The SH-60B's mission equipment included the Texas Instruments AN/APS-124 surface search radar, AN/ASQ-81 magnetic anomaly detector (MAD), Raytheon AN/ALQ-142 electronic support measures (ESM) and the IBM AN/UYS-1 acoustic processor. It also had provision for carrying 25 sonobuoys in pneumatic launchers on the port side of the aircraft. An onboard datalink allowed the sensor data to be transmitted back to the ship's LAMPS operator stations in 'real-time'.

Above: Crewmembers aboard USS* Fletcher *(DD 992) carry out maintenance on an HSL-45 SH-60B. The SH-60B has a towed ASQ-81 MAD 'bird'.

Left: NSH-60B BuNo. 162337, operated by HX-21 at NAS Patuxent River, prepares to land on the deck of USS Ronald Reagan *(CVN 76)* in September 2003. The Seahawk was used as an aerodynamic testbed for the airframe modifications associated with the MH-60R.

Above: An MH-60R fires an AGM-114 Hellfire during tests. At present the MH-60R is restricted to the carriage of four Hellfires on the port weapon pylon, but a modification will add another four missiles on the starboard hardpoint.

Seven additional SH-60Bs were completely remanufactured by Sikorsky as part of the first low-rate initial production (LRIP) batch. Modifications included a series of structural upgrades intended to extend their service lives, allowing them to remain in service through 2020. The airframes also underwent standard depot level maintenance (SDLM) and were equipped with the new H-60 common 'glass' cockpit. The first MH-60R test article made its initial flight at Stratford on 19 July 2001. Following Navy acceptance, the aircraft underwent additional testing in Stratford before being assigned to the NRWTS (redesignated as HX-21 on 1 May 2002). To better reflect its multi-mission capability, the aircraft was redesignated the MH-60R on 25 May 2001.

This view of one of the two NMH-60Rs highlights the asymmetric Hellfire carriage. The trials machines were initially designated YSH-60R, and were converted from SH-60Bs. Production MH-60Rs will be new-build airframes rather than conversions.

Due to the differences between the configurations, the YSH-60Rs were eventually redesignated as NMH-60Rs. Both of the NMH-60Rs are currently stored at Owego. The fourth MH-60R, and the first equipped with a fully operational weapon system, made its initial flight in Owego on 4 April 2002. VX-1 subsequently received the first of two MH-60Rs on 31 March 2003 and the squadron initiated operational testing phase IIA (OT IIA) during the summer of 2003. An Operational Evaluation (OPEVAL) is scheduled to begin in 2004.

Initially the Navy planned to convert 243 SH-60B/Fs to SH-60R configuration but the scope was eventually increased to include 273 conversions of SH-60B/F and HH-60H airframes. The programme was restructured in 2001 and the decision was made to procure new airframes. Sikorsky received a second LRIP contract covering six new-build MH-60Rs that will be delivered beginning in June 2005. The service plans to purchase 270 MH-60Rs and the newest Seahawk model will reach initial operational capability in 2005.

Although the MH-60R shares a common cockpit and -401C engines with the MH-60S, it retains the SH-60 airframe configuration. Besides structural upgrades that provide a 20,000-hour service life, the aircraft features several minor changes over the earlier Seahawk airframes. Rated for a maximum take-off gross weight of 22,500 lb (10204 kg), the MH-60R landing gear has been extended 6 in (15.24 cm) in order to support the increased weight.

Like its predecessor, the MH-60R will be flown by a three-man crew comprising the pilot, co-pilot/ATO and tactical system operator (SO). It features a completely new mission avionics and weapon systems suite. The primary components of the upgrade comprise the Telephonics AN/APS-147 multi-mode radar (MMR) and the Raytheon AN/ASQ-22 airborne low frequency sonar (ALFS). The new multi-mode radar is equipped with Long Range Search (LRS), Target Designate (TD), Periscope Detect (PD) and Inverse Synthetic Aperture (ISAR) modes, as well as air-to-ground track-while-scan (TWS) and coastal mapping modes. Likewise, the ALFS provides expanded deep and shallow-water capabilities.

Although it retains the SH-60B's pneumatic sonobuoy system, the aircraft will not be equipped with a magnetic anomaly detector (MAD). It will, however, feature the new AN/ALQ-210 electronic support measures (ESM), designed by Lockheed Martin. The system will automatically interface with the MH-60R's defensive avionics suite to provide an integrated electronic surveillance, support and radar warning capability. The defensive avionics comprise AN/AVR-2 laser detectors, AN/APR-39A radar warning receiver, AN/ALQ-144 infrared countermeasures, AN/AAR-47 missile plume detectors and AN/ALE-47 chaff/flare dispensers. A Texas Instruments AN/AAS-44 forward looking infrared (FLIR) system, and AN/UYS-2A acoustic processor are also included. Its upgraded communications suite includes AN/ARC-210 radios, and satellite communication (SATCOM) capability. The MH-60R is capa-

ble of deploying Mk 46 and Mk 50 torpedoes, and laser-guided AGM-114 Hellfire missiles.

A Block I update programme, which has already been authorised, will equip the MH-60R with a starboard weapons pylon that will increase its Hellfire capacity to eight missiles. It will also add the Mk 54 antisubmarine torpedo and the updated AN/AAS-44A FLIR. Block I will also provide the MH-60R with a removable infrared suppression shroud similar to HIRSS, as well as the tactical common data link (TCDL). A new air-to-surface missile may also be incorporated.

The USW role remains an important one for the Seahawk, especially since the withdrawal of the S-3 Viking from the mission in 1998. In the MH-60R update USW capabilities are retained and expanded, despite the deletion of the MAD 'bird', which is viewed as largely redundant in modern anti-submarine warfare. The aircraft retains its 25-tube sonobuoy launcher (above), while it has a new dunking sonar in the shape of the ASQ-22 (right, under test at AUTEC/Andros Island) and the latest Mk 54 torpedo.

CONOPS

As the MH-60S and MH-60R are deployed the Navy is undertaking a major restructuring of its rotary-wing organisation. This concept of operations (CONOPS) will see the total number of 'fleet' squadrons increased from 25 to 31 units. Beginning in March 2005 it will also cause the number of type wings to be reduced from six to just four, while the surviving wings will be renamed. The first squadrons will also be renamed in March 2005 when the HC squadrons are redesignated as Helicopter Sea Combat (HSC) squadrons. Under current plans the reorganisation will not be complete until 2015, when the last examples of the MH-60R are delivered.

Currently, each carrier air wing (CVW) deploys with a single HS squadron while surface combatants within the Carrier Strike Group (CVSG) are assigned small detachments from one or more HSL squadrons. Additionally, a logistic support ship attached to the CVSG carries a pair of VERTREP helicopters detached from an HC squadron. The current aircraft mix nominally includes four SH-60Fs, two HH-60Hs, six to eight SH-60Bs and two MH-60S helicopters for a total mix of 14-16 aircraft.

Under CONOPS the structure of the CVW will change dramatically, and the CVSG will be assigned a single Helicopter Maritime Strike Squadron (HSM) with 12 MH-60Rs. Four of these aircraft will operate from the carrier while the bulk will be detached aboard the surface combatants. An HSC squadron will divide its eight MH-60S multi-mission helicopters between the carrier and the logistic support ship, with the former operating six and the latter just two examples of the multi-mission helicopter.

Beginning in 2008, each CVSG will deploy with 18-20 helicopters and 10 HSC and 10 HSM squadrons will be dedicated to the Navy's 10 deployable CVWs. From 2012 six so-called 'expeditionary' HSC squadrons – evenly divided between the Atlantic and Pacific fleets – will detach two to four MH-60Ss with each amphibious group to provide SAR, CSAR and AMCM support. These squadrons will also detach aircraft to deployed combat support vessels. Under the CONOPS, five so-called 'expeditionary' HSM squadrons will detach MH-60Rs aboard surface combatants that are tasked to deploy individually and those assigned to surface strike groups. Although each will operate 9-10 MH-60Rs, three squadrons will support the Pacific Fleet and two will be assigned to the Atlantic Fleet. A sixth squadron will be assigned to the Naval Reserve Force.

Eventually, 38 MH-60S Knighthawks will support SAR duties at 14 Naval Air Stations, replacing the Bell HH-1N and Sikorsky UH-3H. As an interim measure, redundant SH-60Fs will be assigned to these facilities from 2005. Five fleet readiness squadrons (FRS) currently support training, but as part of the CONOPS the FRS community will be reduced to four squadrons comprising individual HSC and HSM training units within both the Atlantic and Pacific Fleets.

Tom Kaminski

Mikoyan MiG-23/27 'Flogger'

Part 1: Fighter versions

Although it was rugged, fast and powerful, the MiG-23 never matched its Western counterparts in terms of handling or systems. However, it was built in vast quantities and became the backbone of the Warsaw Pact fighter forces. It was also issued in number to the Soviet 'client' states, in whose hands it came into conflict with US and Israeli fighters.

The MiG-23's rapid development, extensive production and prompt entry into service in the early and mid-1970s can be viewed as a fairly impressive achievement by the then-mighty Soviet aircraft industry – although the aircraft was considered by many in the West to be low-technology. The 'Flogger' fighter family underwent comprehensive changes over 15 years; it was a prolonged process that called for many evolutionary rather than revolutionary improvements in airframe structure, powerplant, equipment and weapons. After an expensive, slow and often painful development, the MiG-23M of the early/mid 1970s – a crude and disappointingly unreliable 'pure' high-speed fighter-interceptor with markedly limited manoeuvrability – was transformed within a decade into the MiG-23MLD air superiority fighter, a reliable and potent weapons platform.

As the first variable-geometry (VG) wing combat aircraft to be flown in the Soviet Union, the MiG-23 was much more complex in terms of airframe construction, powerplant, equipment, mission avionics and weapons than its nimble and short-legged predecessor, the MiG-21. The MiG-23's much higher level of airframe/powerplant/systems and mission avionics sophistication, compared to its delta-wing predecessor, resulted in a three-fold increase in overall theoretical air combat effectiveness. The trade-off, however, was a huge rise in the aircraft's acquisition price, direct operating costs and maintenance requirements; as might be expected, it created huge difficulties during the type's development, testing and introduction into service with the VVS (Voyenno-Vozdushniye Sily, Soviet Air Forces).

Towards the new fighter generation

Design work on a new fighter-interceptor intended to be the successor to the MiG-21 with the VVS's Frontal Aviation (FA) arm began in accordance with a 3 December 1963 resolution of the government and the Central Committee of the Communist Party of the Soviet Union. The resolution ordered the prototype of the new-generation combat aircraft to be submitted for joint state testing in late 1965.

The Mikoyan OKB, which at the time was the principal design bureau for front-line combat aircraft in the Soviet Union, assembled a pair of teams to work in parallel on two different air superiority fighter designs: the first was conceived with short take-off and landing (STOL) characteristics using lift jets, whereas the second was planned to use the then-revolutionary variable-geometry wing design.

The STOL project was referred to within MMZ Zenith (as the Mikoyan Design Bureau was officially known before 1970) as Aircraft 23-01, and alternatively under the draft designations of MiG-23UVP and MiG-23PD. Under chief designer Vano A. Mikoyan, it took shape as a supersonic design with normal aerodynamic layout, employing a fixed delta wing and conventional all-moving tailplane. Powered by a Tumanskiy R-27-300 main turbojet rated at 51.01 kN (11,470 lb) at military power and 71.61 kN (16,102 lb) with afterburner, it also featured a pair of Kolesov RD-36-35 lift jets rated at 23.03 kN (5,178 lb) each, installed in the centre fuselage in a vertical position and used for take-off and landing. Fed by intakes on the upper fuselage, the lift jets were intended only to reduce take-off and landing speeds, and were never deemed capable of providing a true vertical take-off and landing

First of the breed – the 23-11/1 prototype was recognisably a MiG-23, but the design underwent many significant alterations to make it suitable for large-scale service. Even then, the early 'Floggers' were hampered by poor systems, and some handling and manufacturing idiosyncrasies that would have been deemed unacceptable in a contemporary Western aircraft.

Above: MiG-23s were assembled in the plant at Lukhovitzy. This garish MiG-23ML is on display as a tribute to the efforts of the factory and its workers.

Left: During the 1980s and into the early 1990s the MiG-23 lay at the heart of the defence of the East European countries. Here a Hungarian MiG-23MF cruises with wings fully forward and slats partially deployed.

Below left: Unlike other Soviet types, MiG-23s were encountered by Western aircraft infrequently. This MiG-23M, intercepted by the Swedish Air Force, carries four R-60 missiles on APU-60-2 twin launchers, in addition to the standard R-23R (AA-7 'Apex') long-range radar-guided missile.

Below: For those who appreciate brute strength and purposeful looks in their fighters, the MiG-23 represents true beauty. Although not particularly manoeuvrable by modern fighter standards, the MiG-23's noise, power and dart-like looks made it a firm favourite at air shows. Here a Hungarian pilot pulls hard during a routine (note the amount of taileron deflection), a manoeuvre that was not without its risks due to the MiG-23MF's notoriously poor high-Alpha handling characteristics.

capability. The maximum speed was projected to be between Mach 2.0 and 2.3.

Aircraft 23-01's sole prototype made its maiden flight on 3 April 1967, piloted by Mikoyan OKB test pilot Piotr Ostapenko. After the initial test flights, it became clear that the STOL concept with lift jets suffered from too many design problems ever to satisfy the requirements. Only 14 flights were recorded but results were disappointing due to notably poor stability and controllability during landing. This prompted rapid abandonment of the STOL design in favour of the more promising VG option.

Fashionable variable-geometry

Aircraft 23-11 with VG wings was eventually viewed by both the Mikoyan OKB and the VVS as promising much better performance than the technologically complex STOL design of Aircraft 23-01. Appropriately serialled 231, the first Aircraft 23-11 prototype – designated 23-11/1 – made its maiden flight on 10 June in the capable hands of Alexander Vasilievich Fedotov, the famous Mikoyan OKB chief test pilot; during this flight, the wings remained fixed in the minimum sweep position. On 9 July 1967 Fedotov took the aircraft up again, and this time moved the wings from full forward to full backward position. He was so impressed that he entered the following nota-

Technical description – MiG-23MLA 'Flogger-G'

The **fuselage** of semi-monocoque construction is made of two parts – nose and tail – joined together at frame no. 28. The nose part comprises a pressurised compartment housing the radar and other avionics boxes (extending from frame no. 1 to no. 6), cockpit and nose undercarriage leg niche (from frame no. 6 to no. 11), rear avionics compartment with gun housing below (from frame no. 11 to no. 14), air intakes and fuel tank no. 1 (from frame no. 14 to no. 18), fuel tank no. 2 used as the principal load bearing fuselage element (from frame no. 18 to no. 20), fuel tanks nos 3 and 4 (from upper part of frame no. 22 to no. 28), and finally the engine compartment (from frame no. 20 to no. 28). The tail part of the fuselage comprises afterburner, tailplane mounting and four petal-type ribbed airbrakes.

The **variable-geometry wing** is made of two parts. The fixed wing glove has a sweep of 70°. The movable wing panels feature sweep angles that are fully and continuously variable from 18.5° to 74.5°. The three main positions used in flight for which controllability and stability are fully explored are 16°, 45° and 72° (precise sweep values are 18° 40', 47° 40' and 74° 40'). The wing sweep changes are made by the SPK-1 system, controlled manually by the pilot via a small lever moving forward and aft, which is located beneath the throttle. The wings have a number of high-lift devices: full-span, four-section, trailing-edge flaps deflected at 25° for take-off and 50° for landing, leading-edge slats deflected at 20°, and two-section spoilers.

The **all-moving tailplane** is swept at 55° 40' and is mounted to frame no. 31. Deflection of the tailerons is from -28° 30' to +8° 30' symmetrically for pitch control and +/-10° differentially for roll control. A large folding **ventral fin** improves high-speed yaw stability, and when the undercarriage is down the ventral fin is stowed to port.

The **tailfin** is fastened to fuselage frames nos 29 and 31 and has a rudder mounted on a damping device with deflection of 25° left and right. At the top of the tailfin in a dielectric housing are antennas for the radio, SO-10M RWR and RSBN-6S tactical radio-navigation system. A **brake parachute** container is mounted at the fin base and the parachute itself has an area of 21 m^2 (226 sq ft).

The **landing gear** is of classic three-unit type. The nose unit retracts aft and has a double 520 x 125-mm (20 x 49-in) KT-152 wheel. The main units were designed to fit into a tight space and have a single KT-105 wheel measuring 840 x 290 mm (33 x 114 in). Disc brakes are used and brake control is pneumatic. The hydro-mechanical control system has channels for pitch, yaw, roll, speed brake activation and wing movement.

Two independent **hydraulic systems** – main and booster – have a working pressure of 210 kg/cm^2 (2,986 psi). The booster hydro system feeds the SPK-1 wing sweep system. Two independent **pneumatic systems** are used for brake control, emergency landing gear and fin retraction, canopy operation and pressurisation, brake parachute opening, wing seals, radar compartment pressurisation, and air conditioning of the avionics equipment. The main undercarriage legs are used as pressurised air tanks. AC voltages of 36V, 115V and 380V/400Hz, and DC voltage of 28V, are used in the **electrical system** which is powered by the GSR-ST-18/7KTS starter-generator.

The **powerplant** consists of one Soyuz R-35F-300 turbojet (*izdeliye* 77) with maximum dry thrust of 83.83 kN (18,849 lb) and maximum thrust with afterburning of 128.98 kN (28,800 lb). Engine start-up is provided by an onboard TS-11 turbo-starter.

Fuel is carried in three fuselage tanks and four cells in each movable wing panel, giving a total internal volume of 3000 litres (660 Imp gal). Two additional tanks under the wings holding 800 litres (176 Imp gal) can be used only at wing sweep position 16°, and one more 800-litre (176-Imp gal) drop tank can be carried under the fuselage. There is a centralised pressure refuelling system.

Communications equipment encompasses one R-862 UHF radio set plus one R-855 emergency radio installed in the pilot's survival kit that is housed in the ejection seat. **Navigation** equipment comprises the Polyot-21-23 system incorporating an RSBN-6S tactical radio aid to navigation/automatic landing system, SKV-2NM-2 gyro reference system, DV-30K and DVS-10 air data sensors, ARK-15M ADF and MRP-56 beacon receiver. Other **avionics** include the SOD-59 transponder, SRZO-2 IFF unit (interrogator-transponder), RV-5R Reper-M radar altimeter, SPO-10M Sirena-3 RWR, and the onboard set of the Lasour-SML command datalink for the Vozdukh-1M GCI system.

The KM-1M **ejection seat**, of Mikoyan's own design, enables safe ejection throughout the entire envelope at a minimum speed of 130 km/h (70 kt), including ground level. The cockpit is pressurised and air conditioned above 2000 m (6,561 ft), as is the avionics compartment.

The MiG-23MLD had a typical Soviet cockpit of its era, the 'calming' turquoise background doing little to offset the enormous workload imposed on the pilot.

The MiG-23's canopy offered little drag but also poor outside visibility. The attachment on top housed a rear-view mirror. The pilot above wears standard flight gear, while for operations above 12000 m (39,400 ft) pilots usually wore the bulky VKK-6 partial pressure suit and GSh-6 helmet (above right).

This view shows the crucial wing pivot joint, with locking pin removed and the wing panel moved away.

A brake chute was carried in a fairing above the engine nozzle. Four airbrake sections, each with two actuators, were arranged around the rear fuselage.

Looking aft, this view shows the extended centreline pylon of a MiG-23MLD, containing two six-round KDS-23 chaff/flare dispensers.

Intercepted by the Swedish Air Force over the Baltic, this MiG-23MLD operated with a regiment based at Parnü in Estonia. The unit was one of many which sent aircraft to Afghanistan, this aircraft having seen action there (hence the mission tally on the nose). Upward-firing chaff/flare dispensers were also fitted to aircraft destined for the warzone. The 'Flogger-K' carries the second-generation R-24R missiles, as denoted by the forward sweep on the missile's wing trailing edges. It also has R-60s under the fuselage.

tion in the aircraft's logbook: "A flight with wings being swept at angles ranging from 16° to 72°, for the first time. It's excellent!" During the third test flight of 23-11/1, Fedotov went supersonic, reaching Mach 1.2 with 72° of wing sweep.

The public debut of this revolutionary new Soviet combat aircraft took place during the famous Domodedovo flypast held on 9 July that year, at which both the STOL and VG prototypes were displayed in flight. By that time it was apparent that Aircraft 23-11 would become the basis for development of a high-performance fighter-interceptor to replace the MiG-21. Alexander Fedotov used the flypast – which marked the 13th flight of the first VG prototype – to good effect, demonstrating the wings moving forward and backward at speeds between 400 and 800 km/h (216 and 432 kt) while rolling the aircraft.

A total of nine VG prototypes was built. Seven of them participated in the factory flight testing which commenced in July 1967 at Mikoyan's own flight test facility at the Ministry of Aviation Industry's (Aviaprom) airfield at Zhukovskiy (also known as Ramenskoye) near Moscow.

Design breakthroughs

Five main design requirements were identified in the early/mid-1960s during the project definition phase of Mikoyan's new-generation fighter.

The first of these called for a significant low-level speed increase; the second for an improvement in pilot comfort during high-speed low-level flight; the third for a considerable increase in range; the fourth for a huge advance in take-off/landing performance which would allow the use of shorter runways; and the fifth for integration of a powerful weapons control system that would allow the much-needed use of beyond-visual-range (BVR) missiles (this was considered essential for the air superiority role in the 1970s). Both the McDonnell Douglas F-4 Phantom II and the Dassault Mirage F1 were considered to be the primary future opponents in the air superiority area.

The Aircraft 23-11 design team, led by A. A. Andreev, was given the fairly difficult task of designing a modest fighter-sized aircraft that employed the then-new VG wing concept (known also as swing or variable-sweep wings). In 1964 this approach was considered by the aviation industry and military leadership as a promising new technology that would satisfy a multitude of conflicting main design requirements for the new generation of fighter and strike aircraft planned to enter mass service in the early 1970s. The new Mikoyan fighter was initially conceived as a pure high-speed interceptor only, and would have no particular capability for high-*g* turning air combats, which were out of fashion by that time.

Aircraft 23-11's basic design concept called for a fighter-interceptor type with high, shoulder-mounted wings, and in the event it was strongly influenced by the then-revolutionary General Dynamics F-111 tactical fighter. The fuselage of semi-monocoque construction had a cigar-like shape with pointed nose and rectangular cross-section behind, with rounded angles.

The movable wing panels were hydraulically driven and manually controlled, with three main sweep angles set by the pilot using a small lever located beneath the throttle. Wing position 16° (fully spread) was used for take-

Three generations of fighter 'Floggers'

The MiG-23M was the first mass-produced variant, distinguished by its long dorsal fin.

MiG-23M of the 35th IAP in original grey scheme

MiG-23M, also of the 35th IAP, after unit moved to Germany. Disbanded in 1996

The MiG-23ML had a shorter dorsal fin and numerous other refinements which lightened its structure considerably. The MLA was similar but with improved systems.

MiG-23MLA, Jagdfliegergeschwader 9 'Heinrich Rau', East German air force

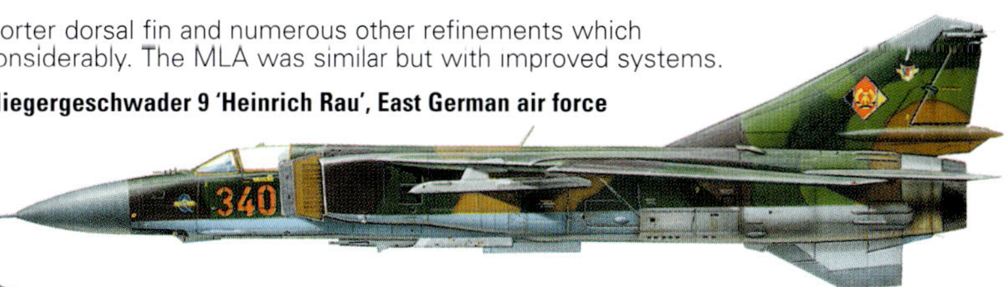

MiG-23MLA, Força Aérea Popular de Angola

The MiG-23MLD was produced by conversion of MiG-23ML/MLAs, the main external feature being a notched leading-edge root extension.

MiG-23MLD, 528th IAP, Smyrnykh air base. Legend reads 'Sakhalin'

MiG-23MLD, 120th IAP, Domna AB. Flew missions in Afghanistan in 1988/89 from Bagram and Shindand

When Pakistan ordered the F-16 in 1980, India decided that it had to urgently acquire BVR fighters. For the longer term the IAF ordered the Mirage 2000, but as an interim step 40 MiG-23MFs were bought for rapid fielding with Nos 223 and 224 Squadrons at Adampur. No. 223 converted to the MiG-29 in 1989, but No. 224 'Warlords' has continued to fly the type to the present day, although it is now used primarily for target facilities and dissimilar air combat training. Here a pair escorts an An-12 over the northern mountains.

off/landing and low-speed cruise (limited to Mach 0.7); mid-sweep of 45° was used for basic fighter manoeuvring, low-level intercepts and high-speed cruise; position 72° (wings fully aft) was used for high-speed dashes at low level and for high-level intercepts. The wings were liberally endowed with high-lift devices such as full-span four-section trailing-edge flaps, leading-edge slats and two-section spoilers, all designed in an effort to provide the STOL performance required by the VVS. There were no ailerons on the wings, in an attempt to prevent twisting the thin movable panels in high-speed flight; instead, spoilers were introduced for roll control at 16° and 45° sweep settings, operating in conjunction with the left and right tailerons (i.e., a tailplane which is used for both pitch and roll control; for the latter task, the tailerons move differentially), whereas at 72° sweep only the tailerons were used for roll control. After landing, the spoilers were also brought into play to increase drag and kill residual lift. A brake parachute further shortened the landing roll.

The tailfin was made fairly large, extending into a long dorsal fillet halfway to the cockpit. A ventral fin was added to improve directional stability at high speeds, and to prevent it scraping the runway during take-off and landing Mikoyan selected a design in which the fin hinged to a root section and folded sideways.

Difficult flight testing

In early May 1969 the first example of the initial production 'Flogger' variant, designated MiG-23S and fitted with the S-21M weapons system borrowed from the MiG-21S/SM, came off the production line at the Znamya Truda factory (the final assembly line was located in the facility at Lukhovitzy near Moscow). It took to the air for the first time on 21 May, piloted by Alexander Fedotov. Later in the same year, a small number of MiG-23Ss were handed over to the two parties which formed the MiG-23's Joint State Test Team – one based at the Aviaprom test centre at Zhukovskiy near Moscow, and the other at the NII VVS (Soviet Air Forces' flight test centre) at Akhtubinsk (also known as Vladimirovka airfield) in the southern part of what is now the Russian Federation. This stage of the MiG-23 test programme commenced in July 1969 and continued until mid-1973: a total of 11 MiG-23s (including some two-seaters) was involved in the protracted joint Aviaprom/VVS testing effort, amassing just over 1,300 sorties.

The flight test programme encountered a number of problems. For example, both Mikoyan and VVS test pilots discovered that the MiG-23S (and, later, its successor the MiG-23M) generally displayed fairly dangerous behaviour when high angles of attack were combined with a certain sideslip, and the aircraft was judged to be prone to spinning in certain circumstances. Such qualities were considered hazardous to flight safety by the time the aircraft entered service in number. In addition, the centre fuselage welding joints and the wings often developed cracks, requiring urgent structural strengthening.

Two MiG-23Ss were lost within a week during the flight test programme in September 1970. The first crashed near Zhukovskiy when the aircraft exceeded the maximum speed limit and disintegrated in the air, killing Mikoyan test pilot Mikhail Komarov; the most probable cause was pilot incapacitation due to interrupted oxygen supply, but no cause was ever confirmed. The second aircraft crashed near the airfield at Vladimirovka after entering a flat spin during low-altitude air combat; it was manoeuvring at high angle of attack (AoA) but supposedly still within design limits. The fatal combination of particular AoA and sideslip caused the departure from controlled flight, killing NII VVS

MiG-23 fighter radars

The MiG-23S variant was equipped with the **RP-22SM** pulse radar featuring a twist-Cassegrain antenna, as found in the MiG-21S. Maximum detection range approached 18 km (9.8 nm) and tracking range was up to 11 km (6 nm). The radar enabled use of the R-3R (AA-2C 'Atoll') SARH missile, which was suitable for tail chase engagements only and had a range of 0.8 to 7 km (0.43 to 3.8 nm). The R-3S (and, later, the improved R-13M) heat-seeking air-to-air missiles were useful only for clear weather visual intercepts. The MiG-23MS exported to a number of Arab states in the 1970s was equipped with a sanitised RP-22 version (from the MiG-21bis) with only one carrier frequency, a design with almost no ECM resistance (in contrast, there were three selectable frequency settings on the RP-22s installed in the MiG-23S and MiG-21bis of the VVS). The RP-22's detection range reached to 25 km (13.4 nm) and its tracking range to 17 km (9.2 nm).

This is the RP-22 radar 'borrowed' from the MiG-21bis for use in the MiG-23MS 'sanitised' version which was sold to several Middle East and North African countries. The radar of the MiG-23S was similar but more capable.

The **Sapfir-23L** was a simplified initial version of the definitive powerful pulse radar set developed for the MiG-23 by OKB-339 (NIIR-Phazotron), initial design work on which dates to 1963. The L version lacked the look-down/shoot-down (LD/SD) capability of the MiG-23's radar, so could accommodate only R-23R SARH missiles used against medium- and high-altitude targets flying above 1000 m (3,280 ft). The radar used a twist-Cassegrain antenna 800 mm (31.5 in) in diameter and had a continuous-wave target illuminator to provide SARH missile guidance in a single target track mode.

The **Sapfir-23D** (known to NATO as 'High Lark II') was the first version with LD/SD capability. It was followed in 1975 by the **Sapfir-23D-III** (*izdeliye* 323D-III) – the definitive version of the first-generation Sapfir-23 family, which was fitted to MiG-23Ms built from 1975. It was considered advanced technology for the Soviet micro-electronics industry of the time, although by Western standards it was somewhat obsolescent. The analog Sapfir-23D-III was a pulse radar working in the 3-cm wavelength (carrier frequency of 9 GHz) and was capable of LD/SD engagements utilising the rather unreliable 'envelope detection' technique, which reportedly was taken by OKB-339 designers from the AWG-10 radar of the F-4J captured in Vietnam in 1967. The radar could not make use of the much more effective pulse-Doppler technique because that required advanced digital processing technology that was unavailable in the Soviet Union at that time. The Coherence On Receive Doppler System (CORDS) of the AWG-10 was a look-down device which was effective for detecting air targets over flat terrain. However, it was of little use in mountainous terrain or urban areas with tall buildings which would generate a large number of false returns, well beyond the capability of the four dozen analog filters whose role was to suppress returns from the earth's surface and project only returns from moving objects onto the ASP-23D sight/HUD (or ASM-17ML, in the case of the MiG-23MLA/MLD).

The maximum detection range of a well-tuned Sapfir-23D-III of fighter-sized targets in look-up mode/head-on engagement approached 45 km (24 nm); in look-down mode/tail-on engagement it was between 10 km (5.4 nm) and 20 km (10.8 nm). Detection range of bomber-sized targets was 55 km (30 nm) in look-up mode/head-on and tail-on engagements, while in look-down mode/tail-on engagement it was between 15 km (8 nm) and 20 km (10.8 nm). The minimal speed of the target in the LD/SD mode was about 60 km/h (32 kt).

Reliability and maintainability problems, especially with fine-tuning the radar's AVM-23 analog computer during regular maintenance sessions, dogged the Sapfir-23's operational employment throughout the entire duration of the MiG-23M's service. For example, in the late 1970s the radars of the MiG-23M-equipped 35th IAP, stationed at Zerbst airfield in East Germany, had maximum detection ranges between 5 km (2.7 nm) and 40 km (21.6 nm), the lower figure being a result of difficulties in fine-tuning the AVM-23.

The **Sapfir-23ML** (**N003**) radar fitted to the MiG-23ML was a considerably

Proudly wearing the chevron national insignia adopted in March 1991 after Hungary had cast aside its Communist shackles, a MiG-23MF lifts off from Pápa. The badge of the 'Sámán' (shaman) squadron depicted an archer on horseback. MiG-23s were withdrawn from Magyar Légierö (Hungarian air force) use in 1997.

test pilot V. Zhukov. Not long after, an experienced VVS instructor pilot from the Lipetsk-based combat training centre was killed when his MiG-23S spun in similar circumstances; in another incident, the VVS's Moscow District Commander, General Odintzov, managed to eject at the last possible moment before his spinning MiG-23S hit the ground.

The first production-standard MiG-23Ss – with airframes cleared for only 4 *g* manoeuvring – entered service with the Lipetsk-based VVS 4th TsBPiPLS (aircrew combat training and conversion centre) and with the 979th IAP (fighter regiment) at Shtutchin, Belarus (other sources give the location as Ross). Interestingly, these moves were executed in a hurry in the mid-1970s, well in advance of completion of the MiG-23's flight test programme.

The MiG-23S was replaced on the production line by another interim variant in 1970-71. Known as the MiG-23 Edition 1971, it featured a considerably redesigned fuselage and dogtoothed Edition 2 wings increased by some 20 per cent in area (due to the increased outer-wing chord) but lacking leading-edge slats. This variant was equipped with the initial production version of the Sapfir-23L, an upgraded version of the Sapfir-23 radar set that enabled the use of the new R-23R BVR missile.

On 14 March 1972, a MiG-23 was lost during testing of the increased-area wings. The cause was traced to structural failure of the fuselage main tank (tank no. 2), which acted as the principal airframe load-bearing element, while the aircraft was manoeuvring at 7.3 *g*. (Some sources say the aircraft had already begun to disintegrate at 4.2 *g*.) The initial design load factor of the MiG-23 had been set at 8 *g*. Luckily, the Mikoyan OKB test pilot, Aviard Fastovetz, managed to eject safely from the disintegrating fighter.

Full-rate production

In 1972 the Znamya Truda facility began production of the definitive first-generation 'Flogger' – the MiG-23M. First flown in June 1972, again by Alexander Fedotov, it was equipped with the improved Sapfir-23D look-down/shoot-down radar and armed with R-23R/T BVR and R-60/R-13M WVR missiles.

Design problems plagued the new type's introduction and employment as an operationally-capable weapons platform for some time. In theory, at least, the MiG-23M was a quantum improvement over the MiG-21PFM/MiG-21SM, which was the mainstay of Frontal Aviation by that time. It came as an unpleasant surprise to the MiG-23 community when the first production MiG-23Ms were pitted against MiG-21M/SMs in practice close air combat at Lipetsk and Vladimorovka – and lost all turning engagements against its nimble predecessor. The result was the same, if not worse, when the MiG-23M was pitted against a Northrop F-5, a trophy aircraft captured in South Vietnam which by that time had been thoroughly evaluated by the NII VVS. The MiG-23 was piloted by one of the most experienced NII VVS test pilots, Vladimir Kondaurov, but lost all turning engagements despite his best efforts to manoeuvre at the maximum permitted AoA (and even beyond it), continuously selecting the best wing sweep (33° to 45°) for each turning speed in order to achieve the maximum possible instantaneous turn rate.

The new fighter-interceptor was officially commissioned into VVS FA service on 4 January 1974, and its mass introduction to the fighter regiments of the VVS's mighty 16th VA (Vozdushnaya Armiya, air army) in East Germany took place between 1974 and 1976. The huge fleet of MiG-23Ms available in East Germany (and also in Czechoslovakia, Poland and Hungary) gave the VVS a temporary quantitative advantage in air superiority in the Central European theatre. No fewer than 450 'Floggers' equipped 33 squadrons in 1976, and by 1979 their numbers had increased to at least 39 squadrons with over 500 aircraft (including

improved and lightened version. It was soon followed by the further improved **Sapfir-23MLA**, which was 170 kg (375 lb) lighter than the Sapfir-23D-III due to the use of new-generation micro-elements and other weight-reducing measures. Its maximum detection range of a typical bomber-sized target in look-up mode approached 65 km (35 nm), reduced in LD/SD mode to below 20 km (11 nm).

The MiG-23P was equipped with the **N006** radar, based on the N003. It had slightly improved LD/SD capability, which was urgently required for countering the cruise missile and low-level strike aircraft operations that gained particular importance in the late 1970s. The new radar was used in conjunction with the ASP-17ML-P sight/HUD.

The MiG-23MLD was equipped with the much-improved **Sapfir-23MLA-II** (known also as **N008**) radar. Fifty per cent of this version's black boxes differed from those in the Sapfir-23MLA and some digital memory was incorporated into the main computer. It weighed 360 kg (794 lb) and its peak emitting power was 60 kW, with an average of 1 kW. The -23MLA-II introduced a useful close air combat mode that enabled a vertical scan in a tall 'window' perpendicular to the aircraft's longitudinal axis, with a field of 45° in elevation (vertical plane) and 6° in azimuth (two-bar scan pattern). In this mode, the radar locked semi-automatically onto the first target to enter the window at distances of 0.3 to 9 km (0.16 to 5 nm), a characteristic that proved advantageous in high-*g* manoeuvres. In the close air combat mode, the N008 could send commands to 'slave' the R-60, R-60M, R-13M and R-73 seeker heads towards the acquired target. This pre-launch acquisition mode extended 12° up/down and left/right relative to the aircraft's nose (up to 45° angle of slew for the R-73).

The N008 had two new look-down/shoot-down modes, which notably expanded the N008's LD/SD capabilities. The first, known as MPKh, assisted the radar to 'see' through the heavy clutter of mountainous terrain, radio-reflective clouds, urban areas with tall buildings, and chaff clouds. The second mode, PSTz, was designed for better discrimination of targets flying as low as 50 m (164 ft); in this mode, detection/tracking range did not exceed 11 km (6 nm) against a target flying at 400 m (1,300 ft) when the interceptor maintained 1000 m (3,200) altitude.

To make these new modes possible, the radar employed 120 Doppler filter channels (compare this to the 49 used in the Sapfir-23D-III). The maximum detection range of a bomber-sized target at medium/high level was 75 km (40.5 nm). In real-world operations, the N008 proved to be moderately reliable and had a mean time between failures of around 60 hours, according to data supplied by NIIR-Phazotron.

Maximum ranges of the N008 radar

	Bomber-sized target	Fighter-sized target
Search range at high altitude	75 km (40.5 nm)	52 km (28 nm)
Tracking range at high altitude	52 km (28 nm)	39 km (21 nm)
Search range at low altitude (look-down/shoot-down mode)		
rear quarter	23 km (12.4 nm)	23 km (12.4 nm)
forward quarter	23 km (12.4 nm)	14 km (7.6 nm)
Tracking range at low altitude (look-down/shoot-down mode)		
rear quarter	23 km (12.4 nm)	15 km (8 nm)
forward quarter	23 km (12.4 nm)	9 km (4.9 nm)
Scan field – azimuth x elevation	60° x 6°	
the scan field of 60° x 6° can be steered up to 60° left and right in azimuth to expand the searched volume		
Autotrack field – azimuth x elevation	112° x -44°/+56°	
Beam width – azimuth x elevation	2.4° x 2.4°	

Seen on a test stand, this is the N003 (Sapfir-23ML) radar used in the MiG-23ML and MLA 'Flogger-G'. This radar had primitive look-down/shoot-down capabilities and was not in the class of contemporary Western radars which employed pulse-Doppler technology for LD/SD. It was, however, a considerable improvement over the Sapfir-23D fitted to the 'Flogger-B' and much lighter.

At the time of German reunification, the East German air force was operating 28 MiG-23MLAs in two squadrons plus nine MiG-23MFs in a third squadron, all assigned to JFG 9 at Peenemünde. The Luftwaffe had no intention of operating the 'Flogger' in the long term, although the aircraft were given Luftwaffe codes in the '20+' series, and this MiG-23MLA received a special scheme in 1991. The badge on the fin is that of JFG 9 'Heinrich Rau'.

the two-seaters). Each VVS fighter 'Flogger' regiment had a squadron specialised in the nuclear strike role (toss bombing) in addition to its primary air defence/air superiority tasking.

This tactical advantage lasted until the McDonnell Douglas F-15A/B joined the tactical fighter wings of the United States Air Forces in Europe in the late 1970s. Nevertheless, the huge numbers of R-23R/T- and R-60-armed MiG-23Ms (which in the 1970s were complemented and then gradually replaced by the refined MiG-23ML/MLA) deployed in Central Europe – the most probable theatre of a conflict between superpowers during the Cold War – ensured that the Soviet armed forces possessed sufficiently potent air superiority and air defence assets.

New capabilities

The MiG-23M offered the VVS a range of combat capabilities that was unthinkable for its MiG-21 predecessor: the 'Flogger' was a high-speed fighter which could engage or disengage at will and possessed genuine head-on, side-on and tail-on BVR intercept capability. The R-23R semi-active radar-homing (SARH) and R-23T infra-red-homing missiles revolutionised the VVS FA's air superiority and defence tactics. 'Silent' (emission-free) tail-on intercepts became possible using the combination of the TP-23 infra-red search and track (IRST) sensor and the R-23T AAM. The TP-23M IRST introduced on the MiG-23M had a maximum range at high altitude against bomber-sized targets of 30 km (16 nm), and fighter-sized targets were detected by well-tuned sensors at up to 23 km (12.4 nm). From 1974, the number of highly-agile R-60 AAMs which could be carried was doubled with the introduction of the APU-60-2 twin-rail launcher units.

At that time, both Soviet pilots and those from client states were still trained mainly in the orthodox and inflexible air intercept tactics of the 1960s which had been mastered during the MiG-21 era. In the 1970s and early 1980s, the VVS and allied air arms flew the MiG-23 the same way as the MiG-21 – as a high-speed point interceptor closely guided and supported by ground control intercept (GCI). It took the VVS 12 years to exploit the MiG-23's capabilities as a true air superiority fighter, after the fleet-wide introduction of the MiG-23MLD.

The initial production MiG-23Ms suffered from numerous teething problems, the most severe of which were the limited manoeuvrability and the vicious handling at high AoA, just like the MiG-23S, as the aircraft was prone to enter an unrecoverable flat spin in such conditions. The difficult and time-consuming process

Air-to-air missiles for the fighter 'Flogger'

R-23R/T (K-23, *izdeliye* 340/360)
The MiG-23M's basic weapons were two R-23R or R-23T beyond-visual-range (BVR) missiles. The aircraft was limited to carrying only one R-23 variant in a single mission because each variant required different black boxes for the fire control system.

The R-23, initially known as the K-23, was developed by Factory No. 134 by a design team led by V.A. Pustyakov. The design requirements called for a missile capable of destroying bomber-sized targets at a maximum distance of 18 km (9.8 nm) and of flying to 4000 m (13,123 ft) above the launch platform (snap-up capability). It was intended to equip the missile with a dual-sensor seeker combining both SARH and IR guidance, but such a device proved difficult to develop. In the event, the K-23 was built in two versions: one with an SARH seeker (*izdeliye* 340) and one with an IR seeker (*izdeliye* 360).

The K-23's first test firings were carried out in early 1967, initially from a specially-modified MiG-21 testbed and later from the 23-11 prototypes, in both guided and unguided (seeker inoperative) modes. Development and testing of the IR version was relatively straightforward, but the SARH version proved quite troublesome and required considerable redesign. The updated K-23 SARH version, tested for the first time in 1970, guided autonomously for three seconds after launch in order to avoid interference from the launch platform's radar side lobes.

Both K-23 variants were fitted with an expanding-rod warhead weighing 25 kg (55 lb) and having a lethal radius of 8 m (26 ft). The SARH version's monopulse seeker gave it fairly good ECM resistance against the F-4E's AIM-7E-2. The IR version employed a seeker with a liquid nitrogen-cooled detector that required lock-on before launch. The IR seeker head could be slaved onto the target by the radar or IRST sensor, or the pilot could use boresight aiming (through the ASM-23M sight/HUD) by manoeuvring the aircraft and positioning the nose toward the target. The missile's probability of kill was rated at 0.8-0.9.

The IR and SARH K-23 variants were commissioned into the VVS with the MiG-23M in January 1974 and were designated R-23T and R-23R, respectively. In real-world combat conditions, the R-23R was reasonably effective against the high-altitude reconnaissance aerostats often encountered over eastern Europe and the Soviet Union, in which cases the missile homed onto the radar-reflective aerostat payload.

Standard BVR weapons for the MiG-23 are the R-23T (IR, left) and R-23R (SARH, right). The improved R-24T/R introduced revised wings.

Beyond-visual-range air-to-air missiles

	R-23R	R-23T	R-24R	R-24T
Maximum launch range at low level				
forward quarter	14 km (7.5 nm)	11 km (6 nm)	17 km (9.2 nm)	11 km (6 nm)
rear quarter	4 km (2.2 nm)	4 km (2.2 nm)	4 km (2.2 nm)	4 km (2.2 nm)
Maximum launch range at high level				
forward quarter	25 km (13.5 nm)	11 km (6 nm)	35 km (19 nm)	12 km (6.5 nm)
rear quarter	8-10 km (4.3-5.4 nm)	8-10 km (4.3-5.4 nm)	20 km (10.8 nm)	20 km (10.8 nm
Minimum launch range at low level				
forward quarter	4-6 km (2.2-3.2 nm)	4 km (2.2 nm)	2.5 km (1.3 nm)	2.5 km (1.3 nm)
rear quarter	1.3 km (0.7 nm)	1.3 km (0.7 nm)	0.5 km (0.3 nm)	0.5 km (0.3 nm)
Max. *g* of launch a/c	4	4	7	7
Max. *g* of target	5	5	7	7
Weight	222 kg (489 lb)	215 kg (474 lb)	243 kg (536 lb)	235 kg (518 lb)
Warhead weight	25 kg (55 lb)	25 kg (55 lb)	35 kg (77 lb)	35 kg (77 lb)

R-24R/T (K-24, *izdeliye* 140/160)
The K-24 programme began in 1975 as an extensive upgrade of the K-23, intended to extend range and increase the effectiveness of seeker heads, radar fuse and warhead. The SARH version, known as the R-24R after it entered service, received the new RGS-24 seeker head with much improved ECM resistance and lock-on after launch capability; the latter not only avoided interference from the Sapfir-23 radar as the missile passed the launch aircraft's nose, but also extended range. The inertial phase of flight was lengthened to 10 seconds, and this, combined with the larger rocket motor, gave the K-24R a range that was 30 per cent longer than its predecessor's. Both the SARH and IR variants were appropriate for use against fighter-sized targets and were able to manoeuvre at up to 8 *g*. The SARH version could even guide onto hovering helicopters and onto high-speed aircraft flying in tight formation. The missile could destroy targets flying at altitudes between 40 and 25000 m (131 to 82,020 ft).

Externally, the R-24 can be distinguished from the R-23 only by its forward-swept wing trailing edge; that of the R-23 is straight. Internally, the new missile has a larger motor, a more powerful expanding-rod warhead weighing 35 kg (77 lb) and having a lethal radius of 10 m (32 ft), and a new, more reliable radar fuse. The minimum launch distance in tailchase engagements was significantly reduced, to 500 m (1,600 ft) from 1300 m (4,160 ft) for the R-23.

The IR variant of the R-24 was equipped with the TGS-23T4 seeker employing a liquid nitrogen-cooled detector with enhanced sensitivity compared to that used in the R-23T; however, just like its predecessor, the R-24T remained a lock-on before launch weapon.

Development and flight testing of both R-24 variants went smoothly, although the usual process of minor improvements and protracted testing delayed their service entry until 1981.

R-60 and R-60M (*izdeliye* 62/62M)
The R-60 and its improved derivative R-60M were the Soviet Air Forces' first dogfight missiles, and from the mid-1970s to late 1980s were the principal VVS close-in weapons. Despite the fleet-wide introduction of the R-73, the R-60Ms retained their worth until the early/mid-1990s. This lightweight, highly-agile missile was capable of

For the most part operators have been content to use Soviet/Russian-supplied weapons on their fighter 'Floggers'. However, in the early 1990s this Czech MiG-23MLA was evaluated with the MATRA Magic II missile.

of expanding the *g* factor and AoA envelope began in the mid-1970s, and required extensive development work by Mikoyan OKB and NII VVS test pilots and engineers to improve the flight control system so that adequate spin-safe handling qualities could be provided. In the event, despite the flight control system improvements and airframe strengthening measures, the improved MiG-23ML and MiG-23MLA were still subject to many manoeuvring restrictions in close air combat. They were removed, and that partially, only with the introduction of the MiG-23MLD variant, a mature tactical fighter with substantial aerodynamic and flight control system improvements.

Notorious in combat

The lack of a history of air-to-air combat success by the 'Flogger', such as that enjoyed by its MiG-21 predecessor, can be attributed to a number of factors. Chief among them were being on the wrong side in local conflicts (in most cases, fighting on the side of the obvious loser), and the aircraft's rather problematic design, which required quite complex handling and maintenance-intensive mission systems that themselves demanded highly-skilled air and ground crews; this was particularly a problem during the early days of 'Flogger' service.

The MiG-23 was rarely employed air-to-air by the VVS during the Cold War. The first combat success of a Soviet-operated 'Flogger' was notched on 21 June 1979, when a MiG-23M of the 152nd IAP controlled by the PVO at Ak-Tepe air base shot down two Imperial Iranian Air Force CH-47C helicopters violating Soviet airspace. Pilot Captain V. Shikinder used two R-60s to down the first helicopter, then strafed the second with the aircraft's GSh-23L two-barrelled 23-mm gun. Seventy-two rounds were spent in a firing pass that damaged the helicopter's starboard engine and forced its crew to attempt an emergency landing. It was the first successful air combat recorded by a swing-wing fighter in the history of the Soviet Air Forces.

However, the 'Flogger' performed less well in the hands of 'client' state export operators. This is especially true of Syrian Air Force MiG-23MS/Ms over Lebanon in 1982 and of Cuban and Angolan MiG-23MLs in southern Africa in the late 1980s.

Perhaps unexpectedly, this dedicated air superiority and air defence VVS FA fighter was often tasked in the ground attack role. Used as medium- and high-altitude 'bomb trucks', the MiG-23MLA/MLDs carried out many combat sorties during the Soviet invasion of Afghanistan. As many as seven 'Flogger-G/K'-

Reverse-engineered from the AIM-9B Sidewinder, the very basic R-3S armed some 'Floggers', notably the 'sanitised' MiG-23MS 'Flogger-E'.

replacing guns in close air combat.

Development work began in 1971 by the Molniya design team, formerly known as OKB-4. The new missile incorporated a simple, compact, uncooled Komar IR-seeker, a radar fuse and a small expanding-rod warhead weighing 3 kg (6.6 lb) and having an effective radius of 2.5 m (8.2 ft). It could destroy a target only in the event of a direct hit. This lock-on before launch weapon imposed few restrictions on the launch platform, as it could reach targets manoeuvring at a maximum of 8 *g* whereas the launch platform was restricted to 7 *g*. The minimum launch range – a significant parameter in a dogfight – was only 300 m (984 ft). Probability of hit of a twin-missile salvo was 0.7-0.8. The pilot could squeeze the trigger during pre-launch manoeuvres, and the R-60 would launch automatically when the missile seeker acquired the target.

The improved R-60M which entered service in the early 1980s had a more-sensitive Komar-M IR seeker with a nitrogen-cooled detector that provided a slightly limited all-aspect acquisition/launch capability against targets with engines in afterburner. It could acquire a target 17° off-boresight (the R-60 was restricted to 12°). Its minimum launch distance was further reduced, to 200 m (656 ft), and the maximum launch distance increased by 500 m (1,640 ft).

R-73 (*izdeliye 72*)

The R-73 was developed by the Vympel Design Bureau in the 1970s to become the principal close air combat weapon of the MiG-29 and Su-27, but later it was identified as a useful addition to the MiG-23MLD's weapons suite. The all-aspect missile employed a compact Mayak-80 IR seeker that made use of digital signal processing for better ECM resistance, and had a detector cooled by liquid nitrogen to 77 K (-196° C). Its off-boresight capability was considerably better than that of the R-60M. The missile had an impressive 60° per second line-of-sight tracking rate. In order to improve probability of kill in a tail-on engagement, 'lead bias' was introduced in the final guidance phase to direct the missile ahead of the target's exhaust plume. The missile's combined aerodynamic/thrust vector control made it remarkably agile. The Kretchet fuse was of radar type and the 7.4-kg (16.3-lb) expanding-rod warhead had a lethal radius of 3.5 m (11.5 ft).

The R-73 was test-fired for the first time from a specially-modified LL-124 testbed and continued its factory test firings from an upgraded MiG-23ML. Its first launch against a target drone – a remotely-piloted MiG-17 fighter – was in 1980.

The R-73 was commissioned into the VVS in June 1984 and the MiG-23MLD could, in theory, carry two missiles on P-72-1D launchers under the wing pivots and two under the fuselage, but in practice only the latter stations were used. The MiG-23MLD's weapons control systems needed only the replacement of three black boxes in order to employ the R-73. A helmet-mounted cueing system such as that designed for the MiG-29 and Su-27 was not available, so the missile was slaved onto the target via either radar or IRST, or the pilot could aim it visually with the ASP-17ML sight/HUD.

An improvement on the R-3S, the R-13M (top) was the main short-range weapon of the MiG-23M/MF until the smaller and more agile R-60 (bottom) became available from the mid-1970s for the VVS, and for export from 1980/81.

Within-visual-range air-to-air missiles

	R-3S	R-13M	R-60	R-60MK	R-73
Max. low level range/rear quarter	2.5 km (1.3 nm)	3.5 km (1.9 nm)	2 km (1.1nm)	2 km (1.1 nm)	2.2 km (1.2 nm)
Max. high level range/rear quarter	7 km (3.8 nm)	13 km (7 nm)	8 km (4.3 nm)	8 km (4.3 nm)	12 km (6.5 nm)
Min. range/rear quarter	1 km (0.54 nm)	0.9 km (0.5 nm)	0.2-0.3 km (0.11- 0.16 nm)	0.2-0.3 km (0.11- 0.16 nm)	0.2 km (0.11 nm)
Maximum *g* of launch aircraft	2	3.7	7	7	9
Maximum *g* of target	3	5	8	8	11
All-aspect capability	no	no	no	limited	yes
Weight	76.5 kg (168.6 lb)	90 kg (198 lb)	43.5 kg (95.9 lb)	44 kg (97 lb)	105 kg (231 lb)
Warhead weight	11.3 kg (24.9 lb)	11.3 kg (24.9 lb)	3 kg (6.6 lb)	3.5 kg (7.7 lb)	7.4 kg (16.3 lb)

In the tussle between the Libyans and the US Sixth Fleet in the 1980s, MiG-23MS 'Flogger-Es' – like this aircraft armed with R-3S missiles – were often encountered by US Navy aircraft. The sparring occasionally turned into hostile action, and on 4 January 1989 two Libyan 'Floggers' were shot down by two F-14s from VF-32.

equipped VVS regiments were deployed there for one year each between 1984 and 1989, based at Bagram, Shindand and Kandahar; their aircraft were much used in the ground attack and escort roles, as well as for quick reaction alert (QRA). The MiG-23MLD crews reported several clashes with Pakistani F-16s around the Afghanistan-Pakistan border, and during one of those engagements, on 12 September 1988, a Bagram-based MiG-23MLD from the 120th IAP was badly damaged by two AIM-9 Sidewinders but managed to get back to base. Russian sources indicate that on 26 September 1988, the Shindand-based pair of QRA MiG-23MLDs from the same IAP shot down a pair of Iranian helicopters flying in Afghan aerospace. As many as 10 VVS MiG-23s are known to have been lost in the Afghanistan war, five downed by enemy anti-aircraft fire, and the rest written off due to hardware failures (two) and pilot error (three).

In the early/mid-1980s, the MiG-23's accident rate with the VVS averaged 12.5 per cent per 100,000 flying hours, equating to one loss per 8,000 flying hours, according to the Soviet Air Forces' official magazine *Aviatsia i Kosmonavtika*, issue 6 of 1990. By that time, the fighter 'Flogger' fleet had built up 1.5 million flying hours.

The East European air arms fared even worse. In Hungarian Air Force service, between 1979 and 1996 the 15 MiG-23MF/UBs recorded a fairly high accident rate of 24.3 major mishaps per 100,000 flying hours (equating to one loss per 4,115 flying hours) and attrition (overall loss) rate was 33 per cent (five aircraft lost from a fleet of 15). This was accompanied by a pilot fatality rate of 71 per cent, five of the seven crew members who had been involved in these accidents having died. In Bulgaria, which operated 92 MiG-23MF/MLA/MLD/BN/UBs between 1976 and 2002, the reported accident rate was 18 losses per 100,000 flying hours (equating to one loss per 5,500 hours), while attrition was 16.3 per cent. Poland's 42 MiG-23MF/UBs (36 MFs and six UBs), delivered between 1979 and 1982 and used until 1999, suffered five losses in 44,200 flying hours, an accident rate of 11.3 losses per 100,000 flying hours (equating to one loss per 8,849 flying hours), while the attrition rate approached 12 per cent. The East German Air Force, which operated a 77-strong 'Flogger' fleet (12 MiG-23MFs, 32 MiG-23MLAs, 22 MiG-23BNs and 11 MiG-23UBs), reported 14 losses in 68,700 flying hours, resulting in an accident rate of 20.4 per 100,000 hours (equating to one loss per 4,901 hours) and an attrition rate of 18.2 per cent.

Fighter 'Floggers' in the 1990s

Between 1969 and 1984, 3,630 fighter 'Floggers' rolled off the production line at Lukhovitzy near Moscow. MiG-23 fighter derivatives were exported to at least 26 countries.

According to official Conventional Forces in Europe Treaty data, in 1990 the Soviet Air Forces had 422 MiG-23s based in the European part of the Soviet Union and in eastern Europe, including a significant number of MiG-23UB/UM conversion and continuation trainers. The Soviet Air Defence Troops' Fighter Aviation branch (IA-PVO) declared an inventory of 749 'Floggers' based in the European part of the Soviet Union. In addition, 400 MiG-23s are believed to have been on strength

Lasour datalink

The onboard Lasour-SMA automated datalink guidance equipment is a follow-on development of the MiG-21's Lasour-1. It was introduced on the MiG-23M and presented in a refined form on the MiG-23ML/MLA/MLD (Lasour-SML). The equipment was integrated with the ground-based Vozdukh-1M automated ground control intercept (GCI) system that equipped each VVS and IA-PVO regimental GCI centre. This system, used in two different versions, was considered useful when operating in friendly airspace, and to a lesser extent in disputed airspace.

The Lasour-1SMA/SML provided the heading and climb commands that the pilot required to build up the intercept geometry. It also supplied the prompts (lights) and audio signals to engage the afterburner, to execute zoom climb for high-altitude missile launch, to switch on the radar in emitting mode for missile firing (this, together with radar antenna cueing and automatic range gate set-up for priority tracking, could also be done remotely by the GCI officer) and, finally, for initiating a timely exit from the attack run. The command messages/prompts were displayed on a special panel located beneath the ASP-23M (or ASP-23ML) HUD/gunsight, or directly on the ASP-17ML HUD in the MiG-23MLD. Heading, attitude and speed commands were displayed through 'bugs' located on the appropriate cockpit instruments.

The pilot followed the cues to fly the designated (covert) intercept profile, in a process known in the VVS and IA-PVO combat manuals as 'instrument-guidance intercept'. In the case of the MiG-23P PVO interceptor, the pilot was required to control only the engine thrust in the fully-automated intercept mode, as the aircraft was under the control of the SAU-23P; in the final stage of the attack run, the pilot was responsible for acquiring the target that had been selected by the GCI officer and executing missile launch. Exiting the attack run and recovery to the home base followed – also flown in automatic mode, including the final landing approach down to 50 m (164 ft).

The chief advantages of the system were that guidance could precisely follow the most favourable intercept geometry profile calculated by the onboard digital computer, and all required operations during the intercept could be completed without voice communication between the pilot and the GCI officer. VVS FA and IA-PVO datalinks for automated fighter guidance are thought to have been made jam-resistant in the 1970s and 1980s.

According to VVS FA guidelines of the 1980s, the GCI officer had near-absolute authority to dictate every action of the fighters in sweep and combat air patrol operations in friendly, disputed or enemy airspace. These could include headings, altitudes, speeds, attack and firing clearances, disengagement, etc. (This process is believed to be followed still by the MiG-29 and Su-27, which use the basic tactics and weapons employment techniques proved by the MiG-23 in the mid-1980s.) Datalink operations of the VVS FA MiG-23s were considered suitable mainly for air defence missions over own or contested territory, because this guidance method's limited range and rigid tactics render it not particularly useful for fighter sweeps where large enemy formations are likely to be encountered deep in enemy airspace.

Air superiority operations with voice or datalink guidance require a well-functioning GCI system, and as a result GCI officers were required to maintain maximum possible situation awareness. This would compensate, to a degree, for the lack of airborne warning and control system assets in the Soviet Union, a situation which continues to the present day.

Both the VVS and IA-PVO mastered this imperfect and rather inflexible technique, which in the 1970s and 1980s was considered useful in the Central European theatre with its fairly dense ground radar coverage. Forty underground and mobile GCI centres fed by 200-250 radars of various types and manned by skilled fighter controllers were available in Central Europe during the Cold War, capable of controlling approximately 160 fighter elements or wings simultaneously. However, it could hardly be said that all export MiG-23 operators have found the Soviet GCI concept to be the most effective command-and-control solution.

In the MiG-23MLD GCI commands were projected on to the ASP-17M head-up display, along with information from the aircraft's own radar.

Bulgarian MiG-23MLD 'Bord 219' was the last of 3,630 MiG-23 fighters to be built. It was delivered in February 1985 and served until Bulgaria withdrew its MiG-23 fleet in November 2002.

with Frontal Aviation and the IA-PVO in the Asian part of the Soviet Union.

The 'Flogger' had a brief front-line service career, compared to the longer-lived MiG-21, a situation that was dictated mainly by the political and military changes that swept across the former Soviet Union and eastern Europe in the early/mid-1990s. The abrupt air force inventory rationalisation moves in Russia (followed several years later by the air arms of Ukraine and Belarus) caused by sharp defence budget reductions led to a rapid withdrawal from use of all single-engined front-line aircraft types; the MiG-23/27 and Su-17 were prime victims of this process. Today, fighter and fighter-bomber 'Floggers' have been entirely withdrawn from use in Russia, most of the CIS republics and eastern Europe. The last MiG-23MLDs serving with the Ukrainian air force flew until 2001, and those in Bulgaria remained in active use until November 2002.

Fighter survivors

The swing-wing fighter, however, is still soldiering on with the air arms of some countries including Syria, North Korea, Libya, India, Angola and Cuba. A substantial number of strike 'Flogger' derivatives are still in a fairly active use in India, Ethiopia, Algeria and Sri Lanka. The service that most recently became a 'Flogger' operator is the embryonic air arm of Ivory Coast, which acquired two ex-Bulgarian MiG-23MLD(Export)s in spring 2003.

RSK MiG (Rossiskaya Samoletostroitelnaya Korporatsiya/Russian Aircraft Corporation MiG, as the organisation is now known) proposed a series of MiG-23 upgrade options in the mid-/late 1990s which included improved weapons system with new or upgraded existing radar, the R-77 (AA-12 'Adder') active radar homing missile, active jammer and 'glass' cockpit. However, this was too expensive for most of the existing operators, which lacked the funds and/or the inclination for such updates and therefore were reluctant to launch such programmes. Most existing 'Flogger-B/G' operators have little or no capability to procure new equipment or launch large-scale upgrade programmes, for reasons that are either economic or political.

They may, however, opt to increase the combat capability of their MiG-23MF/MLA/MLDs at low cost by integrating the R-73 (AA-11 'Archer') high off-boresight missile, which allows the 'Flogger' pilot to acquire and engage an enemy aircraft even at a high angle (up to 45º) off his aircraft's heading. This missile was introduced on VVS MiG-23MLDs in 1983, and its integration with other life-extended variants of the 'Flogger' is described as straightforward. The R-73, with both visual and radar/IRST-slaved aiming modes, would compare well in medium/high-altitude range performance to the aged and patchily-reliable R-23T. Manoeuvrability of its much more effective and sensitive seeker head is considerably greater than that of the R-23T/R-24T, making it a potent weapon in all-aspect near-BVR and in all within-visual-range (WVR) encounters. To exploit the R-73's high off-boresight capabilities fully, however, requires a helmet-mounted cueing system similar to that adopted for the upgraded Indian Air Force MiG-21bis-UPG.

A number of government and private companies in Russia, Ukraine and Belarus – some known to operate in the grey and black weapons markets – could offer new or second-hand R-73s. Kits for the missile's integration, including black boxes, launchers and chaff/flare dispensers, could easily be scavenged from the huge fleet of stored MiG-23MLDs in the CIS republics. Therefore, it can not be ruled out that MiG-23 operators considered hostile to the West – North Korea, Cuba, Libya, Syria – have already chosen to proceed with such a small-scale yet affordable upgrade in order to increase the operational capability of their MiG-23 fleets.

The MiG-23MF/MLA/MLDs sold to export customers were initially slated to serve for only 17 years or 1,500 hours, whichever came first (a main overhaul was required after seven/eight years, or 750 hours). Now these aircraft are being targeted by RSK MiG for its so-called rolling airframe life extension; initially, this would give a life of 20 years or 2,000 hours, then 25 years or 2,400 hours (as the Russian/CIS late-production examples are thought to have been intended to serve). Eventually, service life extensions to 30 years or 3,000 hours may be an option, depending on the airframe's condition in terms of fatigue and corrosion, which is particularly important for the wing pivots, known to be the most sensitive parts of the MiG-23's airframe. This would mean that with the relatively inexpensive airframe/systems service life extensions now being offered by RSK MiG, these MiG-23MLA/MLD(Export)s manufactured in the early 1980s could carry on for 25 years, assuming reliable logistic support was provided.

Alexander Mladenov

MiG-23MLAs were delivered to Angola in about 1985, and were used in the civil war against UNITA rebel forces, being flown and maintained by Cubans and Soviets. During the course of these actions a number were lost to groundfire and sabotage. In the following decade, with Cuban and Soviet 'advisors' gone and South African support for UNITA withdrawn, the MiG-23s were flown on some anti-UNITA missions by South African mercenary pilots working for the Executive Outcomes company.

Mikoyan MiG-23 fighter versions

Aircraft 23-01 ('Faithless')

Aircraft 23-01 was a scaled-up MiG-21 derivative with lateral air intakes and a nose large enough to accommodate a radar with a large-size antenna, as required for BVR intercepts. The wing was fitted with a powerful blown flap system which, in combination with the lift jets, was intended to shorten the take-off run and landing roll considerably. The engine intakes were of semi-circular cross-section, with Mirage III-style shock cones positioned carefully away from the fuselage in order to avoid boundary layer ingestion problems.

The powerplant was an R-27-300 main engine, and two vertically-mounted Kolesov RD-36-35s were used to provide the STOL ability. The lift engines were fed with air by a hinged, shuttered trap door, and were equipped with nozzles with limited movement to deflect the jet up to 5° backward and 10° forward. This provided additional thrust for take-off and braking (i.e., reverse thrust) during landing approach.

Aircraft 23-01 was planned as the prototype of an operational air superiority fighter, and thus was intended to be armed with a built-in GSh-23 twin-barrelled gun and two K-23R SARH missiles or two K-23T heat-seeking missiles under the wings.

The aircraft made its maiden flight on 3 April 1967 but flew only 14 times before its development programme was cancelled. The chief reason was disappointing flight test results, which proved the unsuitability of a STOL design with lift jets in an operational land-based combat aircraft. The complex propulsion system meant that the jet efflux produced by the lift engines interacted with the ground to alter the aircraft's trim during take-off and to cause a suction effect during the final landing approach. This made the pilot's workload excessive, for in order to prevent these potentially dangerous effects he was required to increase the main engine thrust to military power and even use afterburner just before touching down. In addition, the lift jets imposed a significant weight penalty, which reduced the internal fuel and weapons load: understandably, this was deemed unacceptable for the new-generation combat aircraft.

Aircraft 23-01 made its last flight at the Domodedovo air show on 9 July 1962, and soon after was passed to the Moscow Aviation Institute to be used as a ground instruction aircraft. Seen by Western intelligence agencies for the first time at Domodedovo, Aircraft 23-01 was given the NATO code-name 'Faithless'.

The single 23-01 prototype is seen with the intake for its lift engines raised and dummy missiles fitted.

Aircraft 23-11 ('Flogger')

Development of the prototype with VG wing was ordered by the Aviaprom Minister Pyotr Dementiev in 1965. The resultant design, known as Aircraft 23-11, was quite different to its lift jet-equipped STOL forebear. The forward fuselage was of a vertical oval shape, fairing back to a rectangular-shaped cross-section. The low-drag canopy was similar in appearance to that used in the MiG-21 but was hinged at the rear and faired into the fuselage spine, giving the pilot little spare headroom and limiting his field of view down and to the rear. Rear-view mirrors mounted on the central canopy structure partially compensated for this.

The principal load-bearing structural member was the main carry-through box in the top of the fuselage, made by welding thick plates of special VNS-2 steel alloy; it was also used as fuel tank no. 2. The wing pivots were placed near the outer edge of the fixed wing glove and provided a range of sweep angles in an arc of 56°. The three main settings were 16°, 45° and 72°.

Each wing panel featured an extensive array of high-lift devices on the leading and trailing edges. The airbrakes were in four-petal configuration located on the rear fuselage, and moved forward in relation to the tail surfaces. When the landing gear was selected up, the large ventral fin rotated down and locked in a vertical position; when landing gear was selected down, the fin pivoted to the right, to the horizontal (stowed) position.

Aircraft 23-11 was powered by one Tumanskiy R-27F-300 turbojet fed by air entering through rectangular intakes with perforated variable-intake ramps, similar to those used on the American F-4 Phantom II. The R-27F-300 turbojet, rated at 50.88 kN (11,440 lb) dry and 76.31 kN (17,160 lb) at full afterburner, was a scaled-up derivative of the widely-used R-11/13/25-300 turbojet that powered various MiG-21 variants. The R-27F-300's new materials and manufacturing technologies gave it a specific fuel consumption 25 per cent less than that of its predecessor.

Fuel was carried in three fuselage tanks with capacities of 1920, 820 and 710 litres (422, 180 and 156 Imp gal) and three integral tanks in each wing with capacities of 62.5, 137.5 and 200 litres (13.7, 30 and 44 Imp gal).

The undercarriage was very tough and highly effective. Its main units – also used as compressed air bottles – folded into a small space and left the underside of the fuselage free for large centreline stores. However, this design, combined with the intakes which were susceptible to foreign object ingestion, made operations from unpaved, icy or snow-packed airstrips very difficult or almost impossible. The MiG-23 is known to be rather unstable during its landing roll on wet runways.

The first Aircraft 23-11 prototype was intended to have armament initially consisting of four K-23 BVR air-to-air missiles carried on pylons under the wings and under the fuselage.

The first prototype of the new VG fighter, known as 23-11/1 and serialled 231, was rolled out of the assembly hall in May 1967 and transported on 26 May by road to Mikoyan's own flight test facility at Zhukovskiy, where the maiden flight was made on 10 June by Alexander Fedotov.

The 23-11 prototypes were completed with the initial clean wing design and long jetpipe. The no. 1 aircraft (above) survives today in the Monino museum. The second aircraft (below) was the first to carry radar in the form of the Sapfir-23.

After the 45th flight in late 1967, the 25-hour life of the R-27F-300 expired and the aircraft was grounded for installation of a new engine and new equipment such as an AP-155 three-axis autopilot. In April 1968, 23-11/1 was used in firing trials at the NII VVS flight test facility at Vladimirovka. There the aircraft fired 16 missiles at altitudes from 4877 to 16581 m (16,000 to 54,400 ft) and speeds between Mach 0.7 and Mach 1.8, the intention being to investigate engine stability regarding the ingestion of the rocket motor's exhaust gases.

23-11/1's factory flight test programme was completed successfully in July 1968 after 97 sorties, and shortly afterwards the aircraft was passed to the Soviet Air Forces museum at Monino, near Moscow.

23-11/1 was followed in 1969 by two more prototypes, designated 23-11/2 and 23-11/3, which were equipped with the Sapfir-23 radar set and powered by the uprated R-27F2-300 engine. However, the new and rather complex radar, which had been tested on a Tu-104, was still notoriously unreliable and caused many technical problems during the MiG-23's protracted flight test programme. Later, four more prototypes were involved in the trials – 23-11/4 (first flight in 1969; it was used as the prototype for the fighter-bomber variant), 23-11/5, 23-11/6 and 23-11/9.

The 23-11 prototype poses with the main weapon proposed for the type, the K-23 missile (R-23 in service). Early prototypes undertook large numbers of K-23 tests.

MiG-23S (*izdeliye* 23-11S, 'Flogger-A')

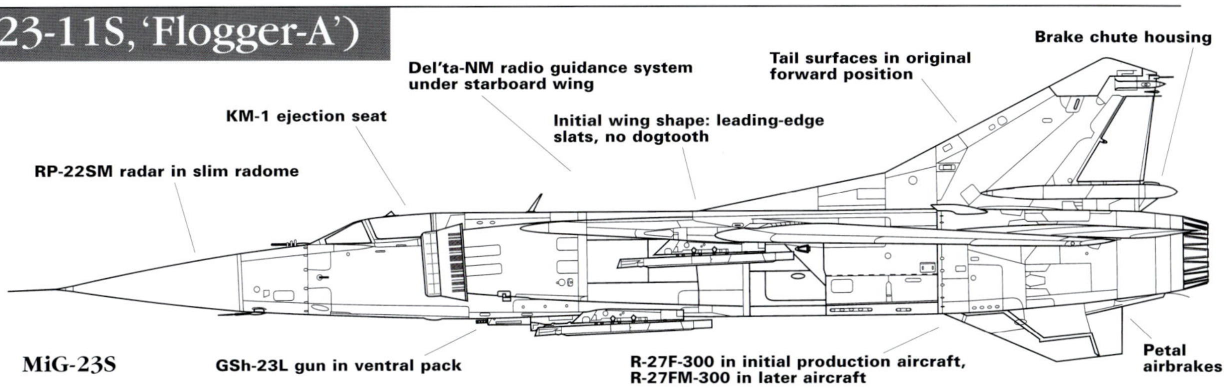

The MiG-23S was an interim variant externally similar to the prototype, but equipped with a weapons control system based on the Sapfir-21 radar (also known as RP-22SM) with an ASP-PFD-21 lead computing gunsight, both taken from the MiG-21S/SM. The Sapfir-23 radar, designed for the 'Flogger' by OKB-339 in Moscow (now known as NIIR-Phazotron), was delayed for some time due to severe development problems, so existing hardware was used as a stop-gap solution. The RP-22SM radar was housed in a purpose-designed conical radome and lacked the IRST sensor. The MiG-23S could carry only four AAMs – usually two SARH R-3Rs and two IR-homing R-3Ss. In addition, the proven GSh-23L 23-mm twin-barrelled gun with 200 rounds was available in a ventral pack.

The Del'ta-NM bullet-shaped pod used to direct the Kh-23 (AS-7 'Kerry') radio-guided air-to-surface missile was built into the root of the starboard underwing pylon. In the ground attack role, the MiG-23S could carry two Kh-23 missiles under the wings, two to four 16-round 57-mm rocket packs, S-24 240-mm rockets or up to 2000 kg (4,409 lb) of iron bombs, napalm tanks or cluster bombs.

The initial production MiG-23Ss were powered by the R-27F-300 turbojet rated at 68.04 kN (15,300 lb) dry and 78.5 kN (17,640 lb) with afterburner. Follow-on aircraft used the uprated R-27F2M-300 derivative rated at 66.71 kN (15,000 lb) dry and 97.84 kN (22,000 lb) with afterburner.

A total of 60 or so MiG-23Ss are reported by Russian sources to have left the Znamya Truda factory in 1969 and early 1970. Their final assembly, ground and flight checks took place at the large facility at Lukhovitzy near Moscow. They had only a brief front-line service career with the 4th TsBPiPLS at Lipetsk and the 979th IAP at Shtutchin. The huge number of problems caused by the unreliable airframe and equipment prevented the MiG-23S being used as a fully-fledged weapons system. Unstable take-off and landing performance – especially on final landing approach – and notoriously poor yaw control were listed among the chief handling problems of the MiG-23S which, in the event, led to considerable attrition.

Notwithstanding the type's poor handling characteristics, a number of surviving MiG-23Ss were transferred in 1976 from the Shtutchin-based regiment to a lead-in fighter training regiment with the Tchernigov-based VVAUL (VVS pilot training school). There the aircraft's radars were replaced with cast iron ballast plates in an attempt to improve their balance, and they remained in active operation for several years as lead-in fighter trainers. The first VVS student pilots soloed on the MiG-23S in 1977, each logging 10 flights on the swing-wing fighter with its demanding handling characteristics.

One of the few MiG-23S interim fighters is seen in a partially disassembled state at Frunze. The long nose of the type is immediately apparent.

This MiG-23S is at Khodinka. The bullet fairing under the starboard wing housed the antenna for the Del'ta-NM radio guidance system for the Kh-23 ASM.

MiG-23 (Edition 1971)

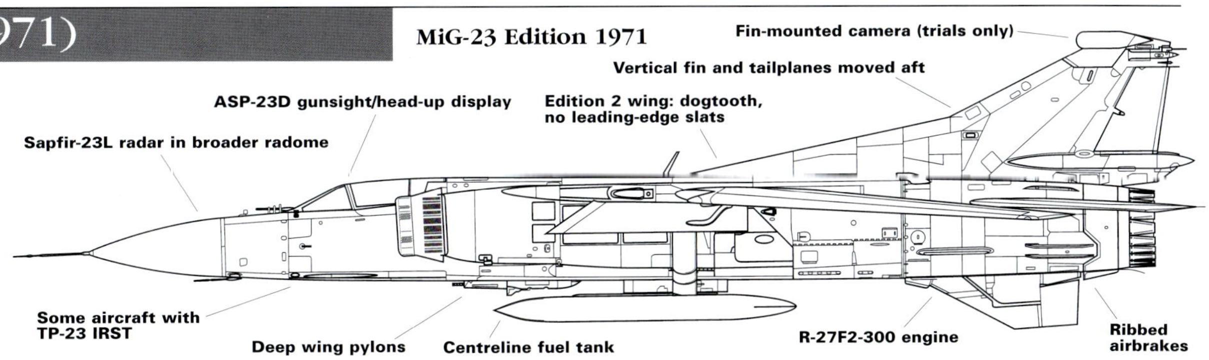

In late 1970 the MiG-23S was swiftly replaced on the production line by another interim 'Flogger' variant, which continued in production in 1971 and was known simply as the MiG-23 (more colloquially as the MiG-23 Edition 1971). It was equipped with the initial production version of the Sapfir-23 radar capable of guiding the new R-23R (AA-7 'Apex') SARH BVR missile. This initial version, the unreliable Sapfir-23L (Lyogkii, lightweight) pulse radar, was an interim solution lacking the definitive look-down/shoot-down capability and originally had been conceived for the MiG-23M. The mission avionics suite incorporated the TP-23 IRST and ASP-23D gunsight/HUD for displaying target data from the radar, replacing the head-down radar scope used on the MiG-23S.

This variant was powered by the R-27F2-300 turbojet, rated at 68.04 kN (15,300 lb) dry and 97.84 kN (22,000 lb) with afterburner. The MiG-23 Edition 1971 introduced a redesigned fuselage in which the tail surfaces were moved aft 86 cm (33.8 in), giving the rear fuselage a very different look. Ribbed airbrakes were introduced and another fuel tank (no. 4) holding 470 litres (104 Imp gal) was inserted into the rear fuselage.

The Edition 1971's improved wing design, known as the Edition 2, had area increased by 20 per cent, or 5 m^2 (54 sq ft). This was urgently needed in order to achieve an acceptable wing loading which, in turn, would affect not only the aircraft's take-off/landing characteristics but also considerably influence the range. The wing area increase led to a change in normal sweep settings of 2° 40', so that the main positions became 18° 40', 47° 40' and 74° 40'. For convenience, however, all cockpit indicators and manuals remained unchanged.

A pronounced leading-edge dogtooth was added but the leading-edge slats were deleted in order to simplify the wing manufacturing process, thus accelerating the launch of production. The dogtooth discontinuity – a jagged forward-pointing portion – created a powerful vortex at high AoA, for example when pulling *g* in a tight turn. Subsequent testing revealed that the Edition 2 wing was a source of potentially dangerous control and stability problems at high AoA, making take-off and landing difficult due to sharply reduced stall margins.

Eighty (or slightly fewer) MiG-23s of this interim variant were manufactured in 1971. They served briefly with several front-line VVS fighter regiments such as the 168th IAP at Starokonstantinov in Ukraine before being assigned to the training role in 1978. The MiG-23 survivors were utilised for advanced and lead-in fighter training with a training regiment of the Tchernigov VVAUL, replacing the MiG-23S in this role at Pevtsi airfield. However, their demanding handling characteristics combined with poor maintenance led to considerable attrition.

Externally similar to the definitive MiG-23M apart from leading-edge slats, the '1971' aircraft introduced the Sapfir-23 radar. This aircraft has a fin-mounted TV camera for trials.

MiG-23M (Type 23-11M, *izdeliye* 2M, 'Flogger-B')

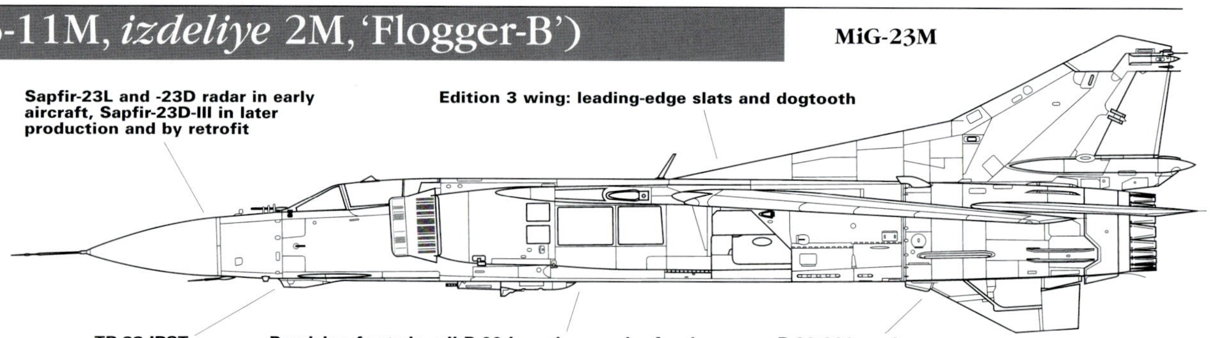

The MiG-23M (NATO 'Flogger-B'), first flown in June 1972, was the first VVS Frontal Aviation fighter that served in large numbers to boast a powerful radar with look-down/shoot-down capability and IRST sensor. This enabled operations in a dense ECM environment where the radar was rendered inoperative. The MiG-23M's weapons suite comprised a pair of R-23R/T SARH/IR-guided BVR air-to-air missiles and a pair of the then-new R-60 (AA-8 'Aphid') dogfight missiles.

This variant was produced in the largest numbers of all the first-generation 'Floggers', and in the mid-1970s became the VVS's most important air superiority fighter. The MiG-23M was powered by a considerably uprated R-27 derivative – the Tumanskiy R-29-300 (*izdeliye* 55A), rated at 81.83 kN (18,400 lb) dry and 113.41 kN (25,500 lb) at full afterburner. It introduced plumbed pylons under movable wing panels for carrying optional 800-litre (176-Imp gal) drop tanks. The tanks could be used only at 16° sweep, and for wing movement to 45° and 72° the pilot was required to jettison the tanks first. The MiG-23M featured the new SAU-23A three-axis automatic flight control system/autopilot and the Polyot-1I-23 navigation system.

The 'Flogger-B' was fitted with the definitive Edition 3 wing, a design that retained the Edition 2's overall dimensions but had leading-edge slats to improve take-off and landing handling characteristics. The type's production at Znamya Truda gathered speed and in 1974 it reached an impressive rate of 30-plus airframes per month, up to as many as 40 airframes at peak times.

The first production MiG-23Ms were equipped with the Sapfir-23L radar, succeeded shortly afterwards by the Sapfir-23D and in 1975 by the definitive Sapfir-23D-III. Later, the Sapfir-23D was brought to the D-III standard during field upgrades by factory teams.

The mass introduction of the MiG-23M was quite a painful undertaking for the VVS, because the new fighter demanded better-educated pilots and technicians than those for the MiG-21. This was due to the complex handling characteristics at the three different wing settings and the highly sophisticated airframe, powerplant, navigation and weapons control system.

The MiG-23M's transformation from limited-manoeuvrability interceptor into (relatively) agile air superiority fighter was not easy. Its strength limits were broadly comparable to those of its MiG-21MF/bis predecessor: 7.5 *g* to 8 *g* at 45° wing sweep (the MiG-23S was restricted to 5 *g* because of concerns regarding airframe and wing pivot strength).

The initial mass production of the wing pivots and no. 2 centre-fuselage steel tank – the MiG-23's principal airframe load-bearing structural elements – was of low quality, and cracks occurred often on the MiG-23S/M/UBs produced in the early and mid-1970s. In addition, there were many wing sweep mechanism failures (mostly pivot structural failures), which prompted VVS command authorities to impose a temporary 5-*g* restriction that effectively prevented combat squadrons from practising basic fighter manoeuvres until 1977. Until clearance was received for high-*g* manoeuvring, the main combat tactic of the MiG-23M-equipped VVS regiments for almost four years was all-aspect BVR intercepts.

Later in the same decade, the welding technology and quality control were considerably improved. The wing pivots were greatly strengthened, allowing the 'Flogger-B'-equipped units to begin practising close air combat in addition to the standard BVR and WVR intercepts. However, the poor high-AoA handling characteristics remained a limiting factor, despite the 1979 fleet-wide installation of an SOUA stick pusher as a crash prevention measure.

Live firings of the R-13M from the underfuselage launchers required skilled pilots in order to avoid engine stalls caused by ingestion of rocket motor exhaust gases, and several aircraft were lost due to this effect. The R-60 had a similar effect, and during the first practice firing at the VVS's 1521st Aviabasa (air base) at Mari in 1976 many inflight engine shutdowns were caused by compressor surge; this necessitated prolonged and expensive operational tests and evaluation firings to be carried out at the NII VVS.

The APU-60-2 double-rail launchers for the R-60 missiles were introduced on aircraft no. 3201. The follow-on production batches (from aircraft no. 3701 onwards) could use the Kh-23 or Kh-23M radio-guided missiles with the associated Del'ta-NG guidance pod carried on the port underfuselage pylon. In addition, all VVS MiG-23Ms were wired to carry a single 10-kT-yield RN-24 nuclear free-fall bomb (later complemented by the 30-kT RN-40) on a special adapter under the fuselage.

The MiG-23M had an important secondary ground attack capability, for which it could carry a warload of up to 3000 kg (6,614 lb) including free-fall and cluster bombs, napalm tanks, S-24 240-mm rockets and 16-round 57-mm rocket packs.

The first production MiG-23M entered service with the 4th TsBPiPLS in 1973, soon followed by front-line fighter regiments. In 1975 the 16th Air Army stationed in East Germany controlled four regiments that converted to the new fighter, a total of 12 squadrons operating some 150 'Floggers'. In the mid-/late 1970s, a small number of IA-PVO regiments also converted to the MiG-23M.

This is the MiG-23M prototype, armed with R-60 (AA-8 'Aphid') missiles which were introduced by this version. Although it looked superficially similar to the '1971' aircraft, it had a new wing which reinstated leading-edge slats.

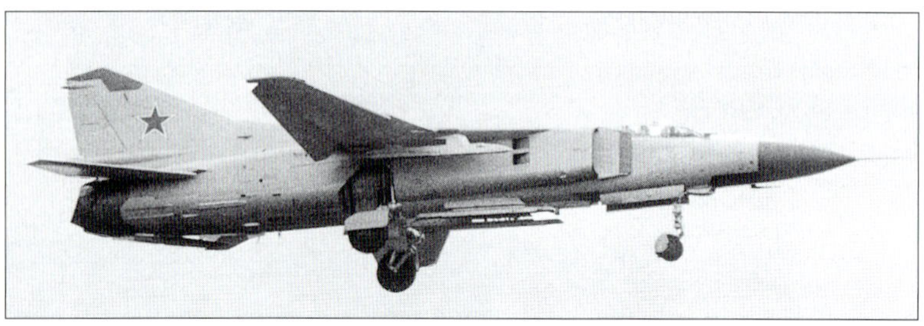

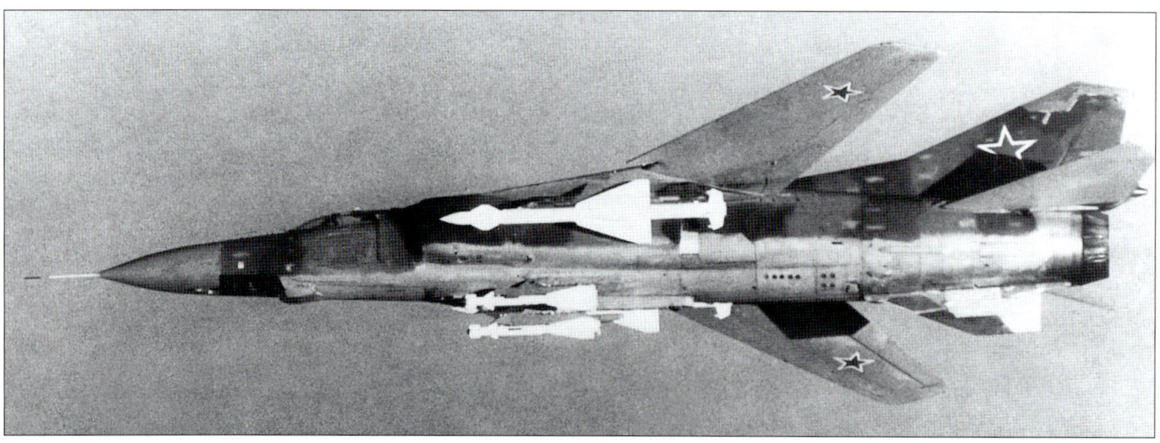

Large numbers of MiG-23Ms were built for the VVS and IA PVO, initially delivered in a light grey scheme (left). The aircraft below carries the standard loadout of two R-23s and two R-60s.

MiG-23MF (*izdeliye* 2A/B, 'Flogger-B')

The MiG-23M's export derivative was designated MiG-23MF and was available in two sub-versions: the first (*izdeliye* 2A or 23-11A) was built for the Warsaw Pact states of Bulgaria, Czechoslovakia, East Germany, Poland, Hungary and Romania, and the second (*izdeliye* 2B or 23-11B) for the client states of Syria, India and Cuba. The *izdeliye* 2A was virtually indistinguishable from the MiG-23M, perhaps with small differences in the communications and IFF (Identification Friend or Foe) equipment only. *Izdeliye* 2B was very different, in both radar (it lacked the ECM-resistance features and often had downgraded performance) and communications equipment (the Lasour-SMA onboard automated datalink was deleted on some aircraft).

All export versions were equipped with the Sapfir-23D-III, redesignated for export customers as Sapfir-23E. Until 1981 the aircraft were delivered with R-13M close air combat missiles instead of the much more agile R-60. The MiG-23MF was in production at Znamya Truda between 1978 and 1983.

MiG-23MF – 'Flogger-Bs' for export

Left: Bulgaria received 12 MiG-23MFs for service with the 1/18 IAP at Dobroslavtsi in October 1978.

Above: Hungary also received 12 MiG-23MFs, from 1979. They served with 1 VRS 'Sámán' of the 47.HRE at Pápa.

Above: Poland's 36 MiG-23MFs served with 28.PLM at Slupsk. This aircraft carries an R-23R missile.

Below: Czechoslovakia received MiG-23MFs from 1979, later bolstering its fleet with MiG-23MLAs.

Above: India bought 40 MiG-23MFs in the early 1980s, mainly as a means of quickly countering Pakistan's F-16s.

Below: The Fortele Aeriene Romane (Romanian air force) received 48 MiG-23MFs, but attrition has been high.

MiG-23MS (*izdeliye* 22MS, *izdeliye* 2MS, 'Flogger-E')

The MiG-23MF was not deemed suitable for export to some non-Warsaw Pact countries during the early 1970s. The Soviet political and military leadership maintained there would be unavoidable leaks of sensitive information that would compromise this new, very promising air defence fighter (as, indeed, was the case with the MiG-25P flown in 1972 to Japan by IA-PVO defector Lieutenant Victor Belenko). In the event, a number of MiG-23s exported to Egypt in 1974 were re-sold to the US and China for technical evaluation after only four years in service.

As a result, a purely political decision was taken in late 1972 or early 1973 by the Aviaprom and VVS leadership to develop a substantially downgraded export version of the MiG-23M. It would employ the same basic airframe and powerplant, but have equipment and weapons system taken from the MiG-21S/SM: the Almaz-23 fire control system (including the radar) and the ASP-PFD-21 gunsight. The resultant MiG-23MS (NATO 'Flogger-E') can easily be distinguished from the MiG-23M/MF by its short nose radome, which houses a downgraded export version of the RP-22SM radar. The undernose IRST was also deleted. The sanitised MiG-23M derivative was armed with up to four R-3S and R-3R air-to-air missiles only, although later the improved R-13M heat-seeking AAM was added.

The MiG-23MS was in production at Znamya Truda between 1973 and 1978. Initially, a small batch was taken on strength by the large conversion training centre for foreign pilots located at Lugovoye/Frunze in Kyrgyzstan. The type was then exported to Syria (up to 54 aircraft, the first delivered in October 1973), Iraq (18), Egypt (eight, delivered in 1974), Algeria (number and delivery timetable unknown) and Libya (54, delivered between 1974 and 1976).

MiG-23MS nose

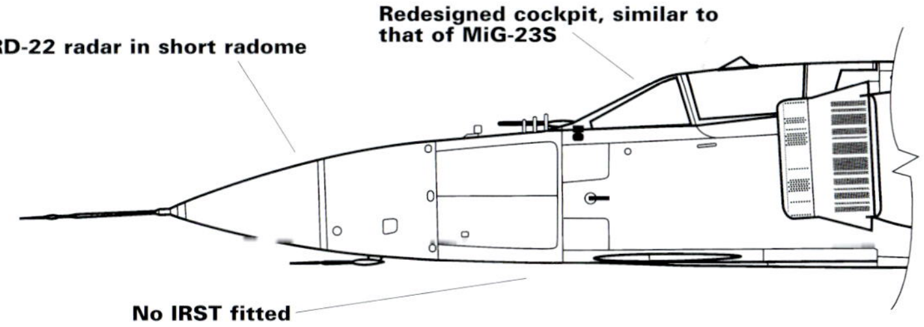

The MiG-23MS was specifically developed for export to countries such as Egypt (right). However, a few served in Soviet colours (above) at the Frunze academy which instructed foreign pilots.

MiG-23ML (Type 23-12, *izdeliye* 3, 'Flogger-G')

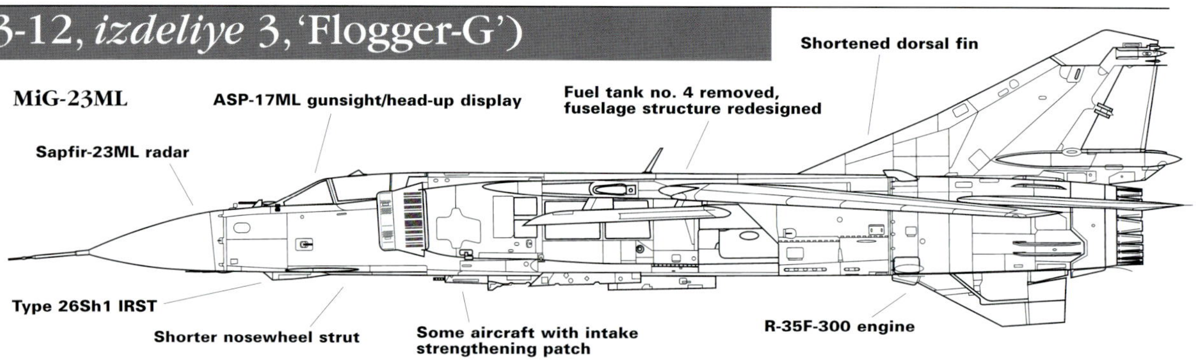

The MiG-23ML (Type 23-12), known to NATO as 'Flogger-G', was a much improved and considerably lightened MiG-23 fighter derivative. It was purposely designed in the mid-1970s in an effort to eliminate the MiG-23M's most obvious design shortcomings, i.e., airframe and powerplant strength and reliability, manoeuvrability, and radar performance.

Removal of the no. 4 fuselage fuel tank and an extensive fuselage redesign made the MiG-23ML (Lyogkii, lightweight) 1250 kg (2,755 lb) lighter than its predecessor. It had somewhat refined aerodynamics to reduce drag, a new flight control system for better handling, an uprated powerplant, and a completely redesigned mission avionics suite (SUV-2ML weapons system) for improved reliability and enhanced combat capability.

Externally, the MiG-23ML is distinguishable from the M variant by its much smaller dorsal fin and the lowered nose attitude when on the ground, which is due to the redesigned main undercarriage units.

The R-35F-300 turbojet was a growth derivative of the R-27F2M-300 and was designed by a team led by K. Khachaturov. It had a dry rating of 83.82 kN (18,849 lb) and maximum afterburner rating of 128.08 kN (28,800 lb). The turbojet featured a compressor with increased airflow (by 242 lb/110 kg per second) and a turbine inlet temperature increased by 70°. They combined to give the engine a lower specific fuel consumption (sfc) than its predecessors: at maximum afterburner, sfc was 1.96 versus 2.09 kg/kgf.h for the R-27F2M-300. The new powerplant enabled the MiG-23ML's theoretical thrust-to-weight ratio to exceed 0.83 (0.77 for the MiG-23M), but in real-world conditions the ratio proved lower due to engine 'detuning' that was needed to reduce thermal loads and extend the time between overhauls; it also increased specific fuel consumption.

During its initial years of service, the R-35F-300 was responsible for many reliability problems due to turbine blade failures and combustion chamber cracking. Reliability subsequently improved, extending time between overhauls for the late-production engines to 450 hours. However, like its R-29-300 and R-27F2M-300 predecessors, it was strictly limited to only 10 hours in total at military power and afterburner settings. The R-35F-300's hydro-mechanical control system (the same as that in the R-27F2-300 and R-29-300) was extremely effective and the pilot could slam the throttle from flight idle to full afterburner at any speed without worrying about engine failure, surge or compressor stall; it took only three to four seconds to change from idle to military power.

The structural weaknesses revealed during the aircraft's initial service were considerably strengthened (particularly the wing pivot mechanism) so that the MiG-23ML could be cleared for manoeuvring with maximum load factor of 8.5 *g* at speeds below Mach 0.85 and 7.5 *g* above Mach 0.85. The SOUA stick pusher permitted the pilot to use the full AoA envelope: at 16° sweep the AoA limit was 20-22°, and at 45° and 72° sweep it increased to 28-30°.

At 1000 m (3,280 ft), a MiG-23ML armed with two R-23 missiles had a maximum rate of climb of 215 m/s (42,323 ft/min), reduced to 145 m/s (28,543 ft/min) at 5000 m (16,404 ft) and to 125 m/s (24,606 ft/min) at 13000 m (42,651 ft) altitude and at a speed of Mach 2. Acceleration time from 600 to 1000 km/h (324 to 540 kt) was 12 seconds at full afterburner at 1000 m (3,280 ft) altitude. Time for 360° turn at 1000 m (3,280 ft) altitude was 27 seconds, at average 6.5*g*, entry speed of 900 km/h (486 kt) and final speed of 540 km/h (292 kt). Maximum rate of turn was 16.7° per second (average rate being 14.1° per second), achieved at the corner speed of 780 km/h (421 kt) and 27° AoA. Time from brake release to 15000 m (49,212 ft) altitude while accelerating to Mach 2.1 at full afterburner was 4.3 minutes. According to RSK MiG, the MiG-23ML/MLA's air combat effectiveness was 20 per cent better than that of the MiG-23M.

The Sapfir-23ML radar, officially known as N003, was among the MiG-23ML's most important new features: it was considerably lighter than the Sapfire-23D-III, was more reliable, had much better look-down/shoot down performance, and introduced the new Type 26Sh IRST. The MiG-23ML's much-improved weapons control system enabled the use of the R-23R and R-23T in the same mission; usually, the former was carried on the starboard wing pylon and the latter to port. The MiG-23ML's weapons suite gained UPK-23-250 23-mm gun pods, two carried on the underwing pylons. The aircraft retained the capability to carry a single nuclear bomb.

The MiG-23ML's navigation suite was the new Polyot-21-23, which was significantly better than its predecessor; the same was true for the Lasour-23SML datalink, SAU-23AM flight control system and the RV-5R Reper-M radar altimeter.

The MiG-23ML prototype, known as 23-12/1, flew for the first time on 21 January 1975 with Mikoyan Design Bureau's test pilot Aviard Fastovetz. The new type swiftly entered mass production the same year, replacing the MiG-23M on the production line. However, export customers continued to receive newly-built MiG-23MFs for seven more years. No fewer than 1,100 copies of the MiG-23ML and MiG-23MLA were built for domestic use and export between 1978 and 1983. The VVS received its last fighter 'Flogger' in 1981.

The MiG-23ML prototype displays the new fin shape and shorter nosewheel strut, but not apparent are the major fuselage redesign, uprated engine and new systems, including the N003 radar.

Above: The MiG-23ML had an important secondary strike/attack capability, with available weapons ranging from unguided rocket pods to nuclear bombs.

Below: In 1978 VVS MiG-23MLs visited France and Finland. The aircraft were considerably 'sanitised', even lacking the IRST sensor from under the nose.

MiG-23MLA (Type 23-12A, 'Flogger-G')

A further improved MiG-23ML derivative utilising the same airframe and powerplant but with a better mission avionics suite and equipment was flown for the first time in 1977. It replaced the MiG-23ML on the production line later in the same year. This aircraft had the Sapfir-23MLA radar, with slightly better range, reliability and ECM resistance. The radar had the frequency spacing feature that enabled group search operations, which the MiG-23M was unable to perform effectively because of interference between the radars within a group. In such operations, the fighters operated in four-ship formations at 7-9 km (3.7-5 nm) lateral separation and 16-17 km (8.6-9.2 nm) distance.

The list of new features introduced on the MiG-23MLA also included the ASP-17ML HUD/gunsight and Type 26Shl IRST. Most important, in 1981 the new R-24R/T air-to-air missile was added, which had much improved range, warhead, fuse and ECM resistance compared to the R-23R/T.

The search performance of the new IRST was limited by its field of scanning, which was restricted compared to the radar – only 60° in azimuth and 15° in elevation – and its maximum tracking rate was 6° per second. The average low-level detection range of receding fighter-sized targets (in tail-on aspect) with afterburner was around 10 km (5.4 nm), and at high level the range could increase to 25 km (13.5 nm). Typical high-level detection range of a fighter-sized target with engine at military power was around 15 km (8 nm) and maximum detection range for bomber-sized targets at high altitude approached 45 km (24.3 nm).

Externally, the MiG-23MLA was virtually identical to the ML variant, and this version, which entered mass production in 1978, was available for sale to foreign customers from 1981. In Bulgaria, Czechoslovakia and East Germany, the newly-exported aircraft complemented the previously-delivered MiG-23MFs. A sanitised MiG-23MLA derivative was subsequently sold to Angola, Cuba, Iraq, North Korea and Yemen, and possibly other countries.

The MiG-23MLA was delivered to the VVS, but is better known as an export version, equipping the second fighter 'Flogger' batches for Bulgaria (above), Czechoslovakia (below left) and East Germany (below).

MiG-23P (Type 23-14, *izdeliye* 6, 'Flogger-G')

The MiG-23P (retaining the NATO 'Flogger-G' designation), formally known within the PVO as *izdeliye* 6, was procured by the IA-PVO as an interim, low-cost solution. At least 500 aircraft are thought have come off the line at Znamya Truda between 1978 and 1981 (no precise figures are available).

It possessed the genuine look-down/shoot down capability deemed necessary to counter the threat posed by NATO's new-generation, low-altitude tactical fighter-bombers (especially the F-111, successfully used at low level in Vietnam between 1968 and 1972), strategic bombers such as the B-1B, and cruise missiles. The new interceptor drew on the basic MiG-23ML fuselage and powerplant but featured a considerably altered mission avionics suite, tailor-made to work with the PVO's remote guidance systems. It replaced both the Su-9/11 and MiG-19P/PM, which operated in relatively large numbers in the late 1970s with the IA-PVO. Externally, it could be distinguished from the MiG-23ML/MLA by the different antennas on the intakes and fin tip.

The MiG-23P (Perekhvatchik, interceptor) was equipped with the much-improved SAU-23P autopilot/flight management system with new-generation digital computer, which provided reliable, fully-automated remote guidance during an optimised flight path intercept using the Lasour-M equipment. The aircraft could automatically follow a heading to its target from the GCI ground station, with no pilot intervention except for engine control. The initial production aircraft were equipped with the ASP-23P sight/HUD, soon replaced in follow-on production MiG-23Ps by the much more modern ASP-23ML-P (DtZM-P). The weapons suite, initially comprising the R-23T/P and R-60, was upgraded in the early 1980s to the R-24R/T and R-60M AAMs.

The MiG-23P proved to be the most popular and numerous PVO interceptor to feature look-down/shoot-down capability, which was urgently needed during the late 1970s and early 1980s. The MiG-23P's Sapfir-23P (N006) radar was an improved Sapfir-23ML derivative, although initially it was unreliable and required much maintenance.

The MiG-23P was the most durable 'Flogger' derivative in Russian service: the last two MiG-23P-equipped IA-PVO fighter-interceptor regiments – one based at Kursk-East (472nd IAP) and the other at Barnaul – operated the type until early 1998. When flown by experienced pilots in BVR combats against the Su-27P, the MiG-23P – with its powerful radar and fairly low frontal-aspect radar cross-section when wings were fully swept (much lower than that of the Su-27P) – performed as well as, if not better than, the newer aircraft. The ready availability of a number of surplus airframes suitable for cannibalisation meant that the combat-capable rate of the last MiG-23 regiments in the 1990s was much better than that of those equipped with the Su-27 and MiG-31.

MiG-23P, 415th IAP, 3rd IA-PVO Corps of the Moscow air defence region, Tunoshna AB, Yaroslavl'

MiG-23P, 28th GvIAP, 2nd IA-PVO Corps of the Moscow air defence region, Andreapol 1992

MiG-23P intake

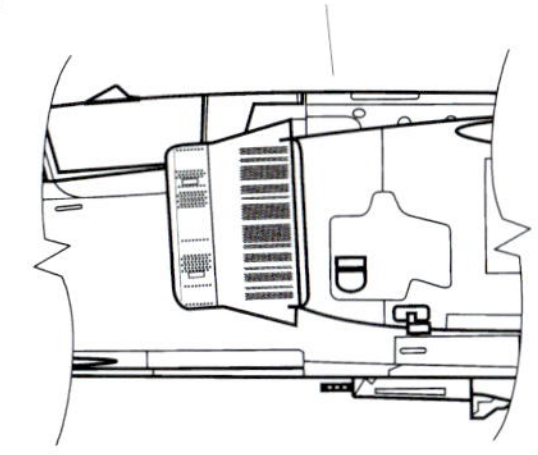

MiG-23MLD (Type 23-18, *izdeliye* 3, 'Flogger-K')

The MiG-23MLD resulted from a comprehensive upgrade programme in 1982 which was aimed at enhancing the MiG-23ML/MLA's overall air combat capability. Initially, the Mikoyan OKB proposed a newly-built MiG-23MLD derivative (Dorabotannyi, upgraded or improved) known as the Type 23-19 (some sources say Type 23-16), lacking the original aerodynamics improvements. In the event, VVS command was reluctant to order newly-built MiG-23s and eventually opted for a large-scale upgrade programme for the majority of the ML/MLA fleet, bringing them to the standard of this enhanced-capability variant with its significant aerodynamic and avionics improvements.

The resolution for commencement of the upgrade was passed in 1981 and approved by the VVS's Chief Engineer in May 1982. The first aircraft (factory number 0390315446) was converted at Znamya Truda, followed by some 560 MiG-23ML/MLAs (including a small number of MiG-23Ps) between May 1982 and May 1985 at three VVS maintenance facilities located in Moscow (Kubinka), Tchuguev and L'vov.

Above: One regiment of MiG-23MLDs (833rd IAP) remained in East Germany (at Altes Lager) until the withdrawal of the 16th Air Army. The 'Floggers' were retired after moving to Russia in 1992.

Below: Russian MLDs had a notched leading-edge root extension. This aircraft was seen at Lipetsk, where many were stored with the 4020th BRS after they had been retired from service.

The primary aim of the 23-18 programme was to enhance control at high AoA, which had been identified as the MiG-23M/ML's chief shortcoming and could be traced as far back as the MiG-23S/M. This prevented the full use of an aircraft's potential manoeuvrability in close air combat. To solve this problem, a host of aerodynamic features was introduced: vortex generators (thin steel plates) were fitted on the pitot boom, and notched leading-edge roots (known as Lambda-shaped leading-edge root extensions) were added to act as vortex generators and energise the airflow over the wings in order to delay the stall. Unfortunately, the LERXes, in conjunction with the large overwing flare/chaff dispensers, generated considerable drag which severely reduced transonic acceleration and sharply increased fuel consumption, which, in turn, restricted range and flight endurance.

The flight control system incorporated the SOS-3-4 synthetic stick-stop device/signals limiter of the MiG-29 to restrict *g*, AoA and pitch rate in order to reduce the probability of departure from controlled flight. The SOS-3-4 also automatically deployed the slats, depending on a combination of AoA, speed and altitude, giving improved stall protection and better turning ability at high AoA; this considerably improved the slow-speed handling and manoeuvring characteristics required for close air combat in the 1980s. The strengthened wing pivots allowed a new sweep position of 33°, which was intended to reduce the turn radius and to facilitate rapid deceleration during close air combat. On the negative side, the 33° wing sweep had a notably poor effect on acceleration characteristics and made handling a great deal more complex. In the event, all combat manuals continued to specify the 45° position, and 33° sweep was used primarily by experienced 'Flogger' pilots.

To enhance survivability, a pair of six-round KDS-23 downward-firing chaff/flare dispensers was installed in the stretched underfuselage centreline pylon. They were complemented on most aircraft by massive, high-drag BVP-50-60 30-round upward-firing chaff/flare dispensers. R-24R/T missiles were added to the MiG-23MLD's weapons suite and the aircraft could employ a pair of B8M1 20-round rocket pods to fire S-8 80-mm rockets fitted with a variety of warheads. The MiG-23MLD had a maximum bomb load of 2000 kg (4,409 lb), usually comprising four FAB-500 500-kg (1,102-lb) GP bombs or ZAB-500 500-kg napalm tanks. Other bombing configurations included up to 16 FAB-100 100-kg (220-lb) GP bombs carried on four multiple ejector bomb racks, up to four FAB-250 250-kg (551-lb) GP bombs, or two RBK-500 cluster bombs. The MiG-23MLD, like its MiG-23M and MiG-23ML/MLA predecessors, could employ Kh-23 and Kh-23M ASMs and a single RN-24 or RN-40 free-fall nuclear bomb.

The MiG-23MLD's appreciably better mission avionics suite centred around the Sapfir-23MLA-II (known also as N008) radar with improved maximum detection range, reliability, ECM resistance and look down/shoot down capability over rough terrain. It also had a newly-added close air combat mode with vertical scan. The new radar's maximum detection range at medium/high level against a bomber-sized target exceeded 70 km (38 nm).

Other important new avionics suite components included the SPO-15L Beryoza radar warning receiver that replaced the old SPO-10, and the A-321 Klystron digital tactical radio navigation/automatic landing system that replaced the RSBN-6S. In addition, the MiG-23MLD received the improved SAU-23-18 automatic flight control system, the all-new MRK-32-25 nose undercarriage leg steering device, and an enhanced-capability SARPP-12-24 crash-resistant flight data recorder.

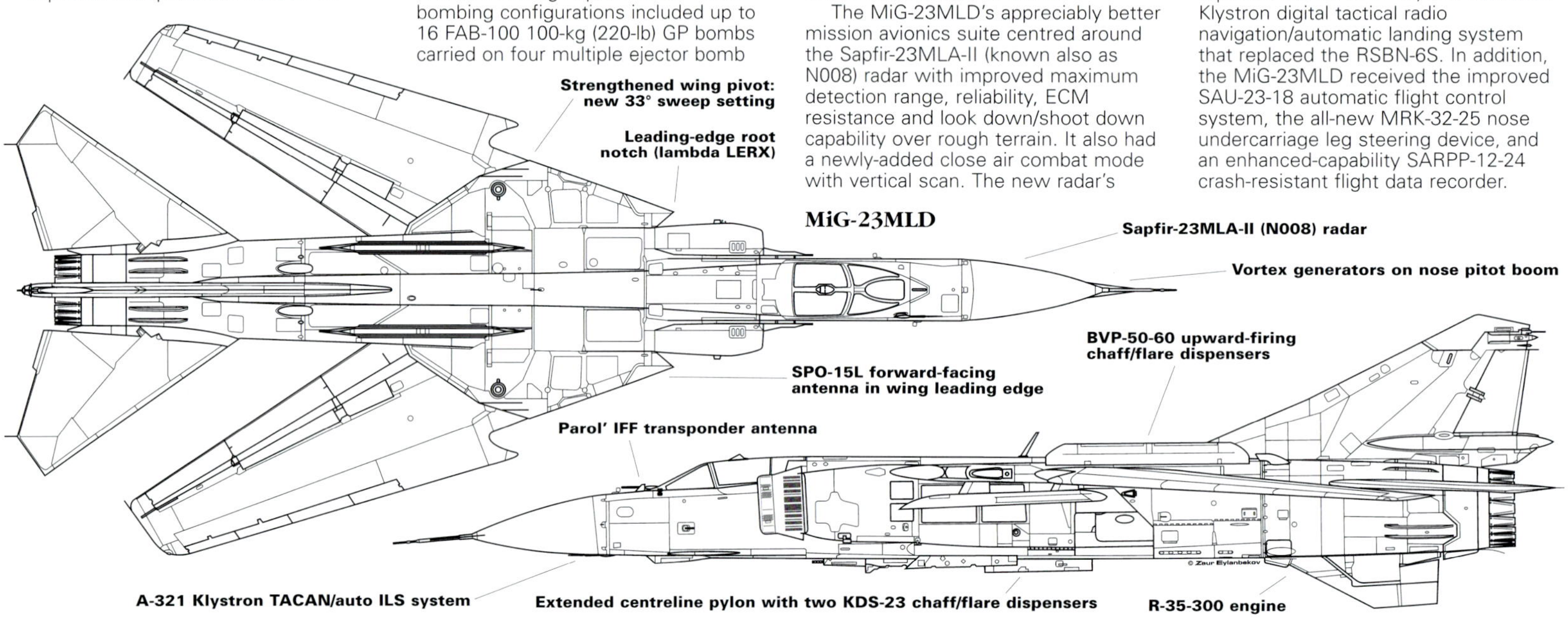

MiG-23MLD (Export) (Type 23-19B, *izdeliye* 3, 'Flogger-G')

Two export MLD derivatives were produced between 1982 and 1984 at Znamya Truda. They featured the new avionics and self-defence systems of the VVS aircraft but lacked the aerodynamic and flight control improvements of the 'true' MiG-23MLD, and had the older-generation IFF systems. These aircraft were the only newly-built MiG-23MLDs, although 66 aircraft in two sub-versions were constructed at Lukhovitzy between 1982 and 1984.

Externally, the MiG-23MLD's export derivatives can be distinguished from the MiG-23MLA/MLD only by the radomes built into the leading edges of the wing gloves, covering the forward-facing wide-band antennas of the new SPO-15L Beryoza RWR (radar warning receiver).

The first derivative, known as the Type 23-19B, was designed specifically for export to Syria and up to 50 aircraft were delivered from late 1982, together with a huge stock of the new R-24R/T BVR missiles. As many as 12 had been taken on strength by the Syrian Air Force by the end of 1982.

The Type 23-19B retained the original MiG-23MLA's Sapfir-23MLAE (N003E) radar but had less capability than the standard VVS version and perhaps lacked ECM resistance (a reduced number of pre-selected frequencies). During main overhauls in the late 1980s, the Syrian 'Flogger-Ks' received fences housing pairs of 32-round ASO-2V chaff/flare dispensers, mounted symmetrically on the rear fuselage beside the fin. According to sources close to RSK MiG, in 2003-04 a number of Syrian MiG-23MLDs were refurbished (or an agreement was reached for them to be refurbished) at the RSK MiG facility at Lukhovitzy, combined with a life extension programme to provide seven to nine more years of useful service.

One Syrian MiG-23MLD defected to Israel in October 1989. The IDF/AF pilots who tested the aircraft found its avionics, including the ASP-17ML HUD symbology, relatively easy to use, and the N003E radar was judged to be significantly more powerful than had been thought. As tests have shown using the MiG-23MLD in Israel, the swing-wing fighter demonstrated – somewhat surprisingly – that it had better acceleration than the escorting F-16 two-seaters.

Now in the IDF/AF museum at Hatzerim, this is the Syrian MiG-23MLD which defected to Israel in 1989. Export MLDs lacked the LERX notch to be found on Russian aircraft.

MiG-23MLD (Export) (Type 23-22A, 'Flogger-G')

A Bulgarian MiG-23MLD taxis in from a mission, still towing its brake chute. It is fitted with upward-firing chaff/flare dispensers.

The second MLD export sub-variant is known under the internal Mikoyan designation of Type 23-22A. Its mission avionics suite was virtually identical to that of the original Type 23-18; no significant differences have yet been discovered, and even the wiring for use of a nuclear bomb is present.

Only 16 aircraft of this sub-version were built in 1983 and 1984, for export to Bulgaria. The second batch of eight MiG-23MLD(Export)s assembled in November and December 1984 represented the very last fighter 'Floggers' to come from Znamya Truda. The aircraft serialled 219 (factory number 29600327219) was the final MiG-23 fighter to be assembled at Lukhovitzy: it underwent functional check flights on 22 December 1984 and was delivered to the Bulgarian Air Force in February 1985.

Improved MiG-23MLDs

Several further improved MiG-23MLD configurations were built in the early to mid-1980s but none reached production. They were designed in an effort to cure a major drawback of the fighter 'Flogger' – the absence of any ECM system (integrated or pod-mounted) for self-protection. As the Beka'a Valley clashes in 1982 clearly demonstrated, such equipment is a vital component of the mission avionics suite of an air superiority fighter.

In 1982, Mikoyan OKB was ordered to develop new MiG-23 versions with self-protection jammers. One proposal was the **MiG-23MLDG** (Type 23-57) with built-in Gardeniya-1 jammer, incorporating rear-facing transmitting antennas in the fin. Another, known as the **MiG-23MLS** (Type 23-47), was suggested to export customers whose aircraft contained the older SPS-141 Siren jammer (note: some sources indicate that the 23-47 was also equipped with the Gardeniya-1). Mikoyan sources refer to another MiG-23 derivative fitted with an ECM system – the **MiG-23MLG** (Type 23-37) – also with the Gardeniya-1 active jammer.

MiG-23 flying testbeds

A number of MiG-23s were used by the Zhukovskiy-based LII (Flight Research Institute) and the Mikoyan Design Bureau as testbeds in various aircraft programmes. The first was the **LLMiG-23** (a converted initial-production MiG-23S) used for flight testing of new automatic landing systems between 1970 and 1975. Later, a two-seat **LLMiG-23UB** was used for this purpose, as well as to conduct research into direct lift control.

Several MiG-23MLs were modified for use in support of test programmes and were used to test the MiG-29's N019 radar (1979) and to carry out the initial test firings of the new R-27 and R-73 air-to-air missiles (1980).

A 'stealth' MiG-23 testbed, employed in the 1980s, had rubber coatings applied to some fuselage areas in an attempt to determine the radar cross-section and thus to decrease detection ranges of enemy radars. This aircraft then went into storage at Mikoyan's own flight test facility and was scrapped at Zhukovskiy in 2002.

MiG-23 upgraded variants

Several ambitious upgrade programmes for the MiG-23M/MLA were loudly announced by Mikoyan in the mid-/late 1990s. At that time, fighter 'Floggers' were still in reasonably active use with over two dozen foreign air arms. Studies had predicted – over-optimistically, as it turned out – that in the mid-1990s a mildly profitable upgrade market would exist for as many as 500 aircraft, with prices ranging from $US300,000 to $US1 million per unit depending on the extent of work. With original service lives to be extended up to 30 years, the upgraded 'Floggers' were advertised as being suitable for safe operations until 2011/2014.

The several MiG-23 upgrades were promoted under the common designation of **MiG-23-98** and took advantage of some design solutions similar to those proven on the upgraded MiG-21-93, which at the time was in development under an order placed by the Indian Air Force. The MiG-23-98's new mission avionics suite made use of a digital weapons system computer and the Russian version of what was known as a 'partial glass cockpit' in which one or two displays replaced some of the original dial instruments. Utilising new aiming modes proven on the MiG-29, the digital computer would provide new capabilities for the gun and unguided weapons. A helmet-mounted cueing system for extending the R-73's cueing capability in close air combat was included in the 'maximum'-standard upgrade package for the 'Flogger-G'. Initially, the NIIR-Phazotron Super Kopyo radar (an enhanced-capability derivative used in upgraded IAF MiG-21bis-UPGs) was proposed, but Mikoyan designers later revealed their new intention to use the more modern radars of the NIIR-Phazotron Moskit family (Moskit-23 or Moskit-21K) in the **MiG-23-98-1**.

New self-defence aids were proposed in the upgrade package, which would probably have been the Gardeniya-1FUE active ECM system. Other features were a new-generation 'solid-state' flight data recorder, all-new communication suite and IFF.

The Mikoyan Design Bureau (now known as Mikoyan Engineering Centre, a design organisation within RSK MiG) proposed a less expensive **MiG-23-98-2** upgrade standard in the late 1990s. It advocated using an improved existing Sapfir-23MLA radar instead of replacing it with a new-generation model, which would give overall air combat effectiveness 2.6 times better than that of the original MiG-23ML.

The upgraded 'Flogger' would be able to fire the new R-77 (AA-12 'Adder') active radar-homing AAM at a maximum range of about 90 km (48 nm) in high-level/tail-on engagements, and also to use the R-27R1 SARH/IR-homing BVR missiles of the MiG-29. The highly-agile R-73 dogfight missile could easily have been integrated into export MiG-23M/MLA/MLDs, working with a helmet-mounted cueing system. For mid-course updates of both the R-27R/T and R-77 BVR AAMs, a fire correction channel datalink system would have been added to the Sapfir-23 radar. Integration of the MiG-29's KOLS-29 IRST/laser rangefinder would have further enhanced the MiG-23-98's air-to-air capability. The air-to-ground weapons suite was to be augmented with the Kh-31A high-speed anti-ship missile (and/or its Kh-31P anti-radar derivative) and the KAB-500Kr TV-guided bomb.

Mikoyan revealed the prototype of the MiG-23-98 in early 1999. It is believed that the aircraft was intended for ground display only, there being no intention to use it in flight trials before an upgrade order was received from a foreign customer; so far, one has not materialised.

NIIR-Phazotron also proposed capability-enhancement options, these being much more expensive and far more capable. They centred on a new radar, the Moskit-23, with a detection range of up to 90 km (48 nm) against fighter-sized targets. The Moskit-23 would enable firing of the R-77 and R-27R/T AAMs as well as the Kh-31A high-speed anti-ship missile. A more affordable alternative would be the Moskit-21 lightweight derivative of the baseline Moskit family, which has a maximum detection range of 45 km (24 nm) against fighter-sized targets.

However, Mikoyan's predictions that there would be a huge potential upgrade market for the 'Flogger' proved to be entirely groundless and no upgrade contracts have yet been announced. The main reason for the lack of strong interest in these comprehensive MiG-23 upgrade proposals is their complexity and costs that would be prohibitively expensive for most of the existing operators. Those air arms still operating the MiG-23M/MLA are all subject to tight budgets, and are more concerned with maintaining airworthiness and extending the life of their aircraft and weapons stocks than with investing a lot of money in costly avionics and weapons upgrades. They would prefer some low-cost maintenance rationalisations, navigation system improvements, and perhaps the integration of R-73 missiles.

Mikoyan's MiG-23-98 'prototype' (ground demonstrator) is seen armed with representative R-73 and R-77 missiles.

The Soccer War of 1969

By 1969 growing resentment in Honduras to the influx of Salvadoran immigrant workers and wealthy entrepreneurs had reached boiling point. That same year the two countries had been brought together to play a series of vital qualifying games for the 1970 Football World Cup finals and, as tensions rose, competition on the pitch led to full-scale conflict on the ground and in the air. The obsolescent equipment of the respective air forces was thrust together in the world's final piston-engined air combat battles.

Above: Undoubtedly the 'star' performers of the Soccer War were the Honduran air force's F4U-4 and -5 (seen here) Corsairs. Not only did the aircraft claim the only air-to-air 'kills' of the war, but also played an important role in providing support for the ground forces delivering bombs, napalm and rockets, as well as conducting devastating strafing attacks.

Below: El Salvador's remaining aircraft were assembled at Ilopango at the end of the war. Here a pair of F-51Ds sits alongside a C-47 and a B-26B Invader.

The 100-hour war, as it is known in El Salvador, Guerra del '69 as it is known in Honduras or the Soccer War as it is known to the rest of the world, is a conflict that has become a part of aviation history as the last instance of piston-engined air-to-air combat. In the late 1960s many Latin American countries had yet to enter the jet age and relied on World War II-era warplanes for the defence of their airspace. Along with obsolescent air forces, neither El Salvador, nor Honduras, possessed air defence radar, or any other detection system – relying on ground observers to locate and track any incursion visually. With few navigation aids available, interception, bad weather and night operations were all verging on impossible, yet both air forces managed to conduct extensive operations during the conflict with these limited resources.

Causes of conflict

The relative economic wealth and density of population were undoubtedly the underlying causes of the dispute between the two countries. In El Salvador overcrowding, caused by 3.3 million people living on just 21,393 km^2 (8,260 sq miles) of land, had led to increasing migration to neighbouring countries. Some 275,000 Salvadorans crossed the border to Honduras, where only 2.6 million people were living in a country almost six-times the size.

The situation was exacerbated by the employment offered by US companies United Fruit and the Standard Fruit Company that cultivated bananas in Honduras. With El Salvador in need of additional space for its growing population and the Honduran people becoming increasingly aggravated by the constant flow of Salvadorans to their country, the potential for conflict was laid. In addition, the Salvadoran economy (together with Guatemala) was a vital member of the Mercado Común Centroamericano (MERCOMUN/Cental American Common Market, created in 1961) and wealthy Salvadoran entrepreneurs had established companies and properties in Honduras and were seen to be flaunting their wealth. Anger in Honduras led to the forma-

Left: Both sides utilised the venerable Texan during the conflict. This aircraft is an SNJ-4 of the FAH and was used for armed reconnaissance and liaison duties. It was especially useful at night as it was the only FAH aircraft equipped with blind-flying instrumentation.

Below: El Salvador possessed only one airworthy B-26B Invader at the outbreak of war and it was little used in the conflict. This example (FAS 602) was one of four examples hastily acquired as the conflict raged and is seen here soon after its arrival in El Salvador, shortly after the ceasefire agreement had been implemented.

tion of civil armed groups, such as 'La Mancha Brava', that began attacks on Salvadorans living in Honduras.

To help keep a lid on the simmering tension the two countries had signed agreements regarding land-use and immigration in 1962, 1965 and 1968 – these, however, did little to pacify the radical groups on either side of the border.

In 1967 isolated border incidents became more serious and included the capture by the Honduran Army of two Salvadoran Army trucks (carrying two officers and 39 soldiers) inside Honduran territory. After hurried negotiations, chaired by ex-US President Lyndon B. Johnson, and the release of a relative of the Honduran president, in jail in El Salvador for common crimes, the captured soldiers were returned.

By 1968 the political and economic situation in Honduras was rapidly deteriorating and many blamed the influx of Salvadoran immigrants. In January 1969 the Honduran government abandoned the bilateral immigration treaty with El Salvador, and deported significant numbers of Salvadorans from the land they occupied. In the spring of that year over 500 families from El Salvador were violently expelled.

Tensions continued to rise as reports of some Salvadorans returning to Honduran land incensed Honduran nationals. The Salvadoran media highlighted the violence involved during deportations. To make matters worse, in June, both countries were to confront each other in an attempt to qualify for the 1970 World Cup finals. During the first of the two matches (at Tegucigalpa, Honduras) violence briefly erupted before order was restored by security forces. However, during the second match at San Salvador stadium, violence erupted on a much greater scale and soon escalated far beyond the confines of the football stadium. Attacks on Salvadoran immigrants in Honduras rapidly rose in frequency and intensity and, on 27 June 1969, the growing problems forced Honduras to break diplomatic relations with El Salvador – the countries were on the brink of war. El Salvador's triumph in a third match, held at a neutral venue (Mexico City) on 28 June incensed the Honduran population further and led to an exodus of 75,000 Salvadorans back to their country.

After an unsuccessful attempt by El Salvador to allow the Organización de Estados Americanos (OEA) to mediate, both countries commenced mobilisation of their armed forces.

First combat

On 3 July 1969 a Douglas DC-3 of SAHSA (Servicio Aéreo de Honduras S.A/Honduran Air Service), took-off from Ocotepeque airport, 8 km (5 miles) from the border with El Salvador, on a routine flight. During the initial climb the aircraft was fired on by anti-aircraft artillery from just inside El Salvador. The pilot advised the control tower, that in turn notified the Fuerza Aérea Hondureña (FAH) – two armed T-28 Trojans were dispatched to investigate. Flying over the border zone, the T-28s entered air combat with a number of Fuerza Aérea Salvadoreña (FAS) Cavalier F-51D Mustangs, however, no aircraft on either side appear to have inflicted any damage on the opposition.

On the following day, a civilian-registered Salvadoran Piper PA-28 Cherokee (YS-234P) crossed the Honduran border, carrying military personnel on observation duties. Between Gualcince and Candelaria, two towns close to the border, a pair of FAH T-28s located the intruder and, as it was over Honduran airspace, ordered the Cherokee to land. The Salvadoran pilot refused and headed for home, the T-28s electing not to open fire.

Meanwhile, border incidents between both armies escalated, with some losses on both sides. To help monitor the deteriorating situation the Salvadorans used a Cessna 310 from the Instituto Cartográfico Nacional and other civilian light aircraft for area reconnaissance and to take photographs of possible targets.

According to El Salvador, on 4 July an SAHSA DC-3 attacked Salvadoran border positions close to El Poy. This would be the location of a further outbreak of ground fighting on

Along with six Goodyear FG-1D Corsairs, the FAS's five Cavalier F-51D Mustang IIs would form the backbone of El Salvador's offensive air operations. FAS 404 and 405 are seen here during delivery from the USA in 1968, the former was shot-down by Capt. Soto in his F4U-5N (FAH 609) on 17 July 1969, killing the pilot, Capt. Varela.

El Salvador launches main attack – 14 July 1969

At 15:00 on 14 July the Salvadoran pilots met to await the order to attack from the country's president and armed forces general commander, General Fidel Sánchez Hernández. Finally receiving the order at 16:30, engines were started and 30 aircraft departed to begin the attack. By 17:00 the C-47s FAS 101, 102 and 104 (configured as bombers with manually dropped 50-kg/110-lb bombs), Corsairs, Mustangs, one B-26B Invader, one Texan and a gaggle of impressed civil aircraft were en route to their targets.

The plan was to attack the airfields at Tocontín with one C-47 and two Mustangs, Catacamas with two Mustangs and the B-26, and La Mesa airport at San Pedro Sula with five Corsairs. In addition the towns of Valladolid and Nueva Ocotepeque would be struck by a single C-47 each, and the 14 civilian aircraft and the Texan were to mount nuisance raids against San Marcos Ocotepeque, Santa Rosa de Copán, Nacaome, Amapala, Quipure, Yoro, Guarita, La Virtud, Jinigual and La Labor.

With the Salvadoran raiders passing overhead, the Hondurans quickly became aware of enemy action on the ground. Radio communications from the army's border posts advised of enemy patrols crossing the border with increasing frequency. In the Ocotepeque district defending troops were coming under artillery fire. The defences were forced to abandon the border city of Guardaraya and retreat further inside the Honduran territory. FAH reconnaissance aircraft reported enemy troop concentrations at many places along the border.... K-2A tower (Tocontín) reported an urgent call from La Paz district at 18:05 stating 'two unidentified aircraft were heading to Tegucigalpa'.

At 18:10, as the light was beginning to fade, the Salvadoran attack force arrived at Tegucigalpa and C-47 FAS 101, commanded by Maj. Jorge Domínguez, proceeded overhead Base Aérea Tocontín at 8,000 ft (2438 m). Through the open door of the C-47 the first bomb was manually dropped. Immediately the first anti-aircraft artillery, a 12.5-mm machine-gun to the south of the base, opened fire, but the C-47 quickly passed out of range. At this point northern area commander Col Mario Chinchilla called the base to inform them about the possible incursion of enemy aircraft, receiving the answer "bombs are falling, my colonel". The colonel ordered all available fighters to scramble and for plans for a counter strike to be immediately formulated.

The Mustangs attacked for 10 minutes without opposition, as the FAH's Corsairs were unable to take off, however, thanks to the aircraft being well dispersed none was lost during the attack. The base also suffered only minor damage because of poor manual aiming from the C-47. On their return, the Mustangs attacked the small towns of Jalteva, El Suyatal and Guaimaca.

Meanwhile, the five Corsairs launched their attack on La Mesa airport and the B-26 bombed Catacamos from medium altitude along with two F-51Ds. The Texan and the civilian aircraft struck the cities of Gracias, Choluteca, Juticalpa, Catacamas, Guavacas, Amapala, Nacaome, Santa Rosa de Copán and Nueva Ocotepeque, and the remaining Corsairs attacked Nueva Ocotepeque and Santa Rosa de Copán.

By 18:20 the raids had ended and by 19:00 the FAS aircraft were landing at their allocated dispersion bases, except the F-51Ds, which returned to Ilopango. As this was happening, the Salvadoran army crossed the border in numbers, beginning a rapid advance across Honduras, occupying El Poy in the North, and the vital Nueva Ocotepequeto Santa Rosa de Copán road. In the south they conquered El Amatillo close to the Fonseca Gulf. The Salvadoran leadership was confident that this 'lightning strike' would force the Honduran to accede to their demands.

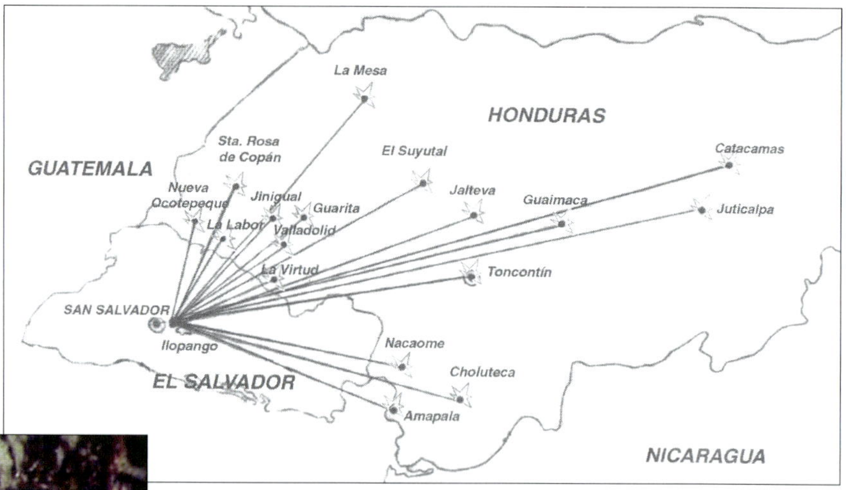

Above: Salvadoran C-47 FAS 104 was used in both the transport and bombing role, with the port cargo door removed (as seen here).

Left: The FAS used a number of forward deployment bases during the conflict, including Carrera airfield, Usutlán, where this FG-1D is seen prior to a mission on 15 July. In the centre is Capt. Guillermo Cortez, who was shot down and killed on 17 July.

13 July that would act as the prelude to full-scale conflict that would begin the next day.

Honduras under fire

The Salvadoran government prepared to take the initiative as full-scale conflict seemed inevitable with its Plan de Campaña Capitán General Gerardo Barrios – a surprise attack designed to give El Salvador an advantage from the beginning. The comparative weakness of the FAS in comparison with the FAH had also dictated that a pre-emptive strike would be necessary. The aircraft of the Club Salvadoreño de Aviación Civil y Reserva (Salvadoran Civil Aviation and Reserve Club), the only civil aeronautical institution in the country, were impressed into use for the operation. These comprised some 13 piston-engined types, mostly agricultural crop sprayers, plus one Beech Mentor, lent by Capt. Adrián Panameño, and armed with 60-mm and 81-mm mortar grenades launched using an improvised mechanism developed by Subteniente Leopoldo Suárez.

The Salvadorans planned to attack the main FAH bases at Toncontín and La Mesa on 14 July to destroy the enemy air force on the ground. To prevent the FAH striking back that day the attack was scheduled for sunset.

El Salvador divided the combat zone into three sections: North (Teatro de Operaciones Norte, TON), Chalatenango (TOCH) and Western (TOOC).

The Salvadoran ground forces were composed of four infantry battalions, one artillery battalion and the Fuerza Expedicionaria. Facing them, the Hondurans had three infantry battalions, six zone battalions, one sapper battalion and two batteries of 75-mm artillery. In response to the Salvadoran threat Honduras had prepared the 'Congolon' plan for the defence of the country that supposed "the enemy will operate offensively to limited targets in the zones of Nueva Ocotepeque, Marcala and Nacaome, with the support of the Navy in the latter zone".

Salvadoran air assets

The FAS's Cavalier Mustangs were to be flown without tip tanks, making them faster and more agile, although during the first oper-

El Salvador's Cavalier F-51Ds were delivered with auxiliary wing-mounted tip tanks, as seen on this example in Panama during its delivery flight in December 1968. The tanks were retained for a number of the early combat missions but were later discarded to improve performance and manoeuvrability.

Above: With the cargo door removed, Honduran C-47 FAH 304 launched on 15 July to bomb Ilopango airfield, but was forced to return with mechanical problems shortly after take-off. It was replaced by FAH 306.

Right: Civilian operator Archie Baldocci provided the FAS with this North American P-51D shortly before the conflict, bringing the FAS's inventory up to six Mustangs.

ations a number of the aircraft still utilised them. The FAS's Corsairs were operating under a number of limitations because of their age. They lacked performance at high altitude because their superchargers were out of service.

After further reconnaissance, including a mission by the Cessna 310 over Tocontín airfield, the order to launch the first air strike of the war was sanctioned and in the late afternoon of 14 July, all available FAS air assets began to take-off from their respective airfields.

The FAH responds

Shocked by the suddenness of the Salvadoran attack, the FAH was not in position to immediately respond. As the last of the attackers were completing their bombing runs FAH Lt Col Serra Hernández managed to get airborne from Tocontín in a F4U-5, determined to intercept the enemy. However, deteriorating weather and the fading light made it impossible to locate the attacking FAS force, so he headed to the Salvadoran harbour of Cutuco, attacking the customs building and piers with cannon fire. He was followed into the air by Capts Fernando Soto Henriquez, Francisco Zepeda Andino and Maj. Oscar Colindres Corrales, also in F4Us, who conducted defensive patrols and, on completion, conducted the first F4U night landings in FAH history. In addition, two F4U-4s from San Pedro Sula, commanded by Teniente Edgardo Mejía and Teniente Marco Tulio Rivera strafed a convoy advancing to Nueva Ocotepeque, one of the aircraft receiving damage from 12.7-mm groundfire.

Honduran bombing

Col Enrique Soto Cano, commander of the FAH, immediately sought authorisation from the Honduran President to launch reprisal attacks in the form of a number of strategic bombing raids inside El Salvador. Only one attack, using a C-47 against Ilopango airport, was authorised as the foreign relations minister, Virgilio Carias, was keen to show to the OEA and the United Nations that El Salvador was the aggressor.

The C-47 (FAH 304) was armed with 18 50-kg bombs that were to be dropped from the open cargo door using a roller rail. At 01:49 the plane departed, however, it almost immediately returned with mechanical problems. After transferring the weapons, it was replaced by FAH 306, commanded by Capt. Rolando Figueroa, and departed at 03:18. Flying almost completely blind, he tracked westwards until he reached the Pacific Ocean and from there headed to Ilopango. With no radar the Salvadorans were unaware of the intruder.

The C-47 continued by dead-reckoning until the crew estimated they were over Ilopango and began to launch the bombs until 14 had

Above: A trio of Honduran F4U-4s formates over the mountains in the west of the country. The superior performance of the FAH's F4Us, in comparison with the FAS's FG-1Ds, gave them a decisive advantage in air combat and allowed them to conduct both air defence and ground attack duties on a single mission. Although a number of the Honduran Corsairs suffered battle damage, none was lost to enemy action.

FAH hits strategic targets – 15 July 1969

After mounting the attack on Ilopango airfield, the first strike package of Corsairs continued to the coast 200 km (124 miles) east of Ilopango for a cannon and rocket (F4U-4s) attack on the Cutuco fuel tanks and La Unión Harbor. The raid was certainly successful and a number of the fuel tanks remained ablaze for days afterwards. Despite coming under intense ground fire all of the Corsairs survived.

The F4U-4s from La Mesa continued to the Acajutla oil refinery, on the Pacific coast. Arriving overhead the Corsairs dropped their bombs over the fuel storage facilities, setting a number alight, before launching low-level rocket and machine-gun strafing attacks.

According to the Salvadoran version of events, one of the Corsairs was intercepted by F-51D FAS 402, flown by Capt. Douglas Vladimir Varela Moreno, who had scrambled during the attack on Ilopango airfield.

Capt. Walter López's F4U-4 FAH 617 was damaged either by ground fire or by the F-51D and, leaking fuel, diverted to Guatemala, landing at El Pilar airfield belonging to the United Fruit Company. The Fuerza Aérea Guatemalteca took the aircraft to La Aurora airport in Guatemala City, escorted by two P-51D Mustangs of Fuerza Aérea Guatemalteca. There the Corsair and pilot were interned until the end of the war. Here, the one remaining rocket on the F4U-4 was subject to particular attention – the Guatemalans discovering that it was a makeshift weapon comprising an 81-mm mortar grenade integrated on an inert practice rocket. The Salvadorans insisted that the F4U had been damaged by the F-51D, with the FAH claiming it was hit by ground fire.

Honduras responded to the Salvadoran attacks on 14 and 15 July by using its Corsairs to inflict serious damage to a number of strategic targets. The fuel storage tanks at Acajutla (right) and installations at Cutuco harbour (below) were both set ablaze by bomb, rocket and strafing attacks by FAH F4Us on 15 July.

Armed with four 20-mm cannon, as opposed to the six machine-gun armament of the FAS's F-51Ds and FG-1Ds (and the FAH's F4U-4s), the FAH's six F4U-5s were the most effective warplane fielded by either side. The more powerful 2,850-hp (2126-kW) Pratt & Whitney R-2800-32W engine also bestowed greater bomb-carrying capacity.

been dropped. Although the crew reported that they were received by anti-aircraft fire, Salvadoran reports stated that the aircraft was never heard at the airport and no bombs fell within the zone. There was only one report of a aircraft flying that night at low altitude near the Zapotitán valley, some 55 km (34 miles) from Ilopango. This could have been the C-47.

With no sign of a ceasefire imminent, the FAH received authorisation for a larger strike. At 04:22 three F4U-5s, armed with four 20-mm cannon and one 250-kg (551-lb) bomb (Maj. Oscar Colindres Corrales, Capt. Francisco Zepeda and Capt. Fernando Soto), together with one F4U-4 (Teniente Santiago A. Perdomo) armed with six 12.7-mm machine-guns, six rockets and one 250-kg bomb, headed for El Salvador at 10,000 ft (3048 m). Elsewhere, at La Mesa, the FAH's sole C-54 took-off for radio liaison with the strike package. It was followed into the air by the F4U-4s of Capt. Carlos Aguirre Corrales, Teniente Edgardo Mejía, Capt. Walter Reyes and Teniente Marco Tulio Rivera, all armed with six machine-guns, one 250-kg and one 50-kg bomb and four 2.75-in rockets.

The Tocontín-based Corsair section appeared over Ilopango just as the sun was rising and began to attack through a defensive shield of anti-aircraft fire. The F4Us dropped their four bombs but failed to destroy any enemy aircraft, although one bomb destroyed a water truck and damaged a hangar. Another fell on the runway without exploding, another on a taxiway causing little damage and the fourth fell on open ground. The reason for the poor aim was explained later by Fernando Soto in an interview with Marco Lavagnino: "when we began the dive, we encountered very intense anti-aircraft fire [Oerlikon 20-mm guns] and of our four Corsairs, only Maj. Colindres descended to the designated altitude, hitting the hangar and the truck. The other three Corsairs dropped the bombs before time at about 9,000 ft [2743 m]". One of the Corsairs was damaged, with parts falling onto the enemy airfield, but managed to return to base. Following the attack at Ilopango, the four Corsairs, along with the second section from La Mesa, attacked various strategic ground installations (see box) before returning to base.

The action intensifies

At 05:00 on 15 July, as the FAH Corsairs were engaged over El Salvador, the FAS launched another airfield attack using three Mustangs, two FG-1D Corsairs and the C-47 FAS 101. the latter had been prepared during the night with the civilian cargo rollers modified to move the 75-kg (165-lb) bombs to the open cargo door.

Expecting further FAS attacks, at 05:31 two

Fuerza Aérea Hondureña – July 1969

At the outbreak of the conflict the FAH possessed some 27 airworthy aircraft, of which by far the most important combat types were the F4U-4 and -5 Corsairs and armed T-28 Trojans. F4U-5 FAH 604 operated throughout the conflict and remained with the FAH until 1978, when it was sold to a warbird collector in the USA. T-28A FAH 213 was one of five purchased from ex-USAF stocks and, armed with two 12.7-mm machine-gun pods and 30-lb (13.6-kg) bombs, flew numerous close air support and armed reconnaissance sorties.

Vought F4U-5 Corsair, FAH 604

North American T-28A, FAH 213

The Salvadoran air force grouped together military and civilian training and liaison aircraft for participation in the conflict. SNJ-5 Texans were used for reconnaissance of the front line and a single example was fitted with two 100-lb (45-kg) bombs and two 7.62-mm machine-guns for close air support duties. FAS 74 (right) is seen here taking off from the main El Salvadoran base at Ilopango.

armed FAH T-28s took-off from Tegucigalpa for a defensive patrol over the city. Meanwhile, on the ground, wrecked Corsairs, T-6 Texans and Vultee BT-13s were dispersed around the airfield to act as decoys.

At 06:03 the F-51s dropped three bombs on the airfield as they commenced the attack. While overhead, FAS Maj. Girón Cortez reported an enemy Corsair departing and dropped his bombs over a hangar. Other machine-gun fire damaged the North American N.A.16 installed as a monument in front of the Escuela de Aviación.

Capt. Guillermo R. Cortez, flying one of the FG-1Ds, then initiated his attack, firing his 12.7-mm machine-guns at various ground installations. Concluding his strafing run, he immediately entered combat with the FAH Corsair that had taken-off to intercept them (FAH 605 flown by Col José Serra Hernández). However, the latter's cannons failed and he was forced to leave the fight. Almost immediately the two patrolling FAH T-28s arrived and, armed only with 7.62-mm machine-gun pods, attempted to intercept the FAS attackers. The F-51 flown by Cortez suffered damage to the rear fuselage, cutting a battery cable, and he was hit in the right leg. According to the official Salvadoran version of events, he separated from the main group and headed to the Honduran President's residence, where he dropped his 75-kg bomb, which failed to explode. He then returned to base, while the remaining Mustangs and the C-47 attacked Nueva Ocotepeque and El Poy in preparation for a ground invasion.

As the C-47 and F-51s began to turn for home they were intercepted by a patrol of FAH F4U-5s. The Corsair pilots claimed the destruction of one Mustang and the C-47, but it appears that both returned to base without damage.

Later in the morning the FAS's sole B-26 attacked, with F-51 support, the city of Choluteca, bombing and strafing the Honduran army positions.

Close air support.

With fighting on the ground well underway, the morning of 15 July marked the final strategic raids of the conflict. From now on the air forces on both sides would be dedicated to the close air support role.

As with the Salvadorans, the FAH looked to boost its capabilities by impressing civilian aircraft into service. Reconnaissance, liaison and medical evacuation were conducted by a fleet of light piston types, whereas SAHSA transport aircraft were used to ferry troops and weapons, LANSA types carried other cargo and provisions and TAN distributed aviation fuel.

FAH T-28s flew reconnaissance and patrol sorties throughout the remainder of the day over Tegucigalpa and Valle de Sula in western Honduras.

At 08.00 three Honduran F4U-5Ns had taken off to attack the bridge at the border town of El Amatillo. Although one of them was to return because of cannon failure (again FAH 605), the other two attacked a Salvadoran troop convoy, one dropping a 250-kg bomb and the other two 50-kg bombs, destroying a machine-gun position.

Meanwhile, two F4U-4s, flown by Capt. Corrales and Teniente Rivera, launched an attack on the frontier post of El Poy, armed

Fuerza Aérea Salvadoreña – July 1969

Apart from the sporadic bombing missions conducted by the FAS's C-47s and the B-26B, the responsibility for offensive operations fell almost entirely to the six F-51Ds and six FG-1Ds. Cavalier F-51D FAS 405 is shown here early in the conflict wearing the USAF SEA-style camouflage adopted by the Mustangs. The tip tanks were soon removed in order to improve manoeuvrability. During the conflict the FAS lost three of its six FG-1Ds, although FAS 220 was one of the survivors and is seen wearing the yellow identification bands added after one of the FG-1Ds (FAS 219) was shot down by friendly ground fire on 17 July.

Cavalier F-51D Mustang II, FAS 405

Goodyear FG-1D Corsair, FAS 220

Above: Wearing the yellow identification marks applied during the conflict, the FAS's surviving FG-1Ds continued to mount armed patrols along the border as the evacuation of Salvadoran nationals was completed.

Right: Cavalier Mustang FAS 405 is seen here undergoing engine repairs at the forward operating airstrip on Madresal Island at the height of the conflict.

with two 50-kg bombs and four 2.75-in rockets. A truck carrying troops was destroyed although Corrales' Corsair suffered damage from ground fire.

By 11.00 Salvadoran troops were advancing along the Pan-American road close to Goascorán River and two FAS FG-1Ds, flown by Maj. Mario Encarnación Echeverría and Maj. Girón Cortez, intercepted a pair of FAH F4U-5s that was preparing to attack. One of the Honduran Corsairs managed to get into position behind Girón Cortez and Echeverría went to his aid, managing to hit the Honduran F4U, which was forced to retire from the fight.

As offensive patrols intensified, the Honduran F4U-5s returned to attack artillery positions at El Amatillo and conducted armed patrols over Tegucigalpa and the region of Las Hormigas, while the F4U-4s continued to patrol over San Pedro Sula and provide close air support to the Honduran Army that was fighting close to the border.

The FAS was also conducting its own offensive missions. Two FAS C-47s were tasked to bomb the defending Honduran ground forces at Nueva Ocotepeque and support Salvadoran troops that were poised to take the city. One of the C-47s (FAS 101) had been equipped with two 12.7-mm machine-guns in the cabin for additional strafing attacks. As the pair of C-47s arrived over the target, four FAH F4U-4s were already overhead, attacking Salvadoran artillery machine-gun positions. C-47 FAS 101, commanded by Maj. Sigfredo Velasco, was attacked by F4U-4 FAH 615, flown by Teniente Marco Rivera. As one of the aircraft mechanics on the C-47 later recalled: "we had just finished launching the bombs and I turned to arrange something on our 12.7-mm guns, when I heard the distinctive sound of machine-guns. When I looked through the cargo door, the first thing I saw was a big oil contrail from the No. 1 engine. The Honduran Corsair had hit the engine, and the propeller blades also suffered multiple hits. The engine lost all its oil from a damaged hose but luckily didn't explode. The pilot shut down the engine. More hits blew one of the tyres on the main landing gear, the Corsair continuing to attack from below until its ammunition was exhausted. We jettisoned everything we could to reduce weight and dived away from the area". The C-47 limped back to Ilopango where it made a single-engined belly landing, sliding through 180°. The aircraft was later repaired and returned to service.

The 15 July also marked the loss of Cavalier F-51D (FAS 400), piloted by Capt. Benjamín Trabanino Santos, which was forced to land at Izábal, in Guatemala. The aircraft was taken to La Aurora airport, Guatemala city where it remained until the end of the war.

By the end of the day, Nueva Ocotepeque was surrounded by the Salvadorans, while the cities of Valladolid, Guarita, Junigual and San Marcos Ocotepeque had been conquered.

The 'kills' of Capitán Fernando Soto

With the Hondurans desperate for revenge following the Salvadoran invasion, FAH air operations commenced early on 17 July with attacks on enemy infantry positions at El Amatillo, just across the border in Salvadoran territory. At 11:30 El Salvador dispatched two F-51Ds on a close air support mission to support the 11th Infantry Battalion who had reported they were under attack from Honduran aircraft. As the battle intensified the FAH sent three Corsairs, commanded by Capts Soto (in FAH 609), Acosta and Zepeda to bomb the same area at 11:55.

Zepeda was forced to return when the cannons on FAH 605 failed again and was pursued by the patrolling FAS Mustangs. The F-51D FAS 404, flown by Capt. Humberto Varela, engaged the fleeing Corsair, Zepeda's attempts to outrun his pursuer being hampered by a mechanical failure preventing him from jettisoning his bombs. Climbing at full power Zepeda requested urgent assistance from Soto and Acosta. They asked for his position and he responded: "Southeast of El Amatillo with two Mustangs on my tail". Sighting the F4U-5 in the process of making violent evasive manoeuvres as the Mustangs closed in for the kill, Soto fired at extreme range to force the Mustangs to leave the unarmed Corsair.

The Mustangs spilt and one dived to the right pursued by Acosta, eventually managing to escape. The other F-51D entered an evading climb, losing speed in the process. Soto positioned himself 800 ft (244 m) and fired a short burst of cannon, damaging the F-51D. The Mustang snapped into a diving turn, its performance hampered by its bombload. Soto followed, firing again and hitting the engine, which began to ignite and belch smoke. With his aircraft on fire Varela opened the canopy and jumped, but because of the low altitude the parachute failed to open completely and Varela was killed.

Capt. Hector Leonel Lobo Perez's F-51D was desperately attempting to get a firing solution on Soto's Corsair and finally managed to hit the F4U-5, however, Soto used his aircraft's superior performance to evade the attack, manoeuvred himself into position behind Capt. Lobo's Mustang, and opened fire. Lobo entered a spiralling climb and, out-turning the F4U-5, managed to hit the Corsair again. Although Lobo claimed the victory, no Corsairs were lost.

On that same day, at 14:00 after a number of tactical support missions, a section of two FAS FG-1D Corsairs was sent to cover the area surrounding the Goascorán Bridge. Capts Guillermo Reynaldo Cortez and Salvador Cezeña Amaya offered support to the troops on the ground without any enemy aircraft present. After accomplishing their mission they returned to refuel.

Meanwhile, at 14:30 in Tocontín, the F4U-5 Corsairs of Soto, Acosta and Zepeda took off again to attack an enemy gun position. Again Zepeda was forced to return with cannon failure. It was later discovered that the problem was the ammunition, which was jamming the gun barrels. The cannons on

The most famous aircraft in the history of the Fuerza Aérea Hondureña is undoubtedly F4U-5N FAH 609, the aircraft flown by Soto when he shot down three FAS aircraft on 17 July 1969. The aircraft has been preserved, with 'kill' markings displayed, at the Museo del Aire de Honduras.

Although both sides claimed the destruction of numerous aircraft, none had actually been shot down.

During the night of the 15/16 July the FAH's AT-6G conducted patrols over Tegucigalpa, as it was the only aircraft equipped to fly on instruments.

An AP agency report reported at the end of the day's fighting that: "In Washington, Jorge Fidel Durón, Honduran representative for the American States Organization [ASO], stated 'that the Salvadoran troops have penetrated 64 km [40 miles] inside the territory of his country, some of the troops using the Pan-American highway. President Nixon had been informed but hadn't made an official response. The ASO was paying attention to the conflict and was working to stop it'."

The fighting continues

At 04:30 on 16 July, following reconnaissance by the Instituto Cartográfico Nacional Cessna 310 the previous day, the FAS pilots assembled for an attack sortie against Puerto Cortez and Goloasón, on the Honduran side of the Golfo de Fonseca. Other attacks were planned at La Presa, Honduras, to prepare the ground for an air assault mission to allow the Salvadoran army to surround the Honduran troops. However, as two of the F-51Ds lined-up on the runway at Ilopango they collided, the wing of one hitting the wing of the other provoking the planes to rotate into each other, damaging both propellers and engine blocks, and the mission had to be cancelled. The F-51s were repaired within 48 hours.

The Hondurans, desperate defence was aided by a supply air bridge using C-47s of the FAH and SAHSA to help in the effort to contain the advancing Salvadoran troops. Altogether the C-47s managed to carry around one thousand troops of the Batallón Guardia de Honor and some journalists to the war zone. FAH F4Us were used to escort the precious cargo aircraft and a Cessna U-17A and Cessna 180 patrolled the border on the look-out for incoming FAS fighters.

The Honduran ground forces were, however, struggling to halt the advance. Following an attack by two FAS Corsairs and two C-47s the Hondurans were forced to abandon the town of Nueva Ocotepeque to Salvadoran forces.

Elsewhere Honduran Corsairs and T-28s conducted close air support missions throughout the day, however, claims by the FAH that they had all but destroyed the FAS were again

Above: Although the F4U-4 lacked the cannon armament of the F4U-5, its 2,450-hp Pratt & Whitney R-2800-18W engine still gave the FAH a decisive advantage over the 2,250-hp R-2800-8-powered FG-1Ds of the FAS.

Left: T-28A FAH 214 was well used during the short conflict, especially in the armed reconnaissance role. Its sister ship FAH 212 even managed to damage an FAS Corsair during air-to-air combat over Tocontín.

this aircraft were new and the ammunition was of incompatible World War II British-vintage. The other Corsairs were armed with older cannon so the problem did not occur. Continuing to El Amatillo, Soto and Acosta spied the two FG-1Ds above them, jettisoned their bombs and climbed into position for an attack. Two F-51Ds had also been spotted and Acosta positioned himself between the two groups of enemy fighters.

After positioning his aircraft Fernando Soto initiated a diving attack, firing first at the FG-1D (FAS 204) of Capt. Cezeña, who, after witnessing the 20-mm cannon rounds hitting his wing, destroying his ammunition boxes and ripping away upper surface wing panels, advised Capt. Cortez of the presence of the enemy planes, before jumping from his stricken aircraft. Soto overshot the second Corsair, convinced that Acosta would shoot it down. However, he discovered that Cortez had followed him and he was taking hits from the FG-1D's machine-guns. Soto accelerated and threw his Corsair into violent manoeuvres to try to shake off his attacker, but his opponent was a talented pilot and managed to again inflict hits on Soto's aircraft. Acosta, at higher altitude, witnessed the pursuit, but could not come to Soto's aid as he was surrounded by the two F-51Ds. Eventually Soto used his aircraft's additional power and superior performance to shake-off Cortez's persistent attacks by entering a full power wing-over and near vertical descent. Cortez followed the dive, but as Soto began to climb the extra power told and Soto snapped into a tight turn to position himself behind the FG-1D. Soto fired five or six bursts without success before a 20-mm shot impacted the left wing, exploding on the aileron and blowing it from the aircraft. The FG-1D was forced to slacken its turn and Soto continued firing, the aircraft eventually exploding in mid-air, killing Cortez immediately. Soto later told that the violence of the explosion surprised him and he was full of admiration for Cortez who had managed to inflict several machine-gun holes on his aircraft's left wing. Witnesses on the ground recalled seeing a ball of fire falling from the sky and, on examining the wreckage, recovered Cortez's badly burned remains. Only his boots and his revolver were saved and preserved.

For his actions Soto became a Honduran hero and the Ace of all Latin America, with his three victories in air combat. Meanwhile, Acosta managed to escape from the two Mustangs and returned to Tocontín.

The day would only get worse for the FAS as shortly after the frenetic air combat FG-1D FS 219, piloted by Capt. Mario Echeverría, was shot down by friendly fire over the Golfo de Fonseca, leaving Salvadoran air assets in a critical state.

Above left: FAH 609 is seen on approach to Tocontín airport. Despite their age the Honduran Corsairs continued to serve into the 1970s, before eventually being replaced by more modern jet aircraft.

Right: Fernando Soto Henríquez became a national hero for his exploits on 17 July 1969 and still regularly conducts interviews today.

The majority of the surviving Honduran Corsairs were sold to American restorers Bob Ferguson and Howard Pardue in the late 1970s. Here the aircraft are seen at Veracruz, Mexico, during the transit flight from Honduras to California. A number of the aircraft remain airworthy in the hands of warbird operators, including FAH 614 that flies from Switzerland sponsored by Red Bull.

rightly discredited by El Salvador, with again no aircraft being lost in air-to-air combat.

During the day the Salvadorans had attempted to deceive the Hondurans into believing that they were invading Amapala Island, on the Golfo de Fonseca, to divert Honduran air attacks. The FAS even carried out a raid against the Honduran 'defenders'. The FAH dispatched Corsairs from Tocontín, but on arrival over the island saw that everything was normal and instead continued to El Amatillo to drop their weapons. There, Capt. Francisco Zepeda, flying an F4U-5, attacked a troop column crossing the bridge, destroying a truck and damaging another two. On this mission a new kind of bomb called 'Piñas' was used for the first time by the F4Us.

During 16 July two FAH Texans participated in their only close air support mission of the war, dropping six 30-lb bombs from 5,000 ft (1524 m) over El Amatillo and making strafing runs with their 12.7-mm machine-guns. In addition, C-47 FAH 306 was used to bomb El Amatillo.

On the ground, the Salvadorans occupied the towns of Caridad, Alianza, Goascorán and Armencina. However, by late afternoon the Hondurans had slowed the advance at Nacaome. As the ground fighting intensified the FAS also used a C-47 and two FG-1Ds to make bombing and strafing attacks at Choluteca.

By the end of the day the FAS had three F-51Ds and one C-47 out of service, which made a significant impact on their ability to conduct offensive operations.

Day of reckoning

Along with the intensive close air support missions and the ensuing air-to-air combat in which Soto claimed his three 'kills', 17 July saw air action all along the front. The fighting was especially fierce as the Salvadorans began to advance to Santa Rosa de Copán. With the Honduran Army adopting defensive positions to protect the town, FAH F4Us set to work attacking enemy troop columns as they advanced from Nueva Octepeque, using 'Piñas' and 2.75-in rockets.

At 17:00 the Honduran army ambushed the advancing forces at San Rafael de las Matarás, and although they managed to halt the advance they failed to push the Salvadorans into a retreat due to a lack of ammunition.

At 17:32 the F4U-4s FAH 612 and 614 took off from La Mesa to attack a convoy travelling from El Portillo to La Labor, two small towns close to the area of the ambush. Locating some enemy trucks, buses and jeeps, they attacked with a 50-kg bomb and four 2.75-in rockets each, destroying some vehicles and causing more than 30 enemy casualties. The Salvadorans retired to Nueva Ocotepeque and adopted defensive positions.

Another strike was conducted by the C-47 FAH 306 on high ground occupied by the Salvadorans near El Ujuste using 52 12.5-kg (27.5-lb) bombs. It was followed by two Corsairs that attacked the same target. In addition, the Texan and T-28s conducted patrols and close support missions, and the Cessna 185 was used for observation duties

By this time the FAS's resources were becoming stretched, with only two Corsairs, five Mustangs (a sixth was interned in Guatemala) and one B-26 available. Desperate attempts were made to acquire more combat aircraft from the United States, especially Mustangs, which were easier to find than Corsairs. Eventually, five North American F-51Ds (FAS 407 to 411) were purchased from US civilians, arriving on 19 July. They were bought alongside four B-26 Invaders (FAS 601 to 605), one of them from the Fuerza Aérea Guatemalteca for use as a spares machine, and a Fairchild Hiller FH-1100, the FAS's first helicopter. Another five Mustangs were seized by the US

Fuerza Aérea Hondureña (FAH) – Order of Battle

Type	Serial (FAH-)	Notes
Cessna U-17B	111	Ex 63-13135
Cessna 185B	112	
Cessna 180	113	
North American BC-1A	200	Ex NC-66142. Out of service
North American SNJ-4 Texan	202	Bu. No. 27491
North American SNJ-4 Texan	205	Bu. No. 09990. Preserved at Museo del Aire de Honduras.
North American AT-6G Texan	206	Ex 51-16075. Out of service
North American AT-6G Texan	208	Ex 51-14916. Out of service
North American AT-6G Texan	211	Ex 49-2998. Out of service
North American T-28A	212	Ex 50-272. Damaged Corsair FAS 220 on 15/7/69 over Tocontín.
North American T-28A	213	Ex 50-267
North American T-28A	214	Ex 50-293
North American T-28A	215	Ex 50-234
North American T-28A	216	
Douglas C-47DL	301	Ex 41-18690
Douglas C-53D-DO (DC-3)	302	Ex 42-68769. Later FAH 321
Douglas C-47A-20-DL	304	Ex 42-93089
Douglas C-47A-75-DL	305	Ex 42-100933
Douglas C-47D	306	Ex 42-93701A. Bombed Ilopango on 15/7/69
Douglas C-47B-15-DK	307	Ex 43-49504. Purchased on 17-6-68. Out of service
Vought F4U-5NL Corsair	600	Bu. No. 124724. Sold to Hollywood Wings, USA in 1979.
Vought F4U-5N Corsair	601	Bu.No. 124560. Sold to B.Ferguson and Howard Pardue in 1979.
Vought F4U-5 Corsair	603	Bu.No. 123168. Out of service from 13/2/67. Used as airfield decoy.
Vought F4U-5 Corsair	604	Bu.No. 122179. Sold to B. Ferguson and Howard Pardue in 1979.
Vought F4U-5N Corsair	605	Bu.No. 122184. Named *Guaguina Fighter* and *Oroquina Kid.* Sold to B. Ferguson and H. Pardue in 1979.
Vought F4U-5N Corsair	606	Bu.No. 124486. Named *El Guajiro* and *El Guardián de Cantarranas.* Sold to B. Ferguson and H. Pardue in 1979.
Vought F4U-5 Corsair	607	Bu.No. 124692. Out of service. Used as airfield decoy. Sold to Hollywood Wings, USA in 1979.
Vought F4U-5N Corsair	609	Bu.No. 124715. Named *Mosquitia Fighter, Snap Shot* and *Sulfato.* Shot down three FAS aircraft on 17/7/69. At Museo del Aire de Honduras
Vought F4U-4 Corsair	611	Bu.No. 97382. Sold to Hollywood Wings, USA in 1979.
Vought F4U-4 Corsair	612	Bu.No. 97288. Sold to B. Ferguson and Howard Pardue in 1979.
Vought F4U-4 Corsair	614	Bu.No. 96995. Named *M-614.* Sold to B. Ferguson and H. Pardue in 1979.
Vought F4U-4 Corsair	615	Bu.No. 97280. Damaged C-47 (FAS 101) on 15/7/69. Sold to B. Ferguson and H. Pardue in 1979.
Vought F4U-4 Corsair	616	Bu.No. 97320. Out of service. Used as airfield decoy. Sold to Hollywood Wings, USA in 1979.
Vought F4U-4 Corsair	617	Bu.No. 97059. Interned in Guatemala on 15-7-69 (Capt. Walter López). Parts sold to Hollywood Wings, USA in 1979.
Douglas C-54B-15-DO	795	Ex 43-17182
Sikorsky H-19		
Sikorsky H-19		

Fuerza Aérea Salvadoreña (FAS) - Order of Battle

Type	Serial (FAS-)	Notes
Beechcraft T-34A Mentor	64	Out of service
North American SNJ-5 Texan	72	Out of service
North American SNJ-5 Texan	76	Out of service
North American SNJ-5 Texan	78	
Cessna U-17A	80	Ex 65-10847
Cessna U-17A	81	Ex 65-10848
Cessna U-17A	82	Ex 65-10849
Cessna U-17A	83	Ex 65-10850
Cessna U-17A	84	Ex 65-10851
Cessna T-41A	90	Ex 65-5132
Cessna T-41A	91	Ex 65-5133
Cessna T-41A	92	Ex 65-5134
Cessna T-41A	93	Ex 65-5226
Cessna T-41C	94	Ex 68-7892
Cessna T-41A	95	
Douglas R4D-1	101	Damaged by a Corsair close to Nueva Ocotepeque on 15-7-69
Douglas R4D-1	102	Out of service
Douglas C-47	104	Ex 43-49772A
Douglas C-47	105	Ex 43-49498
Douglas C-47	106	Ex 42-92104. Out of service
Douglas C-47	107	Out of service
Goodyear FG-1D Corsair	202	Bu. No. 67070. Sold to Charles Osoborn, Sillesburg, IN.
Goodyear FG-1D Corsair	203	Bu. No. 67071. Shot down on 17/7/69 by Capt. Soto. Capt. Cortez killed.
Goodyear FG-1D Corsair	204	Bu. No. 67082. Shot down on 17/7/69 by Capt. Soto. Capt. Cezeña jumped
Goodyear FG-1D Corsair	215	Bu. No. 92629. Sold to Palm Springs Air Museum of California as N62290.
Goodyear FG-1D Corsair	219	Bu. No. 92690. Shot down on 17/7/69 by friendly fire over Golfo de Fonseca
Goodyear FG-1D Corsair	220	Bu. No. 92697. Damaged during an attack on Tocontín on 15-7-69.
Canadair DC-4M	300	Ex YS-27C
Cavalier TF-51D Mustang II	400	Interned in Guatemala on 14-7-69. Capt. Benjamín Trabanino
Cavalier F-51D Mustang II	401	Ex 45-11559
North American P-51D Mustang	402	Ex 44-73350 and YS-210P. Damaged on 16-7-69
Cavalier F-51D Mustang II	403	Lost in an accident in 1973
Cavalier F-51D Mustang II	404	Shot down on 17/7/69 by Capt. Soto. Capt. Varela killed.
Cavalier F-51D Mustang II	405	
North American P-51D Mustang	406	Arrived on 15th of July
North American P-51D Mustang	407	Arrived on 19th of July
North American P-51D Mustang	408	Arrived on 19th of July
North American P-51D Mustang	409	Arrived on 19th of July
North American P-51D Mustang	410	Arrived on 19th of July
North American P-51D Mustang	411	Arrived on 19th of July
Douglas B-26B Invader	600	

government as exportation certification had not been authorised.

Throughout the war, the FAS's C-47s were used to support the troops advancing through Honduras, landing on small runways hastily prepared in the forests, on roads, in fields or on small grass airfields.

The final curtain

18 July began with FAH Corsairs and T-28s attacking Salvadoran troop positions at Llano Largo, Nueva Ocotepeque (TON), Guarita, Cololaca (TOCH) and El Amatillo (TOOC). By this time Honduras had almost complete control of the airspace. Four Corsairs, armed with napalm bombs, attacked El Ujuste hill at El Amatillo, while on the Northern Front F4U-4s attacked with 750-lb (340-kg) napalm bombs.

With the tide turning against the Salvadoran ground forces the FAH planned a large-scale bombing operation near Llano Largon on the Northern Front, however, fearing that this was a trap and the troops could be Honduran, the attack was cancelled after the aircraft had launched.

At 17:00 Honduran ground forces were ordered to advance inside El Salvador through the zone of Arambala, to cut off the Salvadoran forces at El Amatillo, but at 21:30, after advancing with little opposition, a fall back was ordered to coincide with a ceasefire brokered by the OEA, which was to come into effect at 22.00. The terms of the ceasefire ordered both countries to return to their own countries and the borders to be reinstated as they were before the conflict.

The ceasefire brought to an end the major ground battles, however, small scale air and ground operations continued until 20 July. FAS C-47s were used to support the withdrawal of Salvadoran troops from Honduras, under escort from F-51Ds.

Tensions remained high with the Salvadoran army refusing to leave some 1600 km^2 (618 sq miles) of occupied territory, and both armed forces remained on high alert. Finally, on 5 August, the Salvadorans began to return to their country following guarantees made by the American States Organization concerning the treatment of Salvadoran immigrants in Honduras. More than 100,000 Salvadorans eventually were forced to return to their country.

The short war had killed more than 2,000 people, mostly Honduran civilians, and relations between both countries were interrupted for more than ten years, until a peace treaty finally restored diplomacy between the warring states.

In the period that followed both air forces began to rebuild their assets. By the end of the year Honduras had purchased four F-86K Sabres from Venezuela, and in the mid-1970s some 18 Cessna A-37Bs and 21 Dassault Super Mystere B2s were acquired, finally giving the FAH the equipment it needed to be a power force in Central America.

The FAS's Corsairs and Mustangs continued to serve into the mid-1970s, when 12 Fouga Magisters for training and attack, 18 Dassault MD-45 Ouragan fighters and four IAI-101 Arava transports were purchased. The remaining F-51s and FG-1Ds were sold off to private collectors in the USA.

Santiago Rivas and Juan Carlos Cicalesi

Above: The Salvadoran Mustangs continued to serve with the FAS long after the end of the Soccer War. FAS 407 was one of the five P-51Ds acquired from private sources in the USA as the conflict erupted. The aircraft arrived on 19 July, just as the ceasefire was introduced. The aircraft were hurriedly fitted with aiming sights developed by the entrepreneur Archie Baldocchi and fitted with VHF radios purchased from the Guardia Nacional de Panamá for $1,000 apiece.

Right: The Salvadoran pilots that participated in the Soccer War gathered in front of this F-51D at the war's end. The ceasefire marked the final all piston-engined air-to-air combat that the world would see, as both air arms began the transition to more modern jet aircraft.

Dassault Mirage 4000

A lesser-known member of Dassault's stellar Mirage family, the Mirage 4000 was nevertheless a truly remarkable aircraft. Big, powerful and possessing breathtaking performance, it was designed to be the finest air superiority fighter of its generation. Eventually neutered by prohibitive production costs and geo-political circumstances, the Mirage 4000 was destined never to see service. However, the technological progress represented in terms of construction materials and aerodynamics proved invaluable in the design of Dassault's latest combat aircraft – the Rafale.

Above: Designed in parallel with the single-engined Mirage 2000, the Mirage 4000 was optimised as an interceptor. In the mid-1980s the sole example received this 'desert' colour scheme, as Dassault desperately sought orders in the Middle East to enable the type to enter production.

Below: Legendary Dassault test pilot Jean-Marie Saget (right) gives final instructions to Louis Robert of the CEV shortly before he became the second person to fly the Mirage 4000 on 16 May 1979.

As the 1960s drew to a close French fighter production, led by the Dassault company, had never been in a healthier state. That company's Mirage III and Mirage F1 had not only been ordered in large numbers by the Armée de l'Air, but had also garnered appreciable export sales in a market dominated previously by American and, to a lesser extent, British designs. The merger between Dassault and Breguet in 1967 had given an already confident company the size, financial clout and engineering skills base to compete with the American aerospace giants, and finally laid to rest the wasteful and costly in-fighting between the companies for domestic orders. However, to capitalise on the successes of the Mirage series it was necessary to look to the future.

In 1972 the Armée de l'Air launched a new fighter programme under the designation ACF (Avion de Combat du Futur – Future Combat Aircraft). The ACF design emerged as a swept-wing fighter with a conventional tail and powered by two Snecma M53 turbofans, bestowing the aircraft with a planned maximum speed of Mach 2.5. By 1975, however, Dassault had revised its market forecast and elected to pursue two concurrent designs – the single engined Mirage 1000 (which would eventually enter production as the Mirage 2000) and the larger twin-engined Mirage 4000. The ACF programme was abandoned in December 1975 as the Mirage 2000 project was officially launched – to be funded, initially, by the French government.

Nevertheless, Dassault-Breguet was determined to continue development of a twin-engined design as a rival to McDonnell Douglas's F-15 Eagle that was about to enter service with the USAF, and was widely regarded as the most capable fighter in the world. With F-4 Phantom production winding down and, with so many air arms worldwide

likely to be seeking a replacement in the late 1980s and early 1990s, the export potential of the aircraft was obvious. With no state funding available Marcel Dassault took the gamble to develop and build the aircraft as a private venture, as he had previously done with the Mirage III and Mirage F1 – two of the company's more notable success stories.

Sales prospects

Dassault-Breguet had identified opportunities in both domestic and export markets for a large twin-engined multi-role fighter. Although the Armée de l'Air had launched the Mirage 2000 programme in single-seat (for air-to-air missions) and two-seat (precision attack and nuclear strike) variants, the notion of a twin-engined fighter recurred time and again in requests by Armée de l'Air pilots. Additionally, it was abundantly clear that the Mirage 4000, fitted with the Aérospatiale ASMP missile beneath the fuselage, would have been an ideal replacement for the ageing Mirage IV. However, with the Armée de l'Air already committed to acquiring the Mirage 2000 in large numbers, the available budget was small – significant export sales would be needed to place the aircraft into production.

To support export sales, Dassault-Breguet proposed the establishment of an experimental squadron consisting of one company-funded Mirage 4000 prototype, with others funded by the French government, as Bruno Revellin-Falcoz recalls. "The design was well suited for export and if sales were achieved we were confident that France would also order the aircraft. We could build a prototype with our own funds, but to create a series for export at least two other prototypes would be required. That was too much money, out of our range, therefore we made a proposal to the government that they finance the two additional prototypes." The government, however, lacked the political will and certainly the available finance to accede to Dassault's wishes and, instead, lent its support by lending the company Snecma M53 engines initially destined for the Mirage 2000. "We had an agreement with Snecma and the Armée de l'Air to use two M53s from the pool allocated to the Mirage 2000. The frequency of flight tests for the Mirage 4000 was governed by the availability of these engines. When the Mirage 2000 programme needed the engines, the flight tests of the Mirage 4000 had to be stopped."

In the early 1970s a new 'mini arms race' was beginning to gather pace in the Middle East and it was here that Marcel Dassault could see great opportunities for export sales. In June 1974 Iran ordered 40 Grumman F-14 Tomcats, followed by 40 more in January 1975. Saudi Arabia, seeking options to replace its BAC Lightning F.Mk 53s, finally placed a contact for 46 McDD F-15Cs and 16 F-15Ds (contract Peace Sun) from August 1981. However, US sales were restricted by certain conditions in response to concerns from Israel – only 60 F-15s could operate from Saudi territory at any time. These constraints led to Saudi Arabia considering acquiring other fighters from countries that would be less demanding than the Americans.

In addition, Iraq was viewed as a likely prospective customer for the Mirage 4000. Equipped in the early 1970s with Soviet equipment, Saddam Hussein was keen to diversify the supplier base. France won a number of important military contracts in the 1970s, including Aérospatiale Gazelle helicopters, Mirage F1s and Roland missiles, and Dassault heavily promoted the Mirage 4000 as the ideal counter to the Iranian F-14 threat.

In September 1976 the Armée de l'Air finally put pen to paper by ordering the Mirage 2000. This event provided the financial security for Dassault-Breguet to commence construction of the prototype Mirage 4000.

The Mirage 4000 formates with the prototype Rafale in 1987. With interest in the 4000 waning Dassault pinned its hopes on its new design. Flight tests of the Mirage 4000, particularly in the area of canard design, allowed Dassault to steal a march on its European rival – the Eurofighter. The Rafale eventually entered service some four years ahead of the Typhoon.

From the drawing board

With computer-aided design in its infancy, the Mirage 4000 was designed using traditional methods, as Bruno Revellin-Falcoz explains. "We used computers only for aerodynamic and structural calculations, not to design the aircraft

In 1977 Dassault built a full-scale mock-up of the Mirage 4000 at its Saint-Cloud facility. At this time the aircraft was called the Delta Super Mirage and lacked the canards fitted to the prototype. The aircraft would later be renamed as the Super Mirage 4000, before becoming the Mirage 4000 in the mid-1980s.

Dassault's design and engineering team was based at the Saint-Cloud facility south of Paris, where every Dassault military prototype, from the Ouragan of 1948, had been designed and constructed. Here, two views show the Super Mirage 4000 fuselage under construction in the airframe workshop in 1978. Other important structures for the prototype were built at other Dassault facilities before being trucked to Saint-Cloud for final assembly. Ensuring that each section was constructed precisely was a major challenge in an age preceding computer-aided design and production.

itself. The Mirage 4000 inherited many aspects from the Mirage 2000, which had introduced a range of innovations. The Mirage 4000 was initially a very powerful and large fighter with impressive endurance and a very long operating range. We made the choice not only to make it outstanding in the supersonic domain, but also go further in low-level penetration missions. The Snecma M53 has a pretty low compression ratio – appropriate for high speeds at altitude, but a high fuel consumption rate low down. Thus, the Mirage 4000 was literally a 'flying fuel tank' and we made room for fuel in the tail.

"To bestow the aircraft with the desired manoeuvrability we started to introduce 'real' canard surfaces. In fact, we had previously tested various types of flying surfaces, including the folding moustaches on the Mirage Milan S for improved low-speed handling. Various canard designs were analysed and the prototype later conducted many experiments with the canard configuration, this proving valuable in the development of the Rafale. The canards were not just simple control surfaces, they also helped to stabilise the airflow at the apex of the aerofoil – crucial from an aerodynamic standpoint. For us, the Mirage 4000 not only represented a brand new aerodynamic formula [canards and delta wing] but also introduced a host of new technologies, in particular composite materials.

"Along with extensive use of carbon fibre materials, for the first time we introduced optical fibres inside the electronic compartment to check the behaviour *in situ* and for data processing during flight. Along with creating an aircraft vital for experimenting with these new technologies, the tough task given to the design team was to create a prototype that could precede an operational version with world-leading capabilities.

"Following the Dassault tradition, the engineers took many of the successful elements and techniques of previous successful aircraft, while implementing new technologies. Data processing was particularly useful for structural calculations of the airframe. We used composite materials for the rudder, elevons and canards and the Mirage 4000 carried more than 9 tonnes of fuel [including 675 litres in the tail], three times more than the Mirage 2000 could hold. The Messier-Hispano-Bugatti hydraulic system operated at pressure rate of 280 bars [4,000 lb/sq in] and was powered by four pumps [in comparison, the Mirage III hydraulic system uses 207 bars

ACF programme

Current Dassault Vice-President Bruno Revellin-Falcoz here describes the genesis of the ACF project, the technical problems faced by the design team and the programme's eventual evolution into the Mirage 2000 and 4000 fighters.

An artist's impression of the Mirage G8A, displaying the conventional swept wing and tail of the original ACF proposal. Having flight tested the variable-geometry Mirage G8, Dassault worked hard at producing an aircraft to meet Armée de l'Air demands, while continuing to research delta-wing designs. The fixed, conventional swept-wing layout was finally dropped with the cancellation of the ACF requirement in December 1975.

"At the beginning of the 1970s, much of our resources were placed on the continuation and improvement of the Mirage III and Mirage F1 series. In our plans, we had designs for single-engined and twin-jet aircraft. It should be remembered that at this time, we were flying various research aircraft including the single-engined Mirage G and the twin-engined variable-geometry Mirage G8. We were aiming to provide the ultimate design to fulfil a fixed-wing twin-jet programme, launched by the French Defence Ministry, called the ACF [Avion de Combat du Futur].

"At that time, we used the flight tests of our variable-geometry aircraft to analyse the optimal wing sweep to meet the specifications requested by the Armée de l'Air. We launched a first ACF prototype and, in parallel with this twin-jet aircraft, we studied single- and twin-engined delta designs. We wanted to have, like always at Dassault, a reference point and we analysed many formulas. Numerous different alternative aircraft concepts were studied, according to the objectives that we had established ourselves, because during this time there were no precise operational requirements.

"Some models were tested in wind-tunnels and we conducted structural analyses to check that the concept would work in principle. A number of scale and a full-size mock-up were built to check that we could place the engines, landing gear, fuel tanks, pilot's unit and other various equipment within the airframe. The design office for prototypes worked continuously on various configurations and aerodynamic formulas.

"The guideline was to design not only a Mach 2 aircraft, but one that also offered a significant improvement in low-speed handling and performance, allowing short take-offs and landings, and a big improvement in manoeuvrability, particularly at high incidences in combat. Our projects were all based on the Snecma M53 engine.

"This work was undertaken by a team of engineers that produced three-view blueprint airframe arrangements, weight studies and aerodynamic calculations with performance estimations. Thus, for all of the alternative aircraft concepts, we put the engineering data on a *draps de lit* (bed sheet), which filled an entire black board. All the different alternative aircraft concepts were arranged by column and, by row, all engineering data such as mass, fuel, manoeuvrability and performance were detailed. This allowed for an easy comparison of each proposal. Finally the data gathered allowed us to select the correct configuration for the project. However, while the majority of the team worked on perfecting the ACF design a few of us in the design department continued to study alternative concepts, including delta wing designs.

Following its official roll-out at Saint-Cloud in September 1978, the Super Mirage 4000 was trucked to the test centre at Istres for the commencement of flight trials. Here the aircraft is seen in its original bare metal finish following re-assembly at Istres, and shortly before its maiden flight.

and the Rafale 350 bars]. The avionics included a Sagem inertial navigation system and the pilot was provided with the VE 120 head-up-display as used on the Super Etendard. As with the Mirage 2000 the pilot sat in a Martin-Baker Mk 10 ejection seat.

"Although it was never intended to fit a fully-functioning weapons system to the prototype, the nose section offered a diameter of over 1 m [3 ft 3 in] at the antenna station, allowing the installation of a powerful radar. Preliminary studies concluded that the 'Cyrano' 500 radar project, derived from the Thomson-CSF RDM unit fitted to the Mirage 2000, would have been the most likely production option with its ability to detect a small high-flying targets at distances in excess of 120 km [75 miles]."

Prototype takes shape

The Mirage 4000 prototype was assembled in Dassault's Saint-Cloud prototype workshop. The main elements arrived from various Dassault-Breguet facilities: the wings were made in Boulogne; the composite material elevons and canards in Biarritz; Argenteuil and Seclin provided the wing panels, actuators and spars; Argonay delivered the flight command system and the main fuselage was built *in situ* at Saint-Cloud. A number of important subcontractors included Messier-Hispano-Bugatti that designed and manufactured the landing gear, and Auxilec for the design of the electrical power supply generators. With all the various sub-components delivered, final assembly was conducted from the spring of 1978, before the official unveiling of the aircraft, at Saint-Cloud, in September 1978. This site had produced every Dassault military prototype from the Ouragan in 1948.

Named at this stage the Super Mirage 4000, the prototype was moved in early 1979 to Dassault's flight test centre at Istres in southeast France. Engineer Jacques Desmazures was tasked with defining and organising the test programme: "I had just finished the test programme for the Alpha Jet when I was allocated to the Mirage 4000. As with each prototype there was much debate between the engineering and design departments in Saint-Cloud and the guys who were to test the aircraft. We worked together to implement necessary modifications. I was always moving between Istres and Saint-Cloud."

Flight tests and early impressions

Following a medical condition that induced a temporary black-out in flight in April 1979, Dassault Chief Test Pilot Jean Coureau was replaced by Jean-Marie Saget and, in preparation for the Mirage 4000's first flight, the latter conducted fly-by-wire familiarisation flights in Mirage 2000 No. 02 (second prototype) on 12 and 14 February 1979.

After a lengthy preparation period the Super Mirage 4000 was finally ready to take to the air

"We envisaged that new fly-by-wire (FBW) technology would become available, and this would allow us to reduce the 190-kt approach speed of the Mirage III, to 150 kt or a little less, using an FBW delta-wing design.

"As the Armée de l'Air launched the ACF programme, Marcel Dassault stated: 'for export, we will need a single-engined design'. Therefore, we also studied a single-engined project. It was officially named Mirage 1000, but we called it *Mimi* for 'Mirage Mille' ('Mirage Thousand'). The team that worked on the twin-engined design called this one *Mami*.

"In December 1975, when the decision was made to cancel the ACF, Marcel Dassault proposed the alternative Mirage 1000, which was very quickly renamed Mirage 2000. Marcel Dassault added that 'it would be nice to have a twin-engine aircraft for export, particularly to replace the Phantom II'. At the beginning of 1976, we signed our first contract for the Mirage 2000, and Marcel Dassault stated: 'I take the decision to produce a twin-engined aircraft for export that shall offer the best compromise between a superior operating range, Mach 2 performance, a rapid rate of climb, and good low-speed characteristics, all of these appreciably improved in comparison to the Mirage III and Mirage F1 designs'. This signalled the launch of the Mirage 4000.

"I was responsible for the Mirage F1, which was in production, but I desperately wanted to create this new prototype. On the other side, Jean-Paul Emoré, who had just designed the Mirage G prototype, wanted to design a production series aircraft. We went to see Jean Cabrière, the technical director, and asked him to evaluate our various proposals. He agreed, and fortunately I became responsible for the Mirage 4000. I was what we called a *chef d'avion* (aircraft chief), and was project manager in charge of the programme. There were usually three people working with me: one responsible for the structure, another for the systems and a third in charge of weight distribution estimations. I think we made a great team and a great aircraft."

Left: A full-scale mock-up of the ACF proposal (at this stage named Super Mirage) was constructed at Saint-Cloud in 1974. The aircraft clearly drew on the design of the Mirage F1, albeit with a longer thinner wing and enlarged tail.

Above: Dassault began the construction of the prototype ACF Super Mirage in late 1974, and the fuselage is seen here in a semi-complete state in early 1975. Work on the prototype came to a halt in December 1975, when the Mirage 2000 programme was launched. Had the Super Mirage been completed it was slated to make its maiden flight in July 1976.

Right: A fine colour shot of the Super Mirage 4000 during an early test flight with Jean-Marie Saget at the controls. As with all Dassault prototypes the aircraft wore the unit badge of the chief test pilot when serving with the Armée de l'Air – in this case the Chimère of Escadron de Chasse 2/2 'Côte d'Or'.

Left: Jean-Marie Saget prepares to climb aboard the Mirage 4000 to conduct the maiden flight on 9 March 1979. Standing to his right is Dassault's Chief Mechanical Engineer Roger Théron, who was instrumental in transforming the aircraft into a flyable condition.

Below: The entire Mirage 4000 team associated with the design, construction and flight test programme gathered for this group photo to celebrate the type's first flight at Istres.

on 9 March 1979, with Saget at the controls. The maiden flight took place almost exactly one year after the Mirage 2000 (10 March 1978) and lasted 50 minutes. As with the Mirage F1, Mirage 2000 and latterly the Rafale, the aircraft exceeded Mach 1 during the first flight, demonstrating the company's confidence in the aircraft. Jean-Marie Saget remembers the first flight with fondness: "You wait months and months in anticipation and to finally fly the aircraft is a great satisfaction. It was truly an aircraft that performed extremely well thanks to its tested formula, superb thrust to weight ratio, excellent fuel capacity and its ability to carry a wide range of external stores. We took a step into a field that previously had been rather reserved by the Americans". The event marked 30 years of Dassault jet fighter design following the first flight of the Ouragan in February 1949.

On 12 March 1979 the Super Mirage 4000 accomplished its second flight, reaching Mach 1.5 and exploring the aircraft's angle-of-attack up to 20°. The following three test flights were devoted to testing the integrity of the fly-by-wire flight controls as Jacques Desmazures explains: "The tests consisted of voluntarily degrading the capability of the fly-by-wire, to check that in the event of failure the aircraft remained controllable". These tests proved the integrity of the system and posed little problem for Saget: "The fly-by-wire was a great success, however, it took time to adjust. There is a decoupling between what you command and what the control surfaces of the aircraft do. If you ask the aircraft to pull up, you do not know what occurs behind, the computers determine the positions and movements of the control surfaces. It is the computers that fly the aircraft, you order a response from the plane".

Mach 2 was reached on 11 April, during the sixth test flight and, on 5 and 15 May, tests were devoted to the canards. This latter day also saw the maiden flight of the newest iteration of the Mirage III series – the Mirage 50. The following day the second pilot to fly the Super Mirage 4000, Centre d'Essais en Vol (CEV) test pilot Louis Robert, conducted a 1-hour 25-minute research flight followed, on 31 May, by Col Jacques Guillou from the Centre d'Expériences Aériennes Militaires (CEAM). Both pilots were duly impressed by the aircraft's capabilities.

Public debut

The remainder of May 1979 was spent in preparation for the type's public debut at the Paris Air Show, where representatives of a number of important potential export customers would be present. Despite having only flown 15 hours Dassault was desperate to present the aircraft's capabilities in the air and, on 6 June, Jean-Marie Saget conducted two demonstration rehearsals before heading to le Bourget, leaving Istres for the first time. The following day was a special one for Dassault as it presented its two new fighter projects, the Mirage 2000 and the Mirage 4000, to the French public for the first time. Jean-Marie Saget completed four flight demonstrations during the week-long air show, later commenting "At Dassault we always had an aim: to finish our new aircraft just in time for the Paris air show. There were sometimes limi-

As had become a tradition with Dassault fighter prototypes, the Super Mirage 4000 exceeded Mach 1 on its 50-minute maiden flight on 9 March 1979. The aircraft's fly-by-wire controls behaved perfectly during the take-off from Istres (above) and throughout the sortie. On landing back at the test centre Saget deployed the brake parachute (right), housed in a fairing at the base of the tail fin.

Left: With the Mirage 2000 programme progressing well and the Mirage 4000 demonstrating remarkable performance in its initial test flights, the turn of the decade was marked with great pride and optimism at Dassault. Here the 4000, flanked by Mirage 2000 prototypes Nos 1, 2 and 3 (blue camouflage) pose for a company photograph in early 1980.

Below: Resplendent in its new red, white and blue livery, the Super Mirage 4000 was presented to the world at the 1979 Paris air show. Here, French President Valery Giscard d'Estaing (centre) is shown around the new fighter.

Left: To succeed the Mirage 4000 needed export orders and, particularly in the Middle East, potential customers were aggressively courted. Here a delegation from Iraq (in 1979) or possibly from Saudi Arabia (in 1981) take a close-up look at the aircraft at Istres, accompanied by chief test pilot Jean-Marie Saget (far left).

tations imposed on the aircraft, but this was not the case with the Mirage 4000, which was successful from the beginning. This had not been the case with the Mirage G8. In a short time we had succeeded in finding the parameters to best demonstrate the power of the aircraft, without gaining too much speed."

Among the representatives of possible customers King Hussein of Jordan visited the Dassault stand on 12 June and was given a guided tour of both the Mirage 2000 and 4000.

On 18 June the Mirage 4000 returned to Istres to continue the flight test programme. For the remainder of the year the 4000 continued expanding the flight envelope, however, the demand for the Snecma M53 engines for Mirage 2000 development often led to readjustments for the 'lower priority' Mirage 4000. Such was the demand that the Mirage 4000 was flown with two different versions of the M53 installed (M53-2 and M53-5).

Impressing customers

1979 ended with the first flight presentation of the aircraft to an official foreign delegation, in this case Iraq. On 16 December Jean-Marie Saget flew a 30-minute demonstration. The performance of the aircraft took even Saget by surprise: "To demonstrate the 4000's abilities we planned a basic interception mission from brake release to 50,000 ft (15243 m) and Mach 1.8. On this day the head-up-display was not working and at 50,000 ft and Mach 1.8 I cut the afterburner in preparation to return. But I looked down at the instruments to find myself at 61,000 ft (18597 m) and Mach 2, to my great surprise!"

Earlier flight tests had confirmed the Mirage 4000's blistering performance at altitude, with Mach 2.17 having been recorded on 15 May 1979; the aircraft was also able to reach 50,000 ft (15243 m) and Mach 1.9 in 3.45 minutes.

After a short hiatus in early 1980 flight tests resumed on 20 March 1980 and, two months later, the aircraft was presented to French and German military delegations at Bordeaux.

With no orders yet forthcoming the Farnborough air show of 1980 was seen as a good opportunity to present the aircraft's capabilities to a captive audience. Four flight demonstrations were conducted between 31 August and 3 September. A day later Saget returned to Istres to conclude the aircraft's flying programme for the year, during which it had completed 25 flights.

Power from the M53

Key to the Mirage 4000's outstanding performance at altitude was the integration of the Snecma M53 turbofan. Jean Devriese worked on the development of the engine. "We realised in the mid-1960s that a replacement for the Atar engine was becoming necessary. This was a good engine for fighters flying at high altitude,

Few aircraft in the world could match the performance of the Mirage 4000 in its 'clean' state. Had the aircraft garnered any orders, it would have been Europe's fastest production fighter and on interception missions would have been capable of reaching 40,000 ft (12192 m) in two minutes from brake release.

Mirage 4000 walkround

***Clockwise from right:** The air intakes for the M53 engines were mounted on either side of the forward fuselage. The intake cone adjusted its position according to the aircraft's speed.*

The 4000 was equipped with a basic analogue cockpit and central control column. A head-up display was also fitted.

The variable-incidence canards were linked directly to the fly-by-wire control system and added a level of agility not previously seen on a fighter of this size.

Power was provided by two afterburning Snecma M53s. The fairing above the jetpipes contained the brake parachute.

Along with 11 weapons hardpoints, two 30-mm DEFA cannon could be mounted in the forward fuselage.

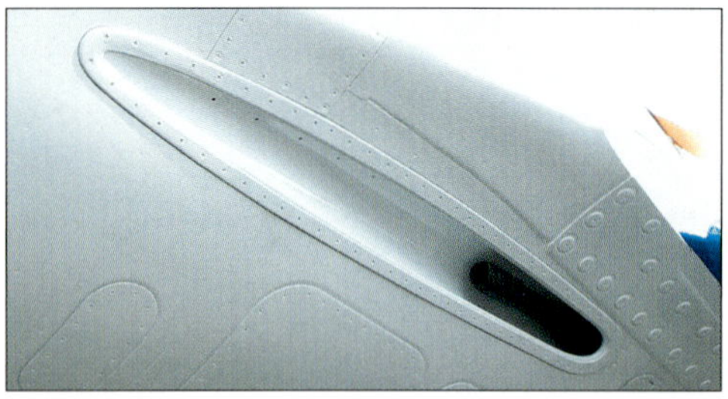

but was less suitable for low-level penetration missions. We wanted a larger engine, capable of high speeds at low altitude using high-pressure ratio modern turbine technology. At this time we had little expertise in this area and reached an agreement with Pratt & Whitney to develop its afterburning JTF10 and TF30 engines as the Snecma TF104 and TF306. We found, however, that these powerplants were prohibitively expensive, very large and complicated to maintain. Their size also dictated an aircraft that was larger than we anticipated and in 1968 the TF306 programme was terminated."

Snecma began studies for a new engine to fulfil the demands Dassault was making for its new generation of fighters. With the Chief of Staff of the Armée de l'Air declaring 'objective Mach 3' for its future fighters, Snecma began working on the compromise between performance and lower fuel consumption at low-altitude and high Mach numbers at higher levels. Jean Devriese describes the task given to the design team: "It is necessary to produce aircraft in the size range of the Mirage III with an engine that is shorter, lighter, more fuel efficient, yet more powerful. We concluded that the project required a common combustion chamber, but with burners differentiated for cold flow and hot flow. With the M53 we wanted to make a simple turbofan, with a low compression ratio so as to ensure that performance at altitude was not compromised. With very high compression ratios the American engines improved fuel consumption at lower altitudes to the detriment of high-altitude performance. They compensated for that by producing larger, heavier and more powerful engines and therefore bigger aircraft."

The M53 did not attempt to address the technical compromises introduced on the Pratt & Whitney F100, which equips the F-15 and F-16, or the Turbo Union RB199 for the Tornado. The M53's pressure ratio of 8.5 compares to 23 for the RB199 and 25 for the F100. At lower altitudes and speeds the F100 and RB199 offered

Dassault Mirage 4000 specification

Powerplant: Two Snecma M53-2, M53-5 or M53-P2 turbofans producing 12,100 lb (53.84-kN), 12,348 lb (54.95-kN) and 14,432 lb (64.22-kN) of dry thrust and 18,700 lb (83.21-kN), 19,800 lb (88.11-kN) and 21,340 lb (94.96-kN) of thrust with afterburning, respectively
Span: 39 ft 4.5 in (12.00 m)
Length: 61 ft 4.5 in (18.70 m)
Wing area: 785.79 sq ft (73.00 m²).
Wing loading: 45 lb per sq ft (1.87 kg/m²)
Max. take-off weight: 70,400 lb (31933 kg)
Internal fuel: 19,899 lb (9026 kg)
Maximum speed: Mach 2.18 at altitude
Approach speed: 162 mph (260 km/h)
Max. rate of climb: more than 60,000 ft (18288 m) per min
Service ceiling: 65,000 ft (19812 m)
Combat radius with external tanks: 1,080 nm (2000 km)

Dassault-Breguet Super Mirage 4000, Prototype No. 1, F-WZRM

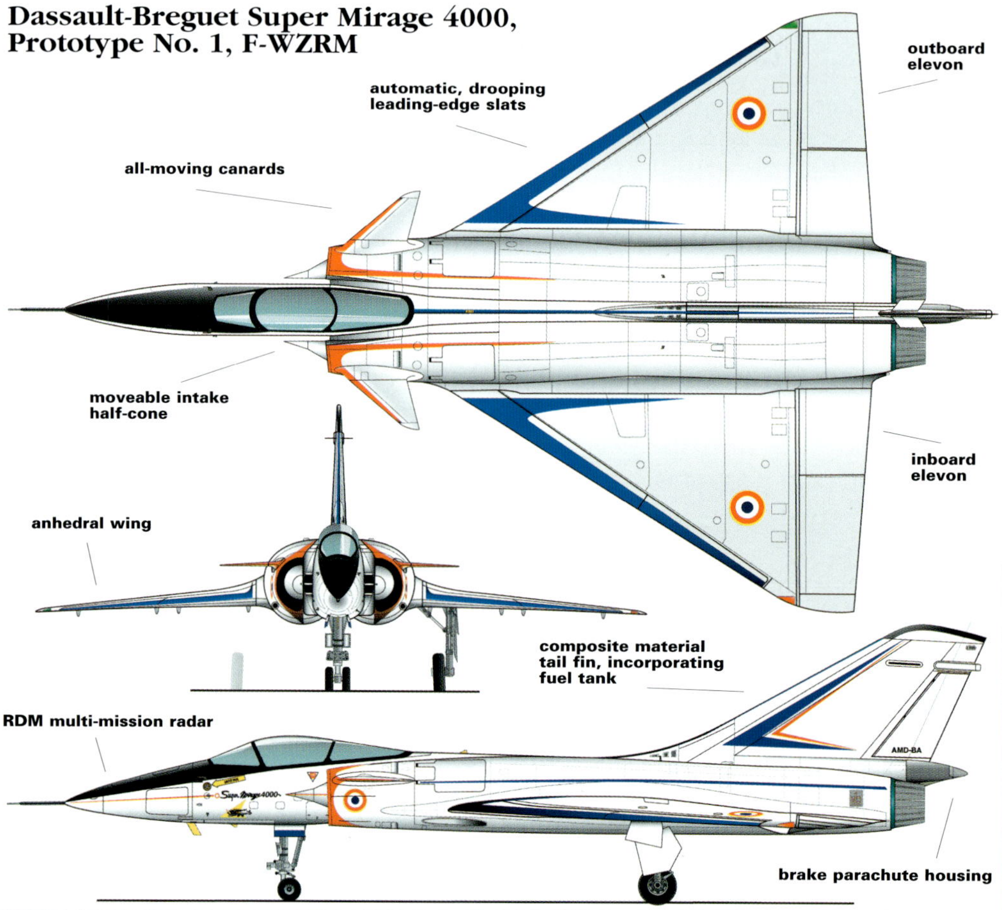

Left: As with all new aircraft, the hours of ground tests far outweighed the test flight programme. Here the Super Mirage 4000 undergoes electro-magnetic and noise vibration tests within the anechoic chamber at Istres in the early 1980s.

Below: After initial handling trials much of the early test flight programme was devoted to testing the effectiveness of the canards. Flights with the canards disabled were compared to the handling of the aircraft with the surfaces operating. The success the canards achieved in improving manoeuvrability led to the technology being integrated into the Rafale.

improved fuel consumption and a better compromise between static power and the weight of the engine. However, above Mach 1.8 the M53 outperforms both rival engines.

An experimental programme to build three complete engines was launched in August 1968, the experience gained during Snecma's participation in the TF306 and Rolls-Royce Olympus programmes being of enormous benefit during the development period. The first prototype engine was installed at the Melun-Villaroche test bench on 26 February 1970, and by August 1971 a power output of 18,740 lb (83.3-kN) had been obtained. Flight tests began in July 1973 aboard a modified Sud Caravelle airliner testbed.

'Magic' at Le Bourget

On 2 February 1981 the flight test programme resumed, with the aim of preparing the aircraft for trials with external loads. On 18 May the first flight with six dummy Matra Magic missiles, installed on wing pylons, was made. It was in this configuration that the aircraft would be demonstrated at that year's Paris Air Show and eight flights were devoted to preparing the most impressive flight demonstration the aircraft could achieve. Arriving at le Bourget on 2 June, the Mirage 4000 conducted 11 flight demonstrations during the show, despite the hard-worked ground crew having to change an engine on 8 June.

The 4000's performance at the show seemed to re-ignite foreign interest and, on 9 July, a new presentation of the aircraft was arranged for Iraqi officials, Jean-Marie Saget making his 120th flight in the aircraft. However, as negotiations with Iraq progressed, that country's declaration of war against Iran in September 1980 had a decisive moderating effect in Iraqi interest. With the country committed to war Iraq had neither the time nor the available finance to procure the Mirage 4000.

Just as war in the Gulf erupted and Iraqi interest waned a prospective customer emerged in the shape of Saudi Arabia. On 16 September a Saudi delegation oversaw flight and ground presentations of the aircraft at Istres, but again the decisive order that would have ensured the aircraft's future was not forthcoming.

Modifications and new equipment

Following the Saudi demonstration the prototype was grounded for a 12-month period, during which it received a new tail. Constructed entirely from composite materials, it made its maiden flight, in the hands of Saget, on 27 October 1981. Following flight tests of the new tail throughout November, the last flight of the year saw the aircraft flying in a new configuration, with underwing external fuel tanks and Magic missiles. On this first flight the external tanks were empty, however, on 25 February 1982 the Mirage 4000 took-off at a weight of 27000 kg (59,524 lb), with the tanks half-filled. During this flight Saget exceeded 10,000 career flight hours and, to mark the occasion, was escorted by four EC 1/2 'Cigognes' Mirage IIIEs, led by Saget's son Claude. Test pilot Guy Mitaux-Maurouard participated in the series of test flights throughout the remainer of the spring of 1982, carrying various external loads.

With the pressure to achieve export sales mounting, the Farnborough air show of 1982 assumed increased importance. The four months preceding the show were devoted to the preparation of the aircraft and perfecting the flight demonstration. Despite conducting seven flight demonstrations, with and without various

Although designed primarily as an interceptor, the Mirage 4000 could be fitted with up to 11 weapons hardpoints, allowing the aircraft to operate swing-role missions. Here the aircraft is seen with six Matra Magic IR-guided air-to-air missiles on the wing pylons.

Weapons options

Dassault always imagined a secondary low-level penetration attack role for the Mirage 4000 and, in addition to the twin 30-mm DEFA cannon mounted in the forward fuselage, the aircraft could be equipped with up to 11 hardpoints (six underwing, four underfuselage and one centreline), allowing carriage of up to 8000 kg (17,637 lb) of stores. The principal air-to-air weapon demonstrated on the prototype was the Matra Magic IR-guided missile, usually carried on the underwing pylons (right).

A wide range of air-to-ground stores was planned for the aircraft including two AS30L laser-guided missiles or two 1000-kg laser-guided bombs (with associated designating pod), the ASMP medium-range nuclear missile, 250- , 500-, or 1000-kg unguided bombs, cluster munitions and rocket pods. For the anti-shipping role the aircraft was to be capable of carrying the Exocet missile, and low- and high-level reconnaissance missions could have been flown with external recce pods. Two large auxiliary fuel tanks could be carried on the inboard wing stations.

During the 1982 Farnborough air show the Mirage 4000 was configured to demonstrate its multi-role capabilities and flew with this heavy load, comprising two Magic air-to-air missiles on the outboard wing pylons, two AS30L laser-guided missiles and two 2500-litre auxiliary fuel tanks inboard of these, plus two BGL 1000 1000-kg laser-guided bombs and two Sycomor ECM pods beneath the fuselage. Beneath the port air intake is a mock-up of the Antilope terrain-following radar.

Seen during ground tests at Istres, the Mirage 4000 is configured for the air superiority role with eight Magic missiles. This included a pair attached to each of the 600-litre RPK (Réservoir Pendulaire Kangourou) auxiliary fuel tanks that were carried in place of the larger 2500-litre tanks for increased manoeuvrability.

external load combinations, the Mirage 4000 returned to Istres on 12 September empty-handed. With the test programme largely complete there was little work for the Mirage 4000 to conduct and the aircraft was grounded for the following nine months.

With Dassault unable to commit significant funds to continue with the Mirage 4000 programme, the Paris air show of 1983 was seen by many as the final chance. The aircraft returned to the air on 4 May. After initial system check flights, the work began in preparation for ground and flight demonstrations. In early June Saget completed 10 flight demonstrations during a week blessed with excellent weather: "I profited from superb visibility. I felt like I was the king of the sky. It was a dream aircraft for a fighter pilot and the demonstrations showed the optimum performance." However, on 6 June the Mirage 4000 returned to Istres and, with no orders forthcoming, the aircraft was relegated to outdoor storage at Istres. The flight test team believed the Mirage 4000 programme was dead.

Mirage 4000 strikes back

On 26 September 1985, Saudi Arabia and the United Kingdom signed the Al Yamamah contract, comprising 74 Panavia Tornados, divided into 48 Tornado IDSs and 24 ADVs (it also included 30 BAe Hawk and 30 Pilatus PC-9s for training). Important though this contract was, its conclusion opened the door for further negotiations for Mirage 4000 procurement. Hopes were raised and, acknowledging that the Tornado ADV could not compete with the Mirage 4000 in terms of air combat performance, the King of Saudi Arabia negotiated a purchase of 100 Mirage 4000s. The programme was estimated at 70 billion francs (100 billion including armament), with the first production examples to be delivered in 1990 – the Mirage 4000 had risen from the dead.

Having been immobilised and placed in outdoor storage for over two years, much work was needed to bring the aircraft back to flight status. This was accomplished remarkably quickly and, on 18 December 1985, the Mirage 4000 returned to the air, again in the hands of Saget, making its 200th flight. Six days later Saget made his final flight in the aircraft, accompanied by a surprise escort of a Mirage 2000N, Mirage 2000C and Mirage IIIB from the CEV. On retirement Saget noted, "I am happy that it was my last fighter. After having known the Mirage 4000 I could not foresee anything better to hope for". Having totalled 129 flight hours in the Mirage 4000 he noted, "Of all the aircraft I flew the Mirage 4000 had the most formidable performance. At the same time it possessed really exceptional acceleration and manoeuvrability. The Mirage G8 was faster, but its higher wing loading limited its abilities". After 34 years as Dassault chief test pilot Saget was replaced

As Dassault's chief test pilot Jean-Marie Saget had played a vital role in the development of both the Mirage 2000 and 4000. To mark his retirement, on 24 December 1985, he made his final Mirage 4000 flight escorted by a Mirage 2000C, a 2000N and a Mirage IIIB belonging to the co-located CEV.

All commercial flights were brought to a halt at Nice airport on 25 April 1987 as Dassault presented both the Mirage 4000 and Rafale A to King Fahd of Saudi Arabia and his air force delegation. Despite Louis Eon putting on a formidable flight demonstration, this last-ditch attempt to secure the Saudi Arabian order was to fail, leaving the Mirage 4000 as a footnote in history.

Right: Mirage 4000 test pilot Louis Eon (right) and Rafale A chief test pilot Guy Mitaux-Maurouard congratulate themselves on a job well done after the flight demonstrations of both aircraft to King Fahd at Nice. Neither aircraft have yet managed to gain export orders.

by Louis Eon, accomplishing his first flight on 20 January 1986. At the end of the month the Mirage 4000 returned to the workshop for important modifications. To achieve success a multi-role capability was needed and the initial very basic cockpit instrumentation and avionics were revised, including the installation of SNA (System of Navigation and Attack), bringing the aircraft closer to operational configuration.

Mirage 4000 and Rafale

In 1978 the Armée de l'Air launched the ACT (Avion de Combat Tactique/Tactical Combat Aircraft) programme. The new aircraft would be twin-engined, but smaller than the Mirage 4000. A demonstrator, the ACX, was ordered in April 1983, and was soon allocated the name Rafale. In December 1985 the aircraft was presented by Marcel Dassault to the world following its roll-out at Saint-Cloud and, transferred to Istres, it made its maiden flight on 4 July 1986. Bruno Revellin-Falcoz describes the important role the Mirage 4000 played in the development of the Rafale. "The Rafale derived directly from the 4000 in many aspects. From a base understanding of the architecture of a twin-jet fighter, through the use of composite materials [paramount in the construction of the Rafale wing], to the aerodynamic properties of canards. Of course we added other things to the Rafale, in particular the air intake that dispensed with moving components and the cockpit, with a seat tilted at a higher angle than in the Mirage 4000 to allow the pilot to endure higher *g* loadings. The Mirage 4000 provided much basic information and great support to the Rafale programme."

Preparing both the Mirage 4000 and the Rafale for the 1986 Farnborough air show assumed great importance in the summer months. The Mirage 4000, still chasing the elusive Saudi order, again conducted a series of impressive flight demonstrations with various external loads. However, the Rafale A was Dassault's new 'star', being presented to the public and potential customers for the first time, and much of Dassault's efforts centred on promoting the new aircraft.

Back at Istres in November 1986 test profiles in the shape of a 'Gothic arch' were conducted to test the Mirage 4000's performance at the subsonic/supersonic boundary, and by the end of 1986 the Mirage 4000 had completed 262 flights, including 60 that year.

From the beginning of 1987 the Mirage 4000 underwent a series of modifications as Jacques Desmazures recalls, "Up until flight 263 [5 February 1987], the aircraft was configured for optimum air-to-air capabilities – not very heavy. From flight 263, we decided to fly with heavy external loads increasing take-off weight to 28-30 tonnes. It was therefore necessary to adapt the fly-by-wire control commands in terms of manoeuvrability to prevent the onset of structural vibrations, which could damage or even destroy the aircraft. This required a series of ground tests to assess the vibrations and determine the parameters of the electronic filters added to prevent any dangerous vibrations."

Saudis lose interest

Meanwhile huge effort was being made to convince the Saudis to order the Mirage 4000 into production. On 12 December 1986 a Saudi delegation had again visited Istres to view the aircraft's new capabilities. To further the cause Dassault elected to present the aircraft in person to King Fahd, during an official visit to France. A first demonstration was arranged at Bretigny, but cancelled at the last minute. The presentation to King Fahd finally occurred, at Nice airport, on 25 April 1987. Serge Dassault, who had succeeded his father in October 1986, proudly showed King Fahd and the staff of the Royal Saudi Air Force around both the Mirage 4000 and Rafale A, before Louis Eon and Guy Mitaux-Maurouard conducted the respective flight demonstrations. All commercial flights into

Although much more than a one-third scaled-up, twin-engined version of the Mirage 2000, the joint heritage of the two types is clearly apparent in this view of the two first prototypes in formation in 1979. Both aircraft were finished in Dassault's smart house colours.

***Left:* Between the presentation to King Fahd at Nice in April 1987 and the Paris air show in June, the aircraft received this 'desert' camouflage in an attempt to woo prospective Middle East customers. It is seen here in multi-role configuration carrying both AS30L laser-guided missiles and R550 Magic AAMs.**

***Below:* Positioned on a Martin-Baker F10R zero-zero ejection seat, the Mirage 4000 pilot was afforded an excellent field-of view thanks to the large transparent blister-type canopy that hinged to starboard on opening.**

***Left:* With the application of the 'desert' scheme the Super Mirage 4000 became the Mirage 4000. The aircraft now wore the badge of Armée de l'Air unit EC 3/13 'Auvergne', in which Louis Eon had served prior to his appointment at Dassault. The aircraft is seen here with dummy Magic missiles on the wing hardpoints.**

the airport had been put on hold. Suitably impressing the visitors with the performance and manoeuvrability of both aircraft, Serge Dassault declared: "It is not only the Americans that can build powerful aircraft". Despite ongoing negotiations, however, the elusive order was still not signed.

In May 1987 the Mirage 4000 received the latest evolution of the M53 turbofan – the M53-P2. Developing almost 10 tonnes of thrust, the Mirage 4000 was bestowed with an even greater thrust-to-weight ratio. Jacques Desmazures noted: "From a performance point of view, the M53-P2 made it possible to increase the maximum take-off weight, increase the rate of climb and allow interceptions at a higher Mach number and greater altitude. Operational range with heavy external loads was also improved." Minor changes to the aircraft's air intakes were the only modifications necessary.

4000 finale

The Paris Air Show of June 1987 was to prove to be Dassault's final major effort in securing the much needed export orders for the Mirage 4000. Appearing in its new 'desert' camouflage and flying with a wide range of external stores, the shows were monitored from a mobile test room, named Amandine. Supervising the parameters of the flights and the aircraft's systems, the team and the pilot, Louis Eon, thrilled the crowds no fewer than 16 times during the show week. It was all, however, to no avail.

In late 1985 Marylène Hermitte had succeeded Jacques Desmzures as the director of the flight test programme – the first woman in military aviation to achieve this role. Her task

Thanks to the introduction of fly-by-wire controls the Mirage 4000 was designed with an aft centre of gravity, improving the overall manoeuvrability. Pitch control was achieved using two-section elevons on the trailing edge of the wing, used in conjunction with the canards.

Left: By the summer of 1987 the resources devoted to the Mirage 4000 project were dwindling. However, the flight test team at Istres continued conducting valuable research work and M53 engine flight tests until the end of the year.

Below: With a digital automated weapons delivery system the Mirage 4000 showed all the potential of being an outstanding swing-role combat aircraft. This capability did, however, come with a prohibitive price tag.

following the 1987 Paris Air Show became increasingly difficult as priority was switched to the Rafale. With a lack of available M53-P2 engines she recalls, "I had to approach the Mirage 2000 team to see if I could acquire any spare engines".

Much of the remainder of the year was devoted to Operation Banzï – the definition and preparation for series production should the hoped-for orders materialise. The outstanding performance of the Mirage 4000 was again highlighted on 22 December 1987, when Louis Eon sustained Mach 2 flight for 18 minutes.

The last mission

On 8 January 1988 Snecma test pilot Alain Rabion flew the 336th and last flight of the Mirage 4000 – the total airframe fight hours adding up to a little over 350 hours.

At the time the flight test team did not know that it would be the aircraft's final sortie, however, with the Saudi contract finally abandoned there was no reason to continue the programme. Ill-feeling remains at Dassault, however, regarding the Saudi Arabian saga, as Bruno Revellin-Falcoz explains: "What we did not realise is what passed with the United States at the time. This became known much later. They considered that the Mirage 4000 was an aircraft that could present a formidable danger to them – they were aware of the aircraft's capabilities and the Americans were worried. There are reports that we later gained knowledge of that highlighted an emergency situation in the Pentagon, stating 'if this aircraft emerges, in performance it will bury our aircraft, it is an aircraft that could be very dangerous in external markets. It is necessary to do everything to kill this programme, so that it does not emerge'."

In January 1988 the French government announced the launch of the Rafale series, which now had priority, and all work on the Mirage 4000 ceased. For Serge Dassault "it was a beautiful but lost opportunity, because this splendid aircraft was the only one that could compete with the American F-15".

The Mirage 4000 was one of the very rare fighter programmes launched and built with company funds, as emphasised by Bruno Revellin-Falcoz. "We did not have administrative constraints, therefore we did what we wanted, at our rhythm, which I think was rather fast. We did not have to justify ourselves to the government or the armed services, or have to produce files on this or that".

Although ultimately a commercial failure, the Mirage 4000 had proved that France was capable of producing a fighter equal in performance to that anywhere in the world and the contribution the programme made to the Rafale should not be underestimated.

The prototype remained in outside storage at Istres before being allocated to the Musée de l'Air at le Bourget, in October 1992. Having spent many years deteriorating outside the aircraft was completely renovated by IT Mercure. Repainted in its trademark white, red and blue scheme, the renovated aircraft was unveiled on 7 October 2002 and stands today as a reminder of a golden era in French jet fighter design.

Alexis Rocher

By 1987 Dassault's future clearly lay with the next-generation Rafale. In its final year of flight tests the Mirage 4000 flew numerous missions in support of the Rafale programme. This included researching the behaviour of canard delta configurations in severe air turbulence and flying 'chase' for the Rafale A prototype.

Lockheed P-38 Lightning
'Fork-tailed Devil'

When it first appeared, Lockheed's classic twin-boom creation was the world's first 400-mph fighter, and its speed, climb performance and hard-hitting armament allowed it to become a cornerstone of USAAF airpower in World War II. It played a key role in North Africa and Italy, and for a brief period was the principal escort fighter for the 8th Air Force's 'heavies', before being overshadowed by the P-51. By contrast, in the Pacific it was 'top dog' right to the end of the war

Painted blue for its photo-reconnaissance role, an unarmed F-5B leads a P-38J fighter during a test flight from Lockheed's Burbank plant. The red border and white bars around the star insignia were applied only between the end of June 1943 and the middle of August, at which time the red border was removed.

267183
267332

The P-38L was the final production version of the Lightning, and also the most numerous with 3,923 built. By the time it entered service in 1944 Lockheed and Allison had ironed out many of the 'bugs' which had plagued the P-38 throughout its early years – many of which had been exposed by operations in the crucible of combat. Following a USAAF edict issued in November 1943, Lightnings were among the first aircraft (along with B-29s) to be delivered without any camouflage paint, apart from the anti-glare panels ahead of the windscreen and on the inner upper surfaces of the engine nacelles.

It is a popular view that the United States and its air forces were ill-prepared when they were pitched into World War II in December 1941. While many instances bear out that belief, there were notable exceptions, among which was the Lockheed P-38 Lightning, which was the only US fighter to be in production and service from the first day of the war to the last. This aircraft notched up a number of 'firsts' that were unmatched by any other type – chief among which was that it was the first fighter capable of reaching the 'magic' 400-mph (644-km/h) mark.

During the course of the war it also became the first US aircraft to shoot down a German aircraft and the first fighter to escort bombers to Berlin. Overshadowed in the glamour stakes by its single-engined contemporaries, the Lightning nevertheless produced the highest-scoring American aces and accounted for more Japanese aircraft than any other USAAF type (estimated at over 1,800). It served in every theatre and, as the war progressed, it embraced a wide range of roles to make it one of the most versatile warplanes of the conflict.

Genesis of the Lightning

In the mid-1930s an air of complacency pervaded the corridors of the United States Army Air Corps. Although development of the Boeing B-17 bomber was well under way, the fighter forces were equipped with poor-performing, short-ranged aircraft and there were no promising designs being worked on to provide any optimism for the future. There were, however, a few enlightened individuals who kept a close eye on events in Europe – events which were unfolding rapidly and which would shape the nature of aerial warfare in the looming global conflict.

With its closely cowled engines and small radiator intakes the XP-38 prototype looked the sleekest of the breed. Its career was cut short by a landing accident at the end of the transcontinental speed record attempt, but by that time it had demonstrated sufficient performance to generate considerable interest in the type.

The 13 YP-38s established the 'look' of the Lightning, which remained virtually unchanged throughout production apart from the introduction of large chin intakes. The most obvious differences between the YPs and the XP were the deeper cowlings and the much larger radiator intakes on the tailbooms. The YP-38s were used extensively for trials. They were armed with two 0.5-in guns with 200 rounds each, two 0.3-in guns with 300 rounds each, and a Browning M9 37-mm cannon with 15 rounds.

In May 1935 the Messerschmitt Bf 109 first took to the air – a sleek monoplane fighter with a powerful inline liquid-cooled engine. The Supermarine Spitfire followed in March 1936. These developments were noted carefully in some quarters in the United States, notably in the Wright Field Development Center, where the earlier complacency was rapidly overtaken by an urgent realisation that the USAAC was not equipped to counter the equipment being fielded in Europe. In July 1936 civil war erupted in Spain, and within weeks bombers, together with fighter escorts, were attacking towns and cities. A new, more sophisticated and terrible form of aerial warfare was being formulated.

Among the Wright staff was Lieutenant Benjamin Kelsey, a visionary young officer who at once identified a pressing need for a new fighter that was much faster and higher-flying than the contemporary US equipment – an aircraft that could climb rapidly to intercept and pursue high-altitude bombers and dispatch them with hard-hitting cannon armament. Kelsey was largely behind a number of requirements that were issued by the Wright Development Center, the first result of which was the development of the Bell Airacuda, which was ordered in June 1936.

False start

Bell's design was designated XFM-1 (Experimental Fighter, Multiplace) and it first flew on 1 September 1937. It was labelled as a 'bomber-destroyer', and it had an unorthodox layout with two wing-mounted gondolas, each of which accommodated a pusher engine in the rear and a gunner in the front, the latter armed with a 37-mm Madsen cannon and a 0.3-in (7.62-mm) machine-gun. Two 0.5-in (12.7-mm) guns were mounted in glazed blisters either side of the main fuselage. The crew numbered five. In flight tests the Airacuda proved to be as cumbersome as it looked, and despite the fact that 12 YFM-1s were procured, the Airacuda was clearly not the answer that Kelsey and his staff were looking for.

While the prototype Airacuda was under construction the Army Air Corps issued two requirements in February 1937 that would form the basis of subsequent US fighter development, Projects X608 and X609. The first was for a twin-engined fighter, and the second for a single-engined aircraft. Especially important to the first requirement was the fact that both were to be single-seaters. It had come to be realised by Kelsey and others that speed, climb and altitude performance were paramount, and to achieve maximum performance on the available power it followed that aircraft size and weight had to be kept to a minimum.

Project X608 was ambitious, specifying a minimum level speed of 360 mph (579 km/h) with 400 mph (644 km/h) as desirable at a critical altitude of 20,000 ft (6096 m), and the ability to reach that altitude in less than six minutes from sea level, where it was required to reach at least 290 mph (467 km/h). Furthermore, the aircraft had to have cannon armament. Five companies submitted proposals for X608: Consolidated, Curtiss, Douglas, Lockheed and Vultee.

Below: The batch of YP-38s was followed three months later by the first P-38s. Despite their designation and the addition of some armour round the cockpit, these aircraft were of little use operationally, and were also destined to be used for trials and training.

On the face of it, Lockheed might have seemed an unlikely competitor, as during its turbulent history it had only produced one fighter, the fast but ultimately unsuccessful P-24. Nevertheless, its successes in the commercial field, centred on the outstanding Model 10 Electra and subsequent twin-engined transports, placed it at the forefront of modern aircraft construction techniques and high-performance twins. Lockheed was also involved in the XC-35 programme, using a modified Model 10 to explore

Above: The P-38D was the first operationally capable version, equipped with the self-sealing tanks that were seen as vital for combat in Europe. This was specified by the USAAC, which also specified that all such fighters should be suffixed 'D'. The P-38B and C designations were therefore skipped.

Below right: Nicknamed the 'Swordfish', this P-38E was modified with a long fuselage nacelle to reduce drag. It also had two cockpits, and was the only Lightning to have full dual controls. The laminar-flow wing sections were part of a later experiment.

Right: This P-38E was modified to test a raised tailplane being considered for a floatplane variant for ferrying to the Philippines. The aircraft would have carried fuel in the floats.

Below: The P-38E had a 20-mm Hispano cannon in place of the 37-mm weapon of the P-38D.

cabin pressurisation and turbo-superchargers for high-altitude flight. Furthermore, the company employed an extremely talented design staff, including Hall M. Hibberd and a young designer of notable engineering flair – Clarence L. 'Kelly' Johnson, who would together work on the X608 proposal and its subsequent developments.

Twin-boom fighter

Lockheed had earlier made proposals under the XFM-2 designation for the competition that was won by the Airacuda. This work was rolled into the X608 proposal, along with new data flowing from the XC-35 programme and from reports emanating from Europe. The internal designation for the new aircraft was Model 22. A number of configurations were studied, including a conventional layout with engines in wing-mounted nacelles. Two potential layouts mounted the two engines in the central fuselage driving wing-mounted propellers (both tractor and pusher) via drive shafts, although this was rejected on the grounds of complexity and cooling difficulties. One layout placed the engines in a push-pull arrangement in a central pod, with the tail surfaces being mounted on twin, wing-mounted booms. This was rejected due to problems associated with cooling the aft engine, and also of pilot egress. The most radical layout comprised twin booms housing the engines with the cockpit situated in the port boom (imagine a single-pilot Twin Mustang), but in the event a twin boom layout with a central fuselage pod was adopted.

This layout provided an elegant solution to a number of problems, as well as achieving optimum use of airframe volume while keeping frontal cross-section to a minimum. The booms provided sufficient space to house (from front to rear) the engines, the superchargers, main landing gear and radiators. In this arrangement the propellers blew air over the rudders to maximise their effectiveness (and therefore kept their size to a minimum) while the viciously hot Prestone liquid used in the cooling system was kept well away from the cockpit area, where it was to cause severe scalding burns to many a pilot of damaged single-engined aircraft in the conflict to come.

The central fuselage nacelle was designed around the need to accommodate the pilot, nose landing gear and armament. By placing all the armament (which initially was to consist of four 0.3-in (7.62-mm) machine-guns and a single 20-mm cannon) in the nose, the Model 22 could dispense with interrupting gear and concentrate its fire-power on the aircraft's centreline, increasing the effectiveness and accuracy of the fire. A heavily glazed canopy sat atop the fuselage nacelle, providing the pilot with a good

Lightning Mk Is for Britain

An order for 150 Lightning Mk Is was placed in May 1940, shortly before the fall of France. The aircraft looked very different from other Lightnings as they lacked turbo-superchargers and had a very different engine cowling and intake/exhaust arrangement. These Lightning Mk Is are seen on test in the US (left), and at Boscombe Down (below), which received three aircraft for evaluation.

Left: Evaluation by the A&AEE revealed that the un-supercharged Lightning Mk I was an unspectacular performer and it was rejected for RAF use. The remainder of the RAF order – both Mk is and IIs – was taken over by the USAAF. The Mk Is were adopted for USAAF use under the non-standard designation P-322 (reflecting the Lockheed model number). They retained RAF serials and were used almost exclusively for training. The P-322 was colloquially known as the 'castrated' Lightning.

all-round view in the upper hemisphere. An unusual feature in the cockpit was the provision of a control wheel rather than the fighter's traditional stick.

In early 1937 the idea of a tricycle undercarriage was extremely novel, although it was supported and encouraged by the Army Air Corps for the new fighter. Drawing on results from tests made with the C-40B (a modified Model 12 Electra Junior) fitted with tricycle gear, Lockheed devised and tested a new undercarriage for the Model 22 by using a rig which was towed at speed behind an automobile over various surfaces. During the course of these tests the characteristics of the gear and its brakes were rigorously examined, clearing the way for its incorporation into the final design.

Power for the Model 22 was to be provided by the only real candidate engine: the Allison V-1710 liquid-cooled Vee-12, which would drive inward-rotating 'handed' three-bladed Curtiss propellers. The powerplant had first been developed in 1931, and initially produced around 750 hp (560 kW). The addition of a General Electric B-1 supercharger and the use of the Army-developed Prestone cooling system had raised power output to around 1,150 hp (858 kW). Provision of supercharging was essential to the Model 22's high-altitude mission. In the low-density air encountered at high altitude power output of a piston engine falls off rapidly, and to restore that power the engine has to be fed with compressed air from a supercharger. The most efficient method is to use the engine's exhaust to drive a turbine – the turbo-supercharger – but this produces extremely high temperatures in the turbine itself and in the ducts which lead to and from it. The lack of suitable materials hindered turbine development, but by the mid-1930s new alloys were being produced which could handle the heat, in turn allowing the development of effective turbo-superchargers.

Contract award

Lockheed submitted its Model 22 proposal for Project X608 in April 1937. Two months later, on 23 June, the company was declared the winner and awarded a contract to construct a prototype, which received the service designation XP-38. Soon after, Bell was awarded the X609 contract for its XP-39 Airacobra.

In the months that followed the contract award, the Lockheed team worked tirelessly to turn their paper proposal into flying hardware. Detailed work began in July, and ran for a year and a half, during which time every element of the overall design was created, tested, refined and tested again. There were many new elements in the XP-38 that were new and untried, requiring many man-hours of work to perfect. Production techniques had to be evolved to produce flush butt joints and flush riveting, which were essential to achieving the design speed. The aerodynamics were based on a high wing loading to achieve the required performance but this, in turn, required high-lift devices to keep landing speeds acceptably low. The result was the fitment of a Fowler-type flap which, as well as drooping, increased wing area when deployed.

By the end of 1938 the XP-38 prototype was complete. It certainly looked stunning – its V-1710-11/-15 engines were cowled extremely closely and its natural metal finish was extraordinarily clean thanks to its flush joints and rivets. Contemporary US service fighters were bulky and powered by radials, and by comparison the XP-38 looked as if it were already travelling at 400 mph while standing still on the ramp.

On 31 December 1938, and like Lockheed's later famous 'black' aircraft, the XP-38 was trucked from the Burbank plant under conditions of great secrecy. The aircraft was taken to March Field, near Riverside, California, where it was reassembled for taxi tests to begin. Lockheed had no fighter test pilots on its books so, appropriately, Lieutenant Benjamin Kelsey – the type's greatest proponent in its formative years – was appointed by the Air Corps as project test pilot.

F-4 - the first photo-reconnaissance Lighting

A PR version of the Lightning was a natural progression, and Lockheed produced the F-4 (left) based on the P-38E and the F-4A (below) based on the P-38F, with cameras mounted in the nose peering through windows in the lower surface. This F-4A (named *Foto Jo*) retains a pair of 0.5-in machine-guns.

This P-38F-13 was one of the aircraft completed from an airframe originally ordered as a Lightning Mk II by the RAF. These aircraft had British-style instrument panels.

Lockheed XP-49 – failed quest for speed and altitude

Although closely based on the P-38, the XP-49's longer engine nacelles and pressurised cockpit gave it a considerably different and more purposeful appearance.

In March 1939, less than two months after the first flight of the XP-38, the USAAC issued Circular Proposal 39-775 calling for a twin-engined high-performance escort fighter. Although the company was heavily tasked with P-38 development, Lockheed submitted an uprated design for the new competition. It was proposed to power the aircraft with Pratt & Whitney's new experimental X-1800 24-cylinder engine, which was expected to deliver 2,000 hp (1492 kW) and propel the aircraft to 473 mph (761 km/h). An alternative engine also under development, the Wright R-2160 42-cylinder six-row radial, was also proposed.

In October 1939 Lockheed's design (Model 522) was declared the winner and the company was awarded a contract for a single XP-49 prototype the following month. At the same time, losing competitor Grumman was also given a prototype contract for its XP-50, which was based on the XF5F-1 Skyrocket.

Although the XP-49 retained the wings of the P-38, it had new engine nacelles to accommodate the P&W engines. The fuselage pod was essentially similar to that of the YP-38, but featured a pressurised cockpit with heavier framing and a much reduced glazed area. Armament was to be two 20-mm cannon plus a quartet of 0.5-in (12.7-mm) machine-guns. The undercarriage was beefed up to handle a maximum take-off weight of 22,000 lb (9979 kg).

Pratt & Whitney abandoned its X-1800 engine in favour of development of its successful radial engines. Lockheed turned to another experimental engine, the Continental XIV-1430-13/-15. This inverted-Vee 12-cylinder liquid-cooled engine developed 1,600 hp (1194 kW), considerably less than the P&W powerplant. As war approached the XP-49 programme languished, and it was not until 11 November 1942 that the prototype, 40-3055, made its first flight. The maiden flight was piloted by Joe Towle from Burbank, and lasted for 35 minutes.

Initial flight trials revealed a pleasant and manoeuvrable aircraft but, as expected, performance on the Continental engines was disappointing. Maximum speed at 15,000 ft (4572 m) was just 406 mph (653 km/h), around the same as most P-38 versions and less than the P-38J/L, which could achieve 414 mph (666 km/h) at 25,000 ft (7620 m). With no obvious benefits to be had from placing the P-49 into production, the project was abandoned. The sole prototype served at Wright Field as an engine testbed, and for tests with pressurised cockpits. It later received a second seat for an observer.

Taxi tests revealed problems with the brakes, but these were solved in time for the aircraft to undertake its first flight on 27 January 1939. Flying chase for the occasion was the unlikely form of a Ford Trimotor – a relic from a bygone age that could barely reach 160 mph at best. On take-off from March, Kelsey could not retract the flaps as a number of actuator rods had broken. Having considered and then rejected the idea of baling out, Kelsey safely brought the XP-38 back to March after 34 minutes in the air. With the flap problem swiftly rectified, the XP-38 returned to the air, exhibiting few problems. Its speed potential was readily apparent from this very early stage of testing.

Transcontinental record attempt

Confident that the aircraft would indeed fly, and that it promised to be fast, the Air Corps was eager to transfer the XP-38 to Wright Field at Dayton, Ohio, for more exhaustive tests. At this time an idea was put forward that the XP-38's ferry flight could be extended to set a transcontinental speed record, which was then held by Howard Hughes in his H-1 racer. Eager to demonstrate to the world that US fighters were not behind those being fielded in Europe, General 'Hap' Arnold approved the idea under the conditions that the ferry flight would be extended to Mitchell Field on Long Island, New York, only if the XP-38 arrived at Wright Field without problems.

On 11 February Kelsey left March in the XP-38. The pairing had just five hours in the air to this point. The flight proceeded well through stops in Amarillo and Dayton. Seven hours and 43 minutes later, of which 7 hours and 2 minutes were airborne, Kelsey arrived in the Mitchell Field pattern, beating Hughes's non-stop time of 7 hours 28 minutes. However, the XP-38 was held behind three landing aircraft, and on the long final approach both engines lost power, a failure later attributed to carburettor icing. The sleek prototype sunk to the earth and ended its days on a golf course just short of the Mitchell threshold. Thankfully Kelsey emerged virtually unscathed.

The loss of a new prototype with so few hours on the clock might have spelled disaster for other projects, but the effect of the unofficial transcontinental record was such that the P-38 was thrust to the top of the Air Corps agenda, at a time when the service was deliberating over an impressive fighter expansion plan. The results were issued on 27 April 1939, and covered a massive order for the Curtiss P-40, plus contracts for development batches of the Lockheed P-38 and Bell P-39 (the Republic P-43, forerunner of the P-47, was added in a matter of days). Lockheed's

Above: Named **Bobbie I**, *the XF-5D was a one-off conversion of an F-5A with vertical cameras worked by an operator lying prone in the nose. Even with the camera and operator, there was still room for two guns.*

Right: The F-5A camera-ship was based on the P-38G, and had a revised nose sensor installation with side-facing cameras. Early PR Lightnings were given a 'haze' paint camouflage, although this was subsequently abandoned.

contract covered 13 aircraft for service test, to be designated YP-38. The price quoted at the time for a production P-38 was $126,000.

Built in the former GGG Distillery next to Lockheed's Burbank works, the 13 YP-38s were known in-house as the Model 122-62-02 and were markedly different from the Model 22-64-01 prototype. Most of the changes were associated with a structural redesign for large-scale production, but there were many other improvements. An improved powerplant, the V-1710-27/-29 with B-2 turbo-supercharger was installed, still producing the same power output but with outward-rotating propellers, and new brakes were fitted. The engine cowlings were deeper, and the radiators on the tailbooms were enlarged considerably. Armament comprised a 37-mm Oldsmobile cannon, two 0.5-in (12.7-mm) machine-guns and two of 0.3-in (7.62-mm) calibre.

First order for the Lightning

On 1 September 1939 the tension in Europe finally broke with the unleashing of the Wehrmacht on Poland. Long before any of the YP-38s were completed, the growing urgency of the world situation led the USAAC to order 66 P-38 aircraft on 20 September 1939. This order was later followed by another contract on 30 August 1940 for a further 410. By this time the P-38 had picked up its first overseas customer. In April 1940 the Anglo-French Purchasing Commission had signed for 150 Model 322-61-03/04 (the -03 for France and the -04 for Britain) aircraft, although France fell in June, leaving the British to fight on alone. The aircraft were based on the YP-38 but with 1,090-hp (813-kW) Allison V-1710-C15 engines, which lacked turbo-supercharging (this technology was not deemed suitable for export by the US authorities). In the Model 322 both propellers turned in the same (right-hand) direction for commonality with the Curtiss P-40 Tomahawk. Having toyed with the name Atalanta for the P-38, the name Lightning was chosen, and the British aircraft were ordered as the Lightning Mk I.

Thus, when the first YP-38 took to the air on 17 September 1940, Lockheed already had a massive order book for the type. However, there were problems to be overcome before the aircraft could be fielded. The first deliveries were made in March 1941, and were soon followed, in June, by the first of 29 P-38s (Model 222-62-02). These first production machines were essentially similar to the YP-38 apart from the addition of some armour around the cockpit and the two 0.3-in guns being replaced by two 0.5-in weapons (for a total of four, plus the 37-mm cannon). A single P-38 (40-762) was modified to become the one-off XP-38A (Model 622-62-10) with a pressurised cockpit. Data from this programme was used in the development of the XP-49.

During tests with the YP-38s pilots began to encounter a new and frightening phenomenon, which only manifested itself at high altitude and at high speed: compressibility. At

speeds above the critical Mach number, especially when in a dive, the aircraft encountered growing buffet and a refusal to pull out. The phenomenon had been forecast to a certain degree by Lockheed engineers, notably 'Kelly' Johnson, but it was not understood in any real way. A long campaign of dive tests gradually revealed the secrets of compressibility and the factors which caused it, and a number of partial fixes were applied. The addition of a faired fillet around the wing/fuselage joint eased buffeting and delayed its onset, but a permanent fix was not introduced to production until 1944, when a dive flap was installed under the wings. This was not intended to slow the aircraft in a dive, but rather to produce a nose-up moment to assist dive recovery at high speeds.

Three P-38s were later modified for research into asymmetric pilot position with a second cockpit mounted in the port tailboom, entailing the removal of the supercharger.

***Above:** The P-38H was an important combat model which had more powerful engines and could carry a greater bombload or larger fuel tanks. It had an M2 cannon in place of the earlier M1. It was the P-38H which began bomber escort missions from England in late 1943, allowing the 8th AF to resume its daylight raids deep into Germany.*

***Left:** The cockpit of the P-38 was quite roomy by the standards of the day, and was relatively conventional apart from one very obvious feature – the wheel in place of the fighter's traditional stick. In practice pilots had little difficulty flying the aircraft, although the wheel may have hampered manoeuvrability in close turning combat.*

Seen with test pilot Jimmy Mattern at the controls,** **Piggie-Back I** **was the first aircraft produced with a second seat shoehorned in behind the pilot. There were no dual controls, but providing prospective P-38 pilots with a demonstration ride just before their first solo in the Lightning greatly reduced the number of training accidents.

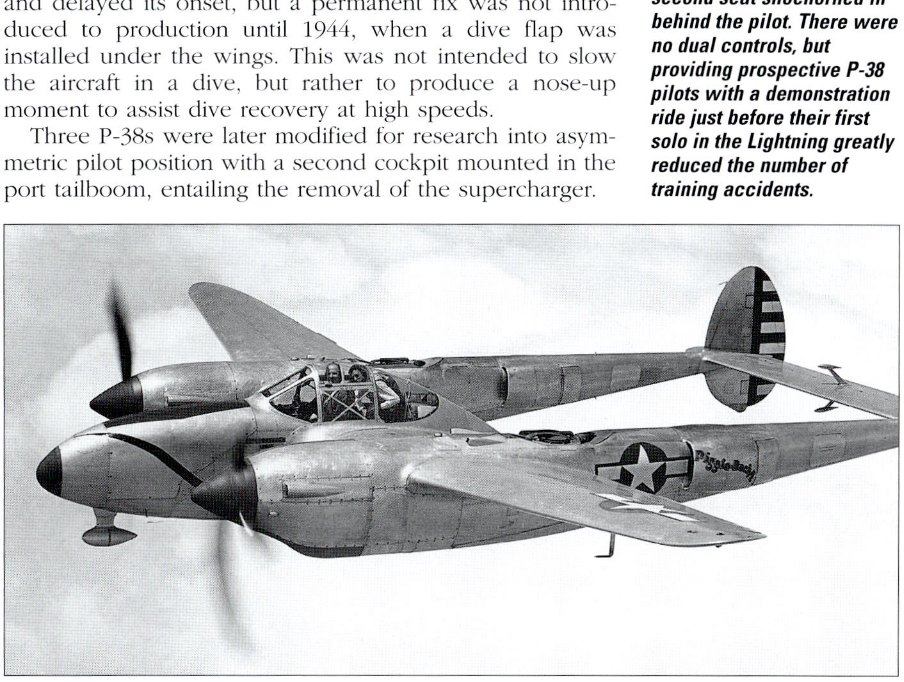

The P-38J introduced a revised propulsion and cooling system. This helped with the overheating problems and also created space in the wings for additional fuel. Although the 'beard' intakes created more drag, the cooling system allowed the engine to run better at altitude, and the J was the fastest of the variants.

Below right: The dive flap was fitted from the P-38J-25 onwards, and assisted the Lightning pilot in pulling out from high-speed dives. The intake on the nacelle side served the carburettor.

Below: The P-38J-10, like this factory-fresh aircraft on an acceptance test, was fitted with a flat-pane bulletproof windscreen.

Reports from Europe showed that the P-38 as first delivered was inadequately equipped for combat operations. The P-38D (Model 222-62-08) set out to address those shortcomings, the main change being the adoption of self-sealing fuel tanks, with an attendant drop in capacity. A low-pressure oxygen system was installed, and the tailplane incidence was altered to cure a low-speed buffet problem. Although only 36 were built, the P-38D was the first true combat-ready version when it appeared in August 1941. The first recipient was the 27th Pursuit Squadron of the 1st Pursuit Group at Selfridge Field, Michigan. Like the P-38s before them, the P-38Ds were principally used for training, preparing crews for the war that would soon arrive on the American doorstep.

By the time of the attack on Pearl Harbor, the US Army Air Forces (having adopted this new name in June) had nearly 50 P-38s and P-38Ds on strength, and production had shifted to the P-38E (Model 222-62-09). In this version the 37-mm cannon had given way to the more reliable 20-mm Hispano cannon (which had 150 rounds instead of just 15), and the hydraulic and electrical systems had been revised. An SCR-274N radio was fitted and the instrument panel had been modified. Later aircraft featured solid Curtiss Electric propellers in place of Hamilton Standard hollow units. P-38E production reached 210 aircraft, and they saw limited service in Alaska.

One P-38E was tested with a tailplane raised by 33 in (84 cm) by means of kinking the rear tailbooms upwards. Rather than being associated with the buffet problem, this aircraft tested the feature for a proposed floatplane Lightning, which was studied as a means of deploying fighters across the Pacific to reinforce the Philippines. The tailplane needed to be raised to provide wave clearance. When the island group was overrun by the Japanese, the need for a floatplane Lightning disappeared and development stopped in favour of more pressing matters.

Another P-38E was modified in 1942 with a much longer fuselage pod ending in a streamlined tail as a means of reducing drag. It was nicknamed the 'Swordfish', and had two cockpits and a full set of dual controls. Late in the war it was fitted with laminar-flow wing sections outboard of the engine nacelles, with a boundary layer control system driven by exhaust bleed air.

Reconnaissance Lightning

During P-38E production Lockheed introduced a photo-reconnaissance version, the F-4 (Model 222-62-13), based on the fighter airframe. This had no armament at all, the guns being replaced by a battery of four K-17 cameras. Ninety-nine were built and they were swiftly committed to action. The 527 P-38Fs that followed were the first Lightning fighters to be deployed into the thick of the fighting in any number.

Burbank production line

Apart from 113 P-38Ls produced near the war's end by Vultee, all Lightnings were built in the Burbank plant (left) by the parent company. Thanks to California's good weather, final assembly could be undertaken outside (below). Peak output was achieved in 1944 when 4,186 were delivered. To keep production flow as great as possible, aircraft were completed to a basic standard before being ferried to modification centres, such as Lockheed's own plant at Dallas or Delta Air Lines at Atlanta, for outfitting with the latest specification equipment or for special requirements.

Meanwhile, a second British order for the P-38 had also been received, covering 524 Lightning Mk IIs with turbo-superchargers, similar to the P-38F but with both propellers rotating to the right. Three Lightning Mk Is from the earlier order were tested at Boscombe Down and Farnborough between January and July 1942. Because of their down-rated powerplants they were considered sluggish and the Lightning was not recommended for further procurement. The remainder of the British Mk Is were retained by the USAAF. Completed Mk Is were designated P-322 in US service and used as trainers, although some were cleared for operations briefly after Pearl Harbor. They retained their RAF serials. One Mk II was completed as such, and was used for torpedo-dropping trials by Lockheed. The remainder of the British Mk II order was completed as P-38F-13s, F-15s or G-13s for operational use by the USAAF.

P-38F sub-variants

In the P-38F the principal change was the adoption of V-1710-49/-53 engines, each producing 1,325 hp (988 kW). The initial P-38F (Model 222-60-09) batch, delivered from March 1942, was followed by the P-38F-1 (222-60-15) which featured two racks under the wings between the engine and fuselage nacelles, each capable of lifting a 1,000-lb (454-kg) bomb, smoke-laying equipment or drop tank. The P-38F-5 (222-60-12) introduced new oxygen equipment, while the P-38F-13 and F-15 (322-60-19) were from the batch ordered for the RAF and had British instrument panels. The F-15 introduced a combat flap setting of 8° droop to enhance manoeuvrability. P-38Fs had SCR-522 and SCR-535 radio.

A number of P-38Fs were modified to have a second seat installed behind the pilot. This entailed the removal of the radio. The second seat was very cramped, and had no controls. The 'piggyback' Lightning was intended primarily to provide air experience to new pilots who had not encountered a tricycle undercarriage before. Twenty F-model airframes were completed as F-4A (Model 222-60-13) photo-reconnaissance aircraft, similar to the F-4 but with the uprated engines.

P-38G production accounted for 1,082 aircraft in six production blocks. The first four blocks (G-1, G-3, G-5 and G-10) were covered by Lockheed Model 222-68-12, whereas the P-38G-13/-15 were taken over from the British order and had the model number 322-68-19. The G differed

Below: Photo-reconnaissance Lightnings, like this F-5B, were produced in large numbers and were the mainstay of the PR effort. They flew with all the main Photo-Reconnaissance Groups: 3rd, 5th, 10th, 67th and 68th in the ETO/MTO; 4th, 6th, 8th and 71st in the PTO/CBI; the 2nd in the US; and with independent units.

By the time the P-38L appeared in 1944 most of the aircraft's deficiencies had been designed out. The more powerful engines were usually derated so that they were operating well within their limits, placing less strain on the cooling and supercharging systems, and they were consequently more reliable. The underwing dive flap allowed a pilot to safely recover even if he reached the critical Mach 0.72 at which compressibility rendered the elevator ineffective. P-38Ls served in Europe with the 9th Air Force, but were mainly deployed to the Pacific Theatre.

only slightly from the F, and had improved V-1710-51/-55 engines, governed to avoid overheating at high altitude and with improved B-13 turbo-superchargers. The radio reverted to the SCR-524N. Late-model P-38Gs could carry three rocket tubes strapped to the sides of the nose.

As with the P-38F, there were also reconnaissance versions, 180 being built with cameras instead of armament as the F-5A (Model 222-68-16). One of these aircraft, 42-12975, was modified to become the sole XF-5D. This was a two-seater, equipped with a glazed nose to carry an observer in a prone position to operate a vertical camera in the nose and two others in the tailbooms. On each side of the glazing was a 0.5-in (12.7-mm) machine-gun.

In September 1942 deliveries began of 601 P-38Hs (Model 222-81-20), which reached the combat zones in May 1943. The chief improvement was the adoption of 1,425-hp (1063-kW) V-1710-89/-91 engines and the increase in maximum take-off weight to 20,300 lb (9208 kg). The maximum underwing load was increased to 3,200 lb (1452 kg). During P-38H production an automatic oil radiator flap was introduced in an attempt to cure the persistent engine overheating problems. The second production block (P-38H-5) employed the General Electric B-33 supercharger for increased boost at high altitude. Top speed at 25,000 ft (7620 m) was 402 mph (647 km/h). P-38Hs were also converted for the photo-reconnaissance role, 123 airframes becoming F-5Cs with cameras.

Revised propulsion system

In August 1943 the first P-38Js began reaching the combat units. This version introduced the first major external change to the look of the Lightning in the form of a deep chin intake under each engine nacelle. Engine overheating had plagued the Lightning throughout its first year of action and the J was an attempt to cure this. The deepened chin housed a new style of intercooler, fed by an intake sandwiched between the oil cooler intakes. From the P-38J-5 the space in the leading edge occupied by the previous duct-style intercooler could now be filled with a 55-US gal (208-litre) fuel tank, raising internal capacity to 410 US gal (1552 litres). Two 360-US gal (1363-litre) drop tanks could also be carried, giving the P-38J a range of around 2,300 miles (3700 km) with 10 minutes fuel for combat.

P-38Js retained the V-1710-89/-91 engines of the H, but a number of developments were introduced during the production run, which accounted for 2,970 aircraft covered by internal model numbers 422-81-14, 422-81-22 and 422-81-23. From the P-38J-10 a flat bullet-proof windscreen was fitted, while the P-38J-25 introduced the dive flaps discussed previously. Hydraulically-boosted ailerons were also installed to cure a roll-reversal problem encountered at high speed, this being one of the first instances of power-actuated controls to be applied to a service aircraft.

A number of P-38Js were converted to TP-38J 'piggy-back' trainers, while 200 P-38J-5s were converted to F-5B reconnaissance aircraft (Model 422-81-21) with a Sperry autopilot. Four of these aircraft were transferred to the US Navy in North Africa, where they were given US Navy

The Lightning's weapons repertoire was expanded in the P-38L by the addition of rockets. The more elegant underwing zero-length launchers (above) were rejected as they entailed too many structural changes in the wings. Instead, the 'tree' installation (right) was adopted which allowed the carriage of 10 5-in (12.7-cm) HVARs (High-Velocity Aircraft Rockets). Bombs of up to 2,000-lb (907 kg) could be carried on the wing pylons.

XP-58 Chain Lightning - escort fighter

As part of the deal which released the P-38 for export to Britain and France, in April 1940 the USAAC instructed Lockheed to proceed with the construction of a company-funded prototype for a two-seat heavy escort fighter. Using the P-38 as a basis, the XP-58 design was considerably scaled up (wing span 72 ft/21.95 m as opposed to 52 ft/15.85 m). The wings had an untapered centre-section, with tapered outer panels being joined some way outboard of the engine nacelles. The central fuselage pod was extended aft of the wing trailing edge, culminating in an aft-facing glazed position for a gunner, who operated turrets situated above and below the rear fuselage. The landing gear was completely redesigned, each unit having two wheels instead of one, and could handle a maximum take-off weight of 43,000 lb (19505 kg), roughly double that of the P-38. Nose armament was proposed as either four 37-mm cannon or a single 75-mm cannon plus two 0.5-in (12.7-mm) machine-guns. Four 0.5-in machine-guns were mounted in the two turrets. The aircraft could also carry up to 4,000 lb (1814 kg) of bombs or fuel.

Initial powerplant choice fell on the 1,600-hp (1194-kW) Continental IV-1430 engine, as eventually used in the XP-49, but this was not powerful enough to achieve the XP-58's desired top speed of 450 mph (724 km/h). Pratt & Whitney cancelled its X-1800 inline engine, ruling that option out, while the radial P&W R-2800 was not compatible with the airframe. Other promising options – the Continental XH-2860 and Lycoming XH-2470 – were also ruled out. Wright's 42-cylinder XR-2160 Tornado engine was the main option until it, too, was cancelled in February 1943. Finally, the 2,600-hp (1940-kW) Allison V-3420-11/-13 was settled on, the seventh engine to be proposed for the Chain Lightning.

Lockheed was also having problems with the customer, too, the USAAF repeatedly changing its requirements. What started out as an escort fighter became an attacker, before being re-roled as an anti-bomber interceptor. These continual changes, both in role and intended powerplant (combined with Lockheed's greater concentration on the more pressing task of producing and improving the P-38), conspired to delay the completion of the first XP-58 (41-2670). It was not flown at Burbank until 6 June 1944. This was also the day of the Normandy landings, and by the time that the XP-58 arrived at Wright Field for evaluation in October, the war in Europe was going well. It was obvious that by the time the Chain Lightning could be readied for service, the war would be long finished. It was also not quite as fast as had once been hoped – maximum speed at 25,000 ft (7620 m) was 436 mph (702 km/h) – and the type had maintenance difficulties. Accordingly, the XP-58 programme was terminated without being fully evaluated by the USAAF, and a second prototype was cancelled.

Two prototypes were planned of the Chain Lighting, but only the first was completed and flown. Visible forward of the aft gunner's compartment is the position of the intended dorsal turret. Another turret was to be mounted below the fuselage. The main armament, however, comprised large-calibre cannon in the nose.

BuAer serials and the designation FO-1. Reconnaissance modifications applied to 105 P-38J-15s and P-38J-25s resulted in the F-5E. Several P-38Js were also converted to 'Droop Snoot' pathfinder configuration with a glazed nose to house a bomb-aimer and a Norden bombsight.

Several P-38Js were used for test work, including one that was used to test an unusual refuelling method, in which a B-24 passed a drop tank to the fighter by means of a cable and hook. The XP-38K was an experimental Lightning that featured 1,425-hp (1063-kW) V-1710-75/-77 engines driving larger-diameter propellers with broad-chord blades. This aircraft (42-13558) was converted from a P-38G-10, and proved to be only marginally faster than a P-38J and was not considered for further development.

Final production version was the P-38L (Model 422-87-23), and it was built in the greatest numbers, accounting for 3,923 aircraft. Again, the main difference was an uprated engine in the form of the V-1710-111/-113, nominally rated at 1,600 hp (1194 kW) although normally derated to 1,475 hp (1100 kW), but was otherwise generally similar to the P-38J. While its bombload was raised to two 2,000-lb (907-kg) weapons, it was the first Lightning to introduce rocket armament, carrying 10 5-in (12.7-cm) rockets on underwing 'tree' fittings. Early trial rocket installations had tested 14 projectiles on zero-length launchers – seven under each wing – but this was rejected as it involved too much alteration to the wing structure and would have slowed production. The landing light was relocated to the port wing leading edge. One was tested as a dedicated strafer, with eight 0.5-in (12.7-mm) guns in the nose and four more in two pods under the wings.

Five hundred P-38L-1s became F-5Es for the reconnaissance role, and some P-38L-5s were converted (along with a single F-5B) with a different camera installation to become F-5Fs. Additional P-38L-5s became F-5Gs, which was the last of the camera versions to be produced. Small numbers of P-38Ls also received the TP-38 'piggyback' modification for training duties.

In June 1944 the USAAF ordered 2,000 P-38Ls to be built by the Consolidated-Vultee Aircraft Corporation at the Vultee works in Nashville, Tennessee. By the war's end only 113 P-38L-5-VNs had been completed and the remainder of the order, along with outstanding orders from the Lockheed plant at Burbank, was cancelled. Total production of the P-38/F-4/F-5 was 10,041.

P-38 at war

The operational P-38D was first allocated to the 1st Pursuit Group in mid-1941, followed by the 14th Pursuit Group at March Field later in the year. In the aftermath of Pearl Harbor the 1st PG was sent to March to bolster the defences of the US West Coast.

Below left: Sizeable numbers of P-38L-5s were converted for reconnaissance roles as either F-5Fs or F-5Gs. This aircraft, seen in India in 1945, has been reported as being a field modification of a P-38L, but may also have been designated as an F-5F.

Below: The F-5G was the ultimate Lightning PR version, featuring a bulbous nose which could house larger cameras of a greater variety. Most also mounted a radio beacon antenna in a teardrop fairing below the nose.

Above: Fresh from the modification centre, a new P-38J is seen running its engines after arriving in England. Assembly of newly delivered aircraft and heavy maintenance was performed at depots at Speke (Liverpool) and Langford Lodge (Northern Ireland).

tionally it converted to the P-47C in February 1943. It later flew Mustangs, becoming unique in the 8th AF in flying all three major USAAF fighter types.

Thus, it was to be another 11 months before the Lightning returned to combat operations over northern Europe. By the middle of 1943 it was becoming plain that operating bombers in daylight without fighter cover was near-suicidal, a fact that was cruelly underlined by the disastrous Schweinfurt raid on 14 October. The heavy losses prompted a temporary halt to bombing operations.

P-38Js were delivered to England from late 1943, the first aircraft retaining an olive drab finish (right) before unpainted aircraft (below) were received from February 1944. Both of these machines served with the 38th FS/55th FG.

In mid-1943 the VIII Fighter Command's main fighter was the P-47 Thunderbolt, which did not have the range to accompany the B-17s on bombing missions deep into Germany. The Lightning certainly did have the range, and so plans were laid to reintroduce the P-38 to the ETO, with the primary task of escorting bombers. Two groups were deployed initially, the 20th and 55th FGs, which were deployed to Kings Cliffe and Nuthampstead, respectively. The 55th FG was the first to go into action: its first mission, a fighter sweep along the Dutch coast, was undertaken on 15 October.

Escorting the daylight 'heavies'

It was not until 3 November that the 8th AF resumed its bombing operations with an attack on Wilhelmshaven. For the first time on a raid into Germany the raiders had fighter cover all the way to the target, provided by P-47s on the transit and by the P-38Hs of the 55th FG during the attack itself. The Luftwaffe's Jagdgeschwader 1 rose to meet the formation and encountered an unpleasant surprise. In the fighting which followed 13 German fighters went down, of which seven probably fell to the guns of the 55th FG (the others were shot down by P-47s). The bomber formation attacked largely untroubled – a new era of daylight bombing had begun.

Over following raids, the presence of fighters ensured that bomber losses were kept down, but the P-38 groups paid for their successes with a steady attrition to the enemy. Although Luftwaffe pilots stated that they would prefer to face the P-38 in combat rather than a Spitfire or P-47, the Lightning nevertheless gained their respect – and the nickname of 'der gabelschwanz Teufel' (fork-tailed devil). Overall, the P-38s came out on top, but a large number were lost in increasingly fierce battles and the margin was small. Lightning losses roughly equalled combat claims, although many of the P-38s were lost to the endless mechanical problems which continued to dog the type.

In December 1943 the 20th FG began combat operations, flying its first mission over enemy territory on the 28th. Bomber escort missions followed soon after. In January 1944 Merlin-engined P-51Bs began escorting

Lightnings of the 367th FG

Little Buckaroo was flown by Major Robert C. Rogers, popular commander of the 392nd FS and inevitably nicknamed 'Buck'. He used this aircraft to achieve five ground strafing kills in an attack on an airfield near Dijon on 25 August 1944.

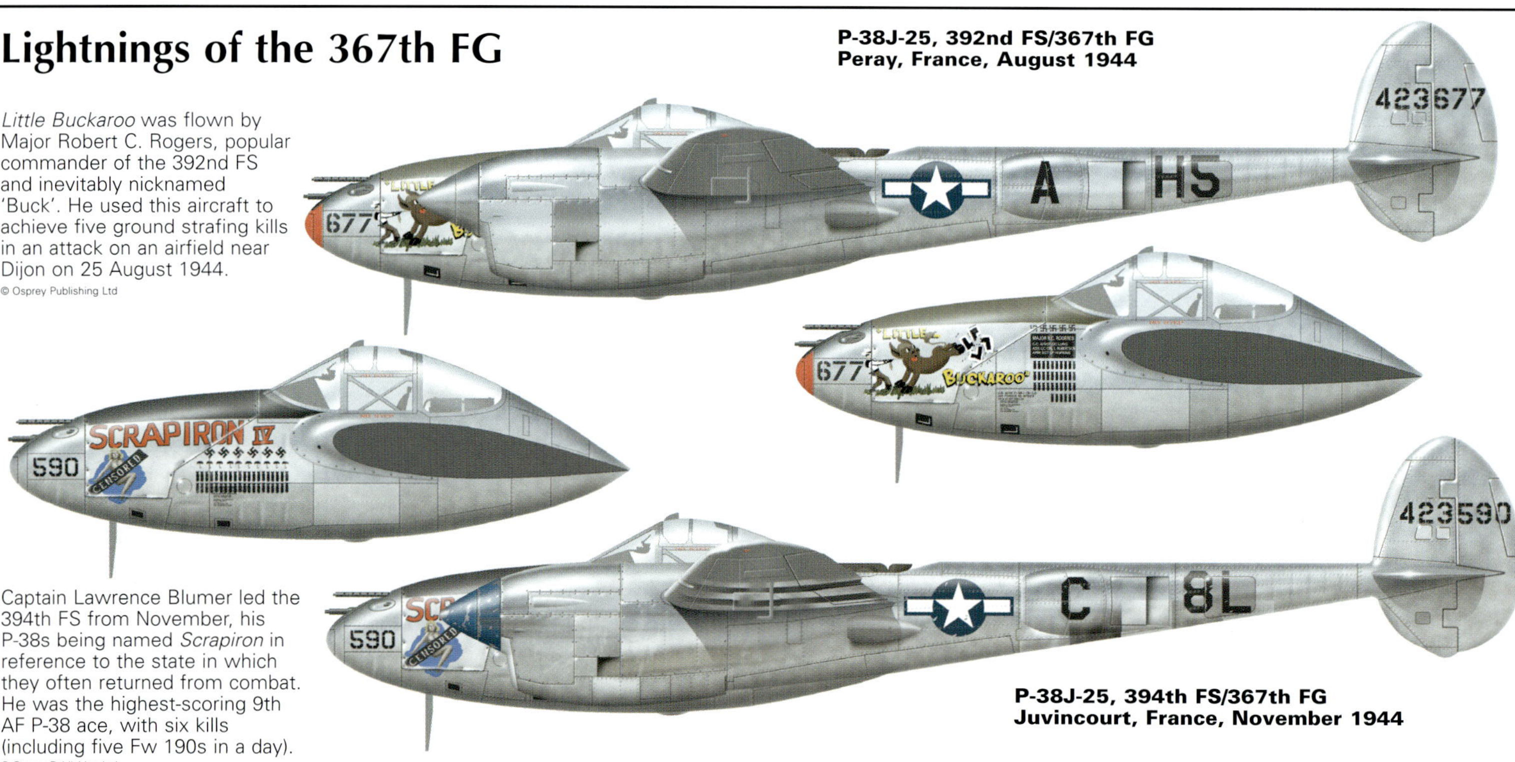

P-38J-25, 392nd FS/367th FG
Peray, France, August 1944

P-38J-25, 394th FS/367th FG
Juvincourt, France, November 1944

Captain Lawrence Blumer led the 394th FS from November, his P-38s being named *Scrapiron* in reference to the state in which they often returned from combat. He was the highest-scoring 9th AF P-38 ace, with six kills (including five Fw 190s in a day).

Lockheed P-38J-15-LO

55th Fighter Squadron
20th Fighter Group
VIII Fighter Command
Kings Cliffe, England, 1944

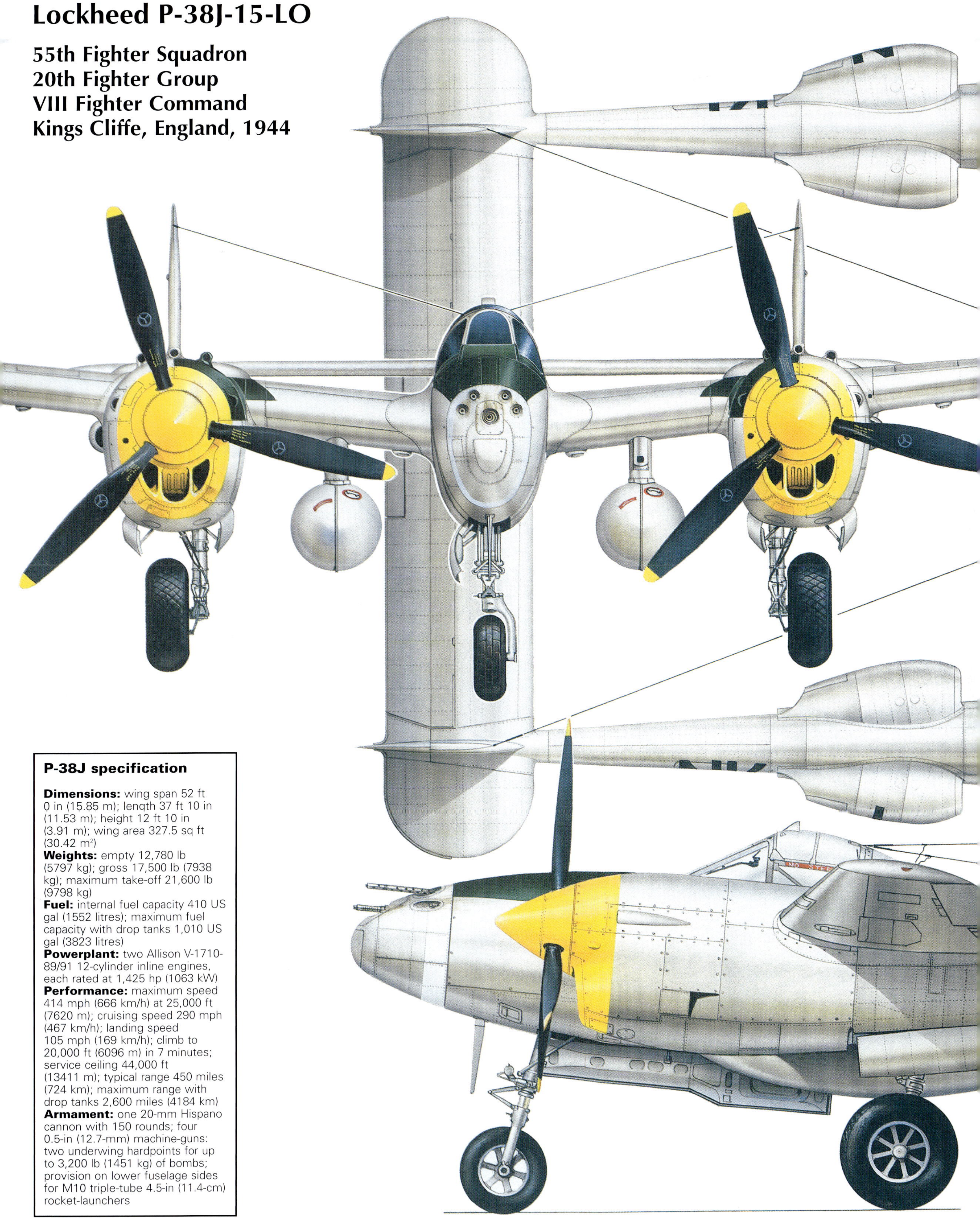

P-38J specification

Dimensions: wing span 52 ft 0 in (15.85 m); length 37 ft 10 in (11.53 m); height 12 ft 10 in (3.91 m); wing area 327.5 sq ft (30.42 m^2)
Weights: empty 12,780 lb (5797 kg); gross 17,500 lb (7938 kg); maximum take-off 21,600 lb (9798 kg)
Fuel: internal fuel capacity 410 US gal (1552 litres); maximum fuel capacity with drop tanks 1,010 US gal (3823 litres)
Powerplant: two Allison V-1710-89/91 12-cylinder inline engines, each rated at 1,425 hp (1063 kW)
Performance: maximum speed 414 mph (666 km/h) at 25,000 ft (7620 m); cruising speed 290 mph (467 km/h); landing speed 105 mph (169 km/h); climb to 20,000 ft (6096 m) in 7 minutes; service ceiling 44,000 ft (13411 m); typical range 450 miles (724 km); maximum range with drop tanks 2,600 miles (4184 km)
Armament: one 20-mm Hispano cannon with 150 rounds; four 0.5-in (12.7-mm) machine-guns; two underwing hardpoints for up to 3,200 lb (1451 kg) of bombs; provision on lower fuselage sides for M10 triple-tube 4.5-in (11.4-cm) rocket-launchers

This P-38J represents those which served with the 8th Air Force in the first half of 1944, before the type was largely phased out of the bomber escort role from July. As well as its escort missions, the 20th FG flew numerous strafing attacks, and gained the nickname of the 'Loco Busters' due to its successes against the rail network. The group received a Distinguished Unit Citation for a fighter sweep on 8 April 1944 led by the CO, Lieutenant Colonel Harold Rau, against German airfields in the Satzwedel area. The sweep accounted for 50 aircraft (including He 177 bombers) destroyed on the ground and in the air, 18 locomotives and numerous trucks, two bridges, and 16 flak towers and gun emplacements.

dge fuel tank
oms and

ap,
erated
rriage leg
g
g strut
el, aft-

ıg strut
cast sub-

ewall
bearer
ure
k

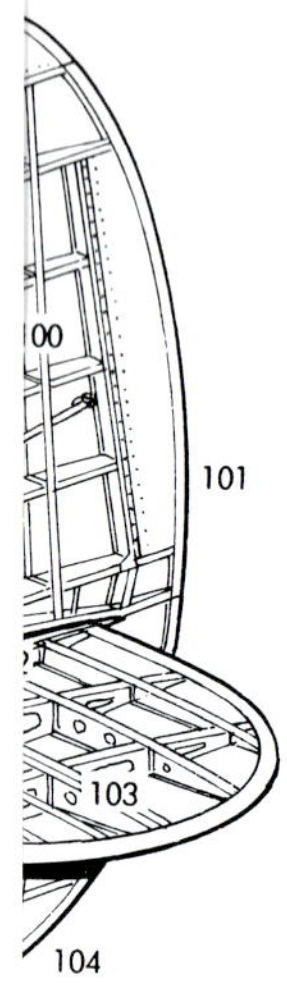

ıtake duct
ıgneto

ort structure
0-89/91
710-111/113
2 engine
manifold
it duct to

let duct
haust flap
ooler

aust
cooler, twin
rd and

ne mounting
er tank
nner

pitch-change

lectric
ed propeller,
tor
l intercooler

spinner
on, mounted
r wing panel
nstallation

68-litre;
external tank
m M8 rocket

be 'Bazooka'
er, fuselage
l on P-38J
g) HE bomb
-kg) HE bomb
m) rocket

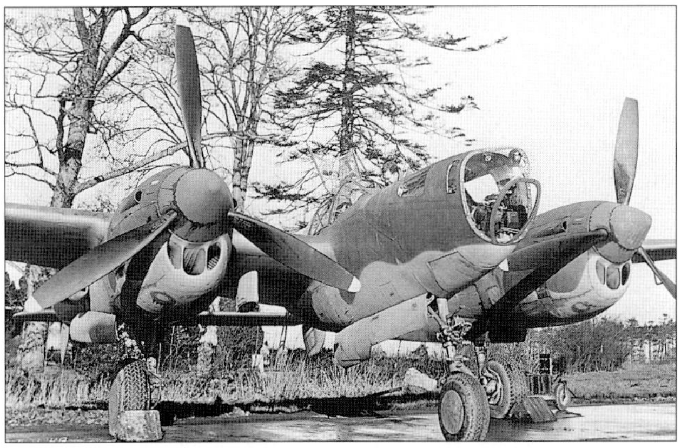

Droop Snoot Lightnings

Using a P-38 as a bombsight-equipped pathfinder followed on from experiments using P-47s as makeshift bombers, dropping their weapons on command from a B-24. As well as supporting P-38 bombing operations, the Droop Snoots also occasionally led P-47 formations. Although pioneered by the 8th AF, the Droop Snoot raids were continued by 9th AF P-38 groups.

bombers into Germany, and the P-38 swiftly lost its importance in the bomber escort role, but there was plenty of work still to do. In early March 1944 a third Lightning group, the 364th FG at Honington, began 8th AF operations with P-38Hs and Js, just as USAAF bombers began to target Berlin. The 55th FG provided the escort for the first USAAF raid on the German capital on 3 March. A fourth P-38 group, the 479th FG, was established in May with P-38Js at Wattisham. On 29 July a P-38 from this group was the first USAAF fighter to tangle with a German 'jet' when it was involved in an indecisive fight with an Me 163 Komet.

By August only the 479th FG was left flying the Lightning in the 8th Air Force as the other units converted to the Mustang. This was somewhat ironic, as the Lightning groups had something of a 'purple patch' in July and August. Among the successful pilots was Captain Robin Olds of the 434th FS/479th FG. His five P-38 kills were followed by eight more in the P-51, and of course four in the Phantom during the Vietnam War. The 479th continued on Lightning operations until September 1944 when it, too, converted to the Mustang. In its last weeks with the P-38 it concentrated on fighter sweeps and airfield strafing attacks.

As well as its fighter groups, the 8th AF also had other Lightning operators. The 7th Photographic Group moved into Mount Farm in Oxfordshire to provide photo-reconnaissance with F-5s from July 1943. The main aim was to photograph targets for the strategic bombing campaign. The F-5 performed satisfactorily, but it was not ideal for long-range, high-altitude missions, which were generally entrusted to the Spitfire PR.Mk XIs that were taken on strength in October 1943. The 8th AF also had two Fighter Training Groups – the 495th at Atcham and the 496th at Goxhill – that flew P-38s and P-47s.

Bombing and attack

Although the Lightning was based in England with the primary aim of fighter duties, it was inevitably used increasingly on attack missions such as strafing raids. It could also be employed as a bomber at low level, although it had no bombsight for medium-/high-level operations. It was reasoned, however, that Lightnings could be used for mass-formation attacks if a means of accurate aiming could be found. The benefits were obvious: fighters could attack at much greater speed than bombers, with greater surprise and survivability.

During the winter of 1943/44, technicians worked in secret at Langford Lodge in Northern Ireland on a special modification of the P-38J that came to be known as the Droop Snoot. The aircraft had its armament removed and a prone position for a bombardier installed. The glazed nose was equipped with a Norden bombsight. Acting as a pathfinder, the Droop Snoot could lead large formations of fighters, which would release their bombs on command from the Droop Snoot.

On 10 April 1944 the first operational use of the Droop Snoot was made by the 55th FG during an attack on Coulommiers airfield in France. After the bombs had been dropped, the formation descended to follow up with a strafing attack. In subsequent operations the concept proved to be successful, and a fair number of Droop Snoot conversions were applied to P-38Js and P-38Ls. One drawback of the system was that if cloud obscured the target the formation could not bomb. Subsequently a few Lightnings were given H_2X bombing-through-overcast radar in an extended nose. They were known as 'Mickeys'.

Resplendent in invasion stripes, 27 Lightning bombers are led by a Droop Snoot aircraft. The photo illustrates the kind of concentration of attack that could be achieved when the formation dropped its bombs simultaneously.

Above left: This was an early Droop Snoot conversion, displaying the glazed nose cone which incorporated a flat pane for the Norden bombsight. The bombardier had a lower escape hatch.

A natural progression from the Droop Snoot was the radar-equipped pathfinder, which housed H_2X (APS-15) radar in a solid nose. Behind the radar sat the operator underneath a glazed panel. The Pathfinders were known as 'Mickey'-ships and were issued in small numbers to the tactical Lightning units.

Three Ninth Air Force groups flew the P-38 from April 1944, employed on tactical operations in the lead-up to the Normandy landings and thereafter as the Allied armies fought through France. This selection shows **Dottie's Great Dane**, ***an aircraft of the 394th FS/367th FG (above), a Droop Snoot from the 428th FS/474th FG (right) and*** **SanDot IV** ***of the 485th FS/370th FG (below) – all in France. The bottom photo shows P-38s from the 430th FS/474th FG taxiing at Strassfeld in Germany.***

In April 1944, as the Allies prepared for the invasion of Europe, 8th AF fighter units embarked on a major strafing campaign against German airfields under Operation Jackpot. In May the German transportation system came under increasing strafe attack under Operation Chattanooga. In the late spring, the 'tactical' Ninth Air Force was preparing its forces for the invasion. Among its array of fighter-bombers were three new groups operating the P-38 – the 367th, 370th and 474th FGs. They joined in the attacks in France in the run-up to D-day.

During the Normandy landings on 6 June the P-38s were mainly tasked with providing low-altitude fighter cover for the ships carrying the invasion force. After the landing, the Lightnings flew attack missions in support of and ahead of the ground forces, including strafing and Droop Snoot attacks. Glide attacks – bombing from low level in a shallow dive – began to be employed, as was the use of finned

fuel tanks and incendiary bombs to create fires. The three Ninth AF units moved to France in July/August, and followed the advancing armies through the country, participating in all the major actions as the Allies closed in on the Reich. In March 1945 the 367th converted to P-47s and the 370th to the P-51, leaving the 474th FG as the last Lightning group in Europe. It ended the war at Langensalza in Germany. The last kill scored by a Lightning in the ETO was a Siebel Si 204 shot down by a 429th FS/474th FG aircraft on 8 May, the last day of the war in Europe.

Combat appraisal

Operations in northern Europe tended to highlight the P-38's weaknesses rather than its capabilities. Among the deficiencies was the poor reliability of the V-1710 engine, which did not work well in the cold and damp of England, especially during the autumn/winter months. In Europe the aircraft was found to be much better suited to combat below 18,000 ft (5486 m), where it suffered less from the effects of the cold, thin air, which played havoc with the

Downning Yamamoto

P-38G, 339th FS/347th FG
Guadalcanal, April 1943

Of all the 10,000-plus Lightnings built, this P-38G-13 was arguably the most important as it was the aircraft in which Lieutenant Rex Barber almost certainly single-handedly shot down the Mitsubishi G4M carrying Admiral Yamamoto on 18 April 1943. Barber normally flew *Diablo* (43-2187) but on the fateful day the aircraft was unserviceable and this aircraft was 'borrowed' from Lieutenant Bob Petit. The ship kill on the nose was for an action by Petit on 29 March.

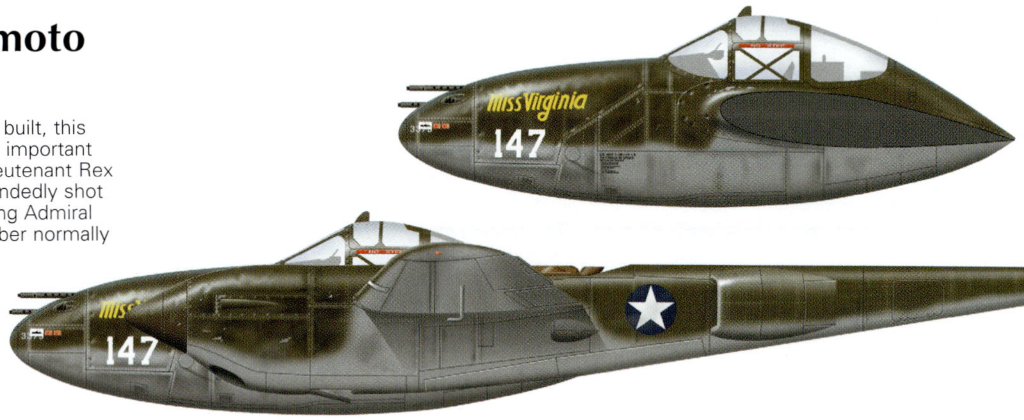

turbo-superchargers. The cabin heater was woefully inadequate, and despite the rapid procurement of heavy flying clothing, Lightning pilots often endured agonising hours of bitter cold. There were only seven groups equipped with the type, a unit strength that was only achieved briefly in mid-1944. Furthermore, each of those units was rarely at full strength, suffering high attrition and difficulty in replacing airframes. As a consequence, P-38s flew far fewer missions than the P-47 and P-51, with a corresponding lower rate of air-to-air and strafing victories.

Yet the effect the P-38 had during the few months that it was the only long-range escort fighter cannot be overlooked, and it allowed the 8th AF to resume its daylight onslaught. It was also very effective as a ground attacker, although 9th AF commanders preferred the immensely strong P-47 to the more delicate Lightning (it is ironic to note that the P-38 was preferred over the P-51 by pilots in the Pacific for exactly the same reason). In the final account, the top Lightning ace in northern Europe was James Morris of the 20th FG who was credited with 7.33 kills. To be accurate, the top 8th AF Lightning ace was another 20th FG pilot, Jack Ilfrey – of Portugal fame – although his score of eight included six from the MTO.

Lightning in the Pacific

P-38 fighters were deployed to the Pacific region from mid-1942. The first deployment comprised P-38Ds and Es to the 343rd FG of the 11th Air Force in Alaska. They were soon replaced by P-38Es modified (as P-38F-1s) to carry drop tanks, which were issued to the 54th FS. After the Japanese capture of Kiska in June 1942, the P-38s were deployed to Unimak. On 4 August 1942 two Kawanishi H6K flying-boats were shot down by two 54th FS Lightnings, the first confirmed victories to fall to the P-38.

Highly prized for their long range and twin-engined configuration, P-38s remained in the Aleutians for the duration of the war, but this theatre did not produce any significant engagements other than the May 1943 recapture of Attu. From their bases in the Aleutians, 11th AF's P-38s nevertheless nibbled away at the northern fringes of the Japanese empire. It should be noted that a P-38J was tested with retractable ski undercarriage for service in the Aleutians, but this remained as an experiment only.

In the south Pacific the Lightning entered the fray in November. P-38s were delivered by ship to Australia, where they were reassembled and issued to squadrons withdrawn from the combat zone for conversion (most had previously flown P-39s). Australia itself operated at least three F-4s, loaned to No. 1 Photographic Reconnaissance Unit from August 1942. One was returned to the USAAF at the end of the year and the other two were subsequently lost in crashes. Two other F-4s were loaned by the 8th PRS to No. 75 Squadron, RAAF, which operated them briefly in New Guinea in the late summer of 1943.

In November 1942 the 339th Fighter Squadron was sent to reinforce the defences of the beleaguered island of Guadalcanal, while the 39th FS was sent back to '14-Mile Drome' at Port Moresby, New Guinea, after having withdrawn to Australia for conversion to the P-38F/G. The first kills were recorded on 18 November by 339th FS Lightnings that were escorting a bomber formation. Three Mitsubishi A6Ms were downed to open the Lightning's amazing account in the Pacific. The first major engagement was on 27 December, when Japanese bombers and their fighter escorts attacked the Allied base at Dobodura, New Guinea. Scrambling from Port Moresby, the 39th FS shot down at least 11 of the attackers for only one loss. Shooting down an Aichi D3A and an A6M that day was then-2nd Lieutenant Richard Ira 'Dick' Bong (temporarily attached to the 39th from the 9th FS), who had claimed the first two of his eventual 40 kills.

Fighting continued over New Guinea into 1943, much of it centred on the Japanese-held harbour at Lae. On 31 December Lieutenant Hoyt Eason became the first P-38 'ace' in the theatre, a status achieved by Bong on 8 January. Soon after, Bong returned to his former unit, the 9th FS, to oversee their conversion to the P-38G and their move to Dobodura. The 80th FS/8th FG followed soon after. Among the 80th pilots was George 'Wheaties' Welch, who had shot down four Japanese aircraft during the Pearl Harbor raid,

P-38J production began with 10 P-38J-1 aircraft for service tests. One of these was used to test a ski undercarriage for potential service in the Aleutians. The ski struts retracted backwards so that the skis lay flush with the underside of the tailbooms and fuselage pod. Here the aircraft is seen undergoing tests at Ladd Field, near Fairbanks, Alaska, in the winter of 1943. By this time the Japanese had been ousted from the Aleutians, and it was unlikely that any Lightning would have to operate from snow. Although the trials were generally successful, the programme was abandoned.

The Royal Australian Air Force operated a handful of F-4 photo-Lightnings which were on loan from the USAAF. Three aircraft were assigned to No. 1 PRU: A55-1 from August 1942, A55-2 from September and A55-3 (illustrated) from February 1943. The second aircraft was given back to the USAAF in December 1942 but the first and third Lightnings were used until lost in crashes in August 1944 and December 1943, respectively. In addition to these aircraft, two further F-4s were briefly loaned to No. 75 Squadron for operations in mid-1943.

This view on Guadalcanal shows the P-38 fighters of the 6th Night-Fighter Squadron, which was the first unit to test the P-38 in nocturnal combat. Their initial exploits used ground searchlights as the primary means of detecting the lone 'Bedcheck Charlie' G4M bombers that plagued the US bases on the island, but later on two aircraft received trial installations of airborne radar in the form of SCR-540.

Right: Armourers load the staggered nose guns of a P-38 at a Pacific base.

Technicians reassemble a newly arrived F-5A on Guadalcanal in 1943. Conditions at many of the bases in the PTO were very primitive, with very little cover for aircraft or personnel. The weather swung between steaming hot and lashing rain. It was a tribute to both men and machine that the Lightning enjoyed such successes in such conditions.

and after the war became the first man to fly the F-86 Sabre (and, controversially, may have become the first man to break the sound barrier) before being killed in an F-100. Half of his 16 confirmed kills were achieved in the P-38.

Many missions at the time were to protect bombers attacking Japanese convoys that sailed from Rabaul to supply the forces at Lae. A large convoy was decimated in February in what became known as the Battle of the Bismarck Sea. Lightnings took centre stage in warding off Japanese fighters, although Hoyt Eason, the first P-38 ace, was lost in the fighting.

In early April Admiral Yamamoto launched Operation I-Go in an attempt to crush the Allied air assets in New Guinea and Guadalcanal, which seriously threatened the Japanese strongpoints at Lae and elsewhere. In heavy fighting the Japanese force came off worst, the P-38 again proving outstanding in the defensive role.

Yamamoto mission

On 18 April 1943, while operating from Guadalcanal, the P-38 flew one of the most famous aerial missions of the war – the assassination of Admiral Isoroku Yamamoto. The admiral was the mastermind behind the Pearl Harbor attack and had become a Japanese national hero. Despite his misfortunes at the Battle of Midway, he remained a very dangerous foe and his national standing and past achievements marked him out as America's 'Most Wanted' in the Pacific. To kill him or capture him would have untold effects on the morale of the Japanese. After the monumental shock of Pearl Harbor, American pride had been partially restored by Doolittle's brave raiders, and the Battle of Midway had proven that the Imperial Japanese Navy was

not as invincible as was once thought. However, true revenge for Pearl Harbor would only be achieved when Yamamoto had been 'taken out'.

The cracking of Japanese naval codes provided the necessary intelligence for the attempt: a signal was intercepted on 14 April detailing the itinerary of a forthcoming morale-building tour by the admiral to forces on Bougainville. Yamamoto would be travelling in a Mitsubishi G4M2 'Betty', with a standard fighter escort of six A6M Zeros. The Japanese believed that the route was well outside the range of US fighters.

The task of intercepting the flight was entrusted to the P-38Gs of the 339th Fighter Squadron, which were based at 'Fighter Two' airstrip (Henderson Field) on Guadalcanal. Eighteen aircraft were prepared for the mission with one 165-US gal (625-litre) and one 310-US gal (1173-litre) tank each. The mission was to be led by Major John Mitchell, whose aircraft was fitted with a naval compass to aid navigation. Four P-38s, flown by Captain Tom Lanphier, and Lieutenants Rex Barber, Jim McLanahan and Joe Moore, made up a 'killer' section, while the remainder would form a fighter cover screen. Both Moore and McLanahan had to abort though problems, and their place in the 'killer' section was taken by Lieutenants Ray Hine and Besby Holmes.

Just before 8 am on 18 April the Lightning force left Guadalcanal. Flying at very low altitude, the formation skirted islands so that no alarm could be raised, and flew a circuitous route so that the attack would appear to the Japanese as a chance encounter by a routine patrol, rather than an obvious interception attempt. This step was necessary to avoid compromising the intelligence 'gold mine' of the code-crackers.

At around 9.30 the formation arrived at the intercept point over Bougainville. The timeliness and accuracy of the rendezvous – and the ultimate success of the mission itself – was entirely due to the superb navigation work by Major Mitchell. Pretty much on cue, one of the covering pilots, Lieutenant Doug Canning, spotted the Japanese formation. There were two G4M2s, creating some confusion.

However, the 'killer' section went into the attack. Holmes could not jettison his tanks, and pulled away to tackle the problem. His wingman, Hine, went with him, leaving Lanphier and Barber to go after the admiral's formation.

What happened next has been the subject of confusion and controversy for many years. As the two Lightnings climbed to attack the formation, Lanphier pulled away from the bombers to tackle three of the Zero escorts, while Barber forged ahead towards the first 'Betty'. With Zeros closing in, he fired from astern, observed pieces falling off and then the bomber snapped over and disappeared beneath his aircraft. The bomber had not fired back, a fact which puzzled Barber (later, examination of the wreckage revealed that the 'Betty' had been stripped of all armament). Lanphier, meanwhile, had claimed to have shot down a Zero and then dived back down to make, what he claimed, was the final *coup de grace* on the damaged bomber. Inside, Yamamoto was already dead, having been hit by bullets from the 'first' P-38 attack.

Immediately after the attack on Yamamoto's aircraft, Barber began to take hits from the Zeros, a total of 104 holes being counted later when he limped back to Guadalcanal. Successfully evading the fighters, he then joined the search for the second bomber (carrying Yamamoto's Chief of Staff). Holmes was also looking for it, having successfully shaken off his hung-up tanks, and he got in a first attack. The inexperienced Hine also attacked, without effect, and Barber finished the job before the Zeros caught up with them again. Holmes and Barber both claimed Zeros as they attempted to disengage, but Japanese accounts suggest that no A6Ms were lost in either this engagement or to Lanphier's earlier attack. In fact, the only fighter that was lost that day was the P-38 of the unfortunate Hine, which was last seen trailing smoke before disappearing.

Post-mission controversy

The Yamamoto mission almost certainly changed the course of the war, and saved many Allied lives. It also became highly controversial on a number of levels. The morality of targeting an individual was subsequently called into question, while others later became incensed at the lack of recognition for the pilots involved. At a more personal level, the debate over who actually shot down Yamamoto's aircraft blew up almost immediately after the P-38s returned to Guadalcanal. Initially the USAAF accepted Lanphier's version of events, in which he had dived back down to administer the fatal blow against the G4M. Barber contested, but Lanphier outranked him. Some

This P-38L of the 18th FG is seen at Zamboanga in the Philippines, in early 1945. The fighting to retake the island group had involved the P-38 in some of the fiercest action it had yet encountered, although it was a prosperous time for the Lightning aces.

Far left: Napalm became a favoured weapon in the latter part of the Pacific War. The inflammable gel was pumped into standard 165-US gal (625-litre) drop tanks (bottom) and then dropped in low-level attacks. The photo at the top shows a demonstration in Hawaii in October 1944 of a finned tank, which was much more accurate as it did not tumble when released.

Below: Photographed from another Lightning, a P-38J is seen providing fighter cover for the invasion convoy heading for Balikpapan in eastern Borneo.

Ace in the CBI theatre

P-38J-10, 459th FS/80th FG
Chittagong, India, May 1944

The leading ace in the CBI theatre was Captain Walter Duke, who flew this flamboyantly marked P-38 to gain the majority of his 10 victories. He was also flying the aircraft when he was shot down by a Nakajima Ki-43 and listed as MIA on 6 June 1944. The 459th FS 'Twin-tailed Dragons' was the only Lightning squadron in the southern area of the CBI theatre, with operations mainly being conducted over Burma.

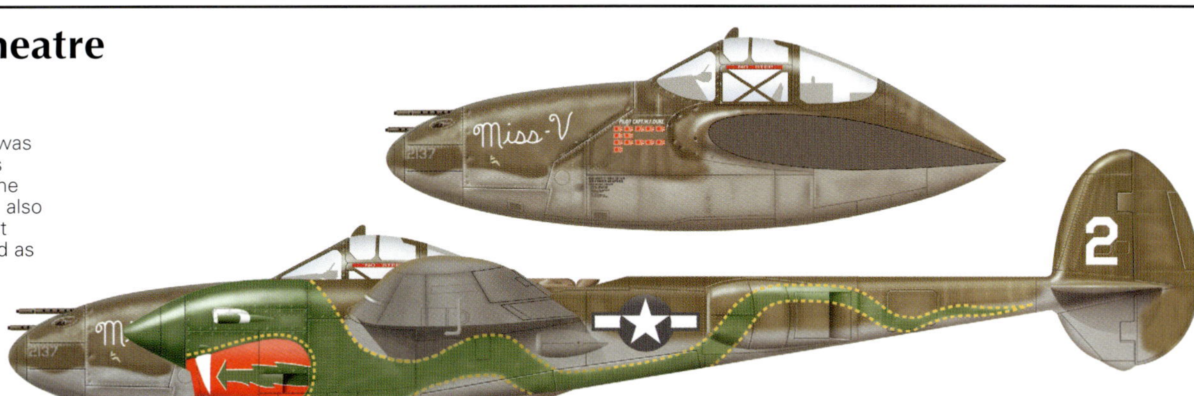

America's top two

Tom McGuire, America's second ranking ace, named his Lightnings *Pudgy*. This aircraft was the second of four to bear the name. He was killed just short of being returned to the US, crashing while helping a squadron-mate in combat, and perhaps striving to eclipse Bong's 40-kill score before his tour ended.

P-38H-5, 431st FS/475th FG
Dobodura, New Guinea, December 1943

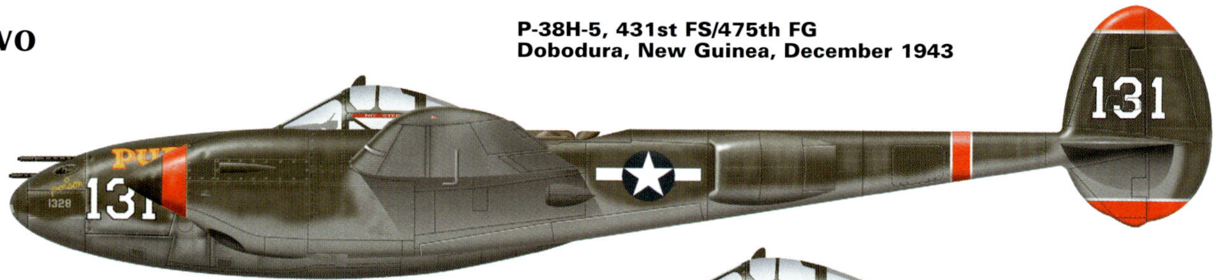

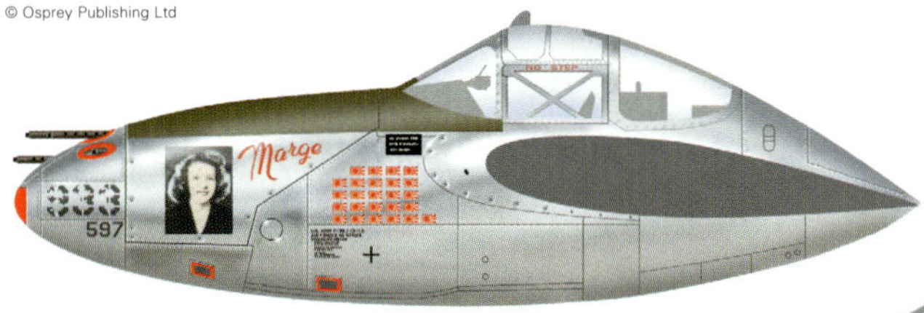

America's 'ace of aces', Dick Bong named his Lightnings after sweetheart Marge Vattendahl, whom he married in February 1945. On the ground Bong was quiet and self-effacing, but underwent a transformation in the air into an aggressive warrior.

P-38J-15, 5th Fighter Command
Cape Gloucester, New Britain, March 1944

years later, after Japanese survivors from the battle had been interviewed, the claim was reinvestigated and both Lanphier and Barber were officially credited with sharing the kill. Examination of all the evidence would suggest that it was Barber alone that shot down Yamamoto's 'Betty', and several august bodies have publicly stated as such.

CBI Lightnings

With the end of the North African campaign in early 1943, the USAAF brass decided that it could spare some precious P-38s from the MTO for services elsewhere. The loudest clamouring came from General Claire Chennault, former leader of the 'Flying Tigers' and now heading the air campaign against the Japanese in China. Rather than getting new aircraft and new pilots, most of the Lightnings and their crews destined for China were drawn from units in North Africa. Most arrived at Kunming in late July, where the Lightnings were initially organised as Squadron X. This subsequently became the 449th Fighter Squadron.

Operations began in late July, but it was not until 21 August that the first kills were achieved. Many of the early missions involved dive-bombing attacks, especially against Japanese shipping in Chinese ports. During one of these missions on 30 October, former American football star Tom Harmon was shot down over Kiukiang. He trekked for 32 days to get back to his unit at Ling-Ling, a remarkable demonstration of his determination to survive.

A second P-38 squadron, the 459th FS 'Twin-tailed Dragons', was established in India for service over Burma. When it disbanded in China soon after the end of the war, it achieved the unique position of being the only USAAF squadron to have never served in the US. The 459th was heavily involved in both air fighting and in strafing/dive-bombing attacks against airfields until mid-1944, after which it was principally involved in ground attack duties. Some P-38s operated with clusters of three Bazooka tubes strapped to either side of the fuselage. Air-to-air victories for both the China-based and India-based Lightnings had largely dried up by the end of 1944.

As in other theatres, the F-5 was the backbone of the PR effort. By the end of the war many squadrons had the F-5G on strength, this version being instantly recognisable thanks to its bulbous nose. This example is seen in China in 1945. The recce group in the CBI theatre was the 8th PRG, headquartered at Bally in India.

Right: Interested groundcrew examine two F-5Bs at Isley Field, Saipan, in the Marianas Islands. The date was 10 July 1944 and the aircraft were the first Lightnings to arrive at the base. On its engine nacelle Stinky carries the badge of the 28th Photo-Reconnaissance Squadron, which was directly assigned to the Seventh Air Force. A detachment of the 28th PRS operated from Saipan on bomber support recce tasks until May 1945, when the whole squadron regrouped on Okinawa. A deployment was then sent forward to Ie Shima.

A number of Droop Snoot aircraft were sent to the CBI theatre, these aircraft being seen at Bangalore in India in 1945. At least one was used as the personal transport of General Stratemeyer.

Above: At the end of the war many P-38s were dismantled and eventually scrapped locally. This scavenged 475th FG aircraft was at Kimpo, Korea, in late 1945.

Throughout the CBI campaign the P-38's twin-engined layout meant that many a lucky pilot limped home safely with one engine. Although there were only two squadrons deployed, they achieved enormous success and earned the greatest respect from their opponents. Around 15 P-38s and 15 F-5s were handed over to the Chinese.

On the offensive

At around the time the P-38 was arriving in the CBI theatre, the Allies in the south Pacific were seizing the initiative in the late spring of 1943, spurred on by the removal of Yamamoto. Forces began to range far and wide from Guadalcanal, preparing for the invasion of the Solomons. The first attempts at using the P-38 as a night-fighter to counter nocturnal G4M raids were trialled over Guadalcanal in January 1943, but they were unsuccessful. However, a group of P-38 pilots was seconded to the 6th Night-Fighter Squadron and continued to experiment with the night-fighter role. On 15 August Lieutenant Henry Meigs Jr scored a kill in a P-38G over a 'Betty' which had been caught in the base's searchlights. He went on to down two more on 20 September. P-38s were used with some success in this role.

Meanwhile, in New Guinea major attacks began on the Japanese bases at Wewak in August. By this time the three P-38-equipped squadrons (431st, 432nd, 433rd FS) of the 475th Fighter Group had joined the fray, and participated in a massive dogfight over Wewak on 18 August. During this fight three fighters became the first kills for the 431st FS's then-1st Lieutenant Thomas Buchanan McGuire Jr.

While continued attacks managed to neutralise Wewak without a costly invasion, it was a war of attrition, and the enemy did not depart the base complex until November, following the fall of Finschhafen. By this time, Allied attention was firmly on the Japanese 'fortress' at Rabaul, New Britain. The air campaign against Rabaul began on

Left and below: As many as 30 Lightnings may have been given to the Chinese air force towards the war's end. These aircraft are an F-5E (below), which was based on a P-38L-1 airframe, and an F-5G (left) based on the P-38L-5.

Late-war Lightnings in the Pacific

P-38L-5, 475th FG
Clark Field, Philippines, March 1945

Colonel Charles H. 'Mac' MacDonald was commander of the 475th FG, and with 27 victories was the third highest-scoring Lightning ace. His P-38, *Putt Putt Maru*, was decorated with the colours of the three squadrons under his command.

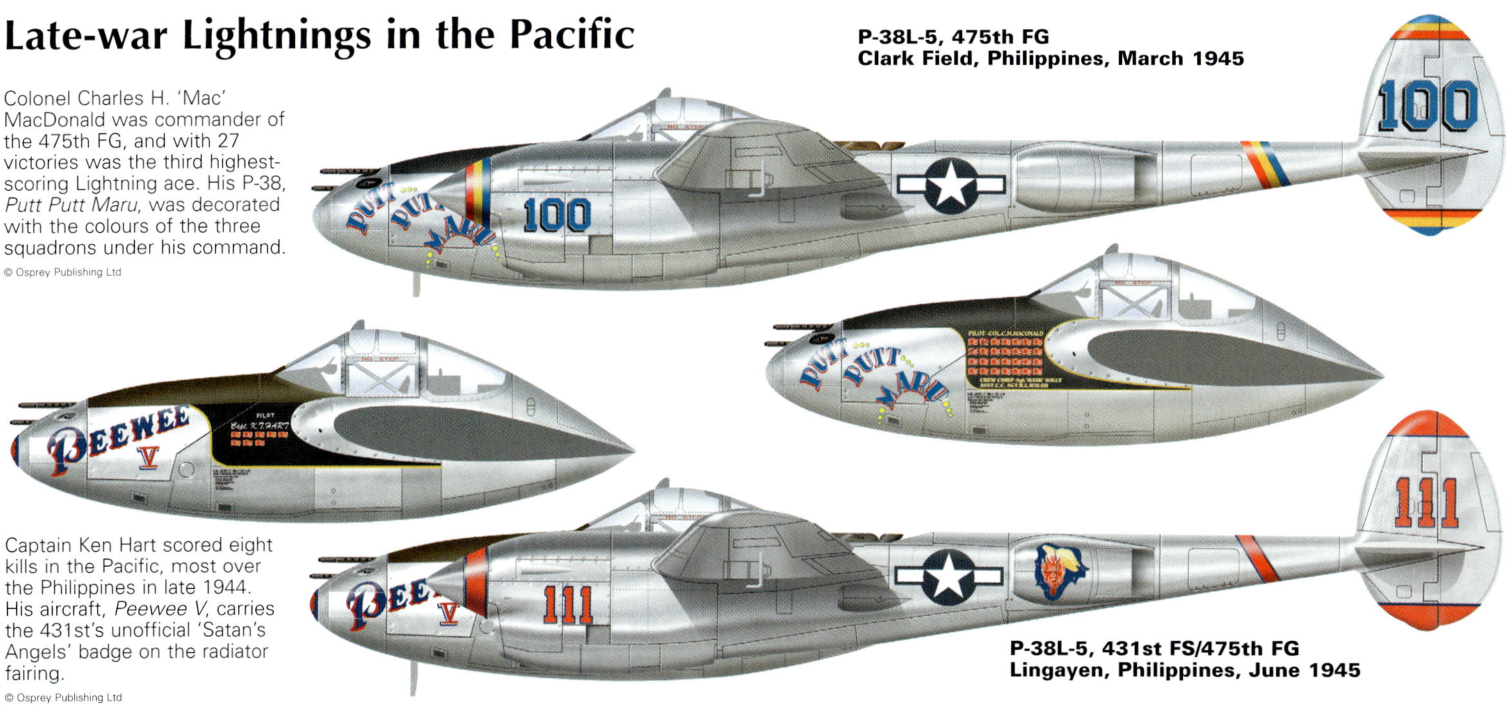

Captain Ken Hart scored eight kills in the Pacific, most over the Philippines in late 1944. His aircraft, *Peewee V*, carries the 431st's unofficial 'Satan's Angels' badge on the radiator fairing.

P-38L-5, 431st FS/475th FG
Lingayen, Philippines, June 1945

44-27234 was a much-photographed P-38M, and shows the gloss black finish adopted by USAAF night-fighters in the latter part of the war. It also carries the 'tree' fittings for 5-in rockets. The definitive night-fighter Lightning was produced following experimental field modifications, none of which had been particularly successful due to their single-pilot operation.

Below: This view gives an idea of just how cramped the radar operator's position was. The radar scope was mounted on top of the instrument panel and was fitted with a hood so that the operator could work it without any light interference.

12 October, when P-38s escorted B-24s and B-25s in attacking airfields in the area. Bad weather hampered operations over Rabaul, but the raids continued sporadically, and the 5th AF P-38 tally rose rapidly. On 15 October Major Charles MacDonald opened his account: the future leader of the 475th Fighter Group would score 27 in P-38s (all named *Putt-Putt Maru*) to become the third-ranking Lightning ace.

On 2 November P-38s from the 9th, 39th, 80th, 431st, 432nd and 433rd FSs provided cover for a low-level B-25 raid. The ensuing fight, which encountered stiff Japanese opposition, came to be known as 'Bloody Tuesday', with several P-38 pilots achieving ace status that day. In November the 13th AF's P-38s, including aircraft from the 18th Fighter Group, also began escorting raids over Rabaul, by which time the attacks had largely switched to low-level missions. These were both hazardous and rewarding for the P-38 units. By March 1944 the threat of Rabaul had been negated, and the fortress was bypassed as the war swept on towards the Philippines.

There was still, however, fighting to be done in New Guinea, where the Japanese were still holding on in Hollandia. Factory-fresh P-38Js arrived in time for the brief assault on Hollandia, a welcome influx of aircraft (P-38 deliveries had hitherto been prioritised for the European theatre). The first raid got under way on 30 March, with 80th FS and 475th FG P-38s covering the B-24 bombers. Compared with the fierce opposition over Rabaul, the Japanese defences were ill-organised and kill tallies rose rapidly over Hollandia. Now a major and assigned to Fifth Fighter Command, Dick Bong notched up his 27th kill during the air battles of 12 April. This surpassed the score of 26 by Eddie Rickenbacker in World War I. By mid-April the Japanese forces at Hollandia were all but beaten and New Guinea was, to all intents, in Allied hands.

Operations carried on relentlessly through the summer of 1944 as the Americans closed in on the Philippines. During this period the 475th FG played host to Charles Lindbergh in his capacity as Tech Rep of the United Aircraft Corporation. The famous aviator instructed the group on how to get the best endurance from their aircraft, and flew a few missions. On 28 July 1944 he shot down a Mitsubishi Ki-51.

Philippine battles

Air combat in the Philippines reached a crescendo on 7 December, when the US landed at Ormoc Bay. Although the landings were virtually unopposed on the ground, there were massive air battles, in which 50 Japanese aircraft were claimed for the loss of just one P-38. Attention soon turned to Clark Field, and at the end of the month P-38s scored many kills during attacks on the former US base.

The fight for the Philippines saw the Lightning involved in intense battles, illustrated by the fact that over 400 kills were achieved during the last two months of 1944. By this time the P-38L had been issued to many units. On 17 December Major Dick Bong scored his 40th and last victory over Mindoro to top the pile of US aces. He was awarded the Medal of Honor and posted back to the US where his successes could be exploited for propaganda. Sadly, soon after he had married his sweetheart, Marge, he was killed in a flying accident in a Lockheed P-80 on 6 August 1945, the day of the Hiroshima bomb.

The 431st FS/475th FG was also in the thick of the action over the Philippines, and the unit's commander, Major Tom McGuire, scored 14 victories in this period. The last four of these, on 26 December, took his tally to 38, making him the

second highest-scoring US pilot of the war. He was killed in action on 7 January 1945 when his P-38L stalled at low level while he was attempting to clear a Japanese fighter from the tail of a squadron-mate.

By early January air action over the Philippines had subsided, and P-38s occupied the captured bases. From them they could range out over Southeast Asia, and a few kills were scored along the coast of Vietnam and over Taiwan. Near the end of the conflict Lightnings were operating from Ie Shima, and could reach as far as Korea and the Kurile islands. However, the great air battles of New Guinea, Rabaul and the Philippines were now behind the Lightning, and the last months of the war were relatively quiet for the P-38 force as the focus shifted to the attack on the Japanese homeland. The recce F-5s, however, were heavily involved in the attacks on Japan, flying missions to provide pre- and post-raid images.

Ten days after the Japanese surrender, a P-38 flown by Colonel Clay Tice landed at Nittagahara airfield in Japan on 25 August 1945. Although the landing was unplanned – Tice landed with engine problems – it was significant as it was the first American aircraft to land in Japan following the end of the war. It was especially symbolic that it was a P-38, for that aircraft had done more than most to defeat the Japanese air forces.

The Lightning night-fighter

Throughout the war in the Pacific high-flying Mitsubishi G4M 'Betty' bombers had been a constant thorn in the Allied side. The hastily prepared Douglas P-70 night-fighter could rarely reach the bombers and, as noted earlier, the P-38 was used as a makeshift night-fighter, albeit without radar. Nevertheless, the success it achieved in this role, together with its excellent high-altitude performance, made the P-38 a natural choice for a dedicated night-fighter, especially as the Northrop P-61 Black Widow was suffering from numerous problems and deficiencies.

Trials of radar-equipped night-fighter versions began in 1943 when a pair of P-38Fs were tested by the 6th NFS in the Pacific with SCR-540 radar, and continued with three P-38Js of the 547th NFS which were fitted with APS-4. One of the latter scored a kill in the Leyte region in January 1945. However, the extra work of operating the radar was found to place a heavy burden on the pilot, and so a two-seat version was formulated. It was the P-38L-5 which was to form the basis of the operational night-fighter, which was designated P-38M-5 (Model 522-87-23). The conversion consisted of squeezing a second cockpit for the radar operator behind the pilot. This second station was extremely cramped, and it was recommended that operators should be no more than 5 ft 6 in (1.68 m) tall. Other features were the fitment of flash suppressors on the gun muzzles to avoid blinding the pilot, and the installation of AN/APS-4, or ASH, radar in a pod, or 'bomb'. At first the ASH was installed under the centre of the fuselage pod, where it was found to be in the way of spent cartridge cases when the guns were fired. It was then moved to the starboard wing, where it created asymmetric handling problems, and finally to a position under the nose.

***Above:** The P-38M was an effective night-fighter that was just too late to see action. The radar, second cockpit and gun modifications had only a marginal effect on aircraft performance.*

***Above left:** This close-up shows the ASH (APS-4) radar 'bomb' under the P-38M's nose. Using a centimetric wavelength, the ASH was developed from an air-to-surface radar. It was very good at low level due to its clutter rejection properties, but its antenna sweep was slower than other night-fighter radars. The view also highlights the flash suppressors fitted to the guns. Thought was given to covering the turbo-supercharger outlets to further enhance visibility, but at night they glowed with heat anyway, and covering the exhausts would have had a negligible effect in cutting light emission.*

***Left:** Two pilots from the 418th Night-Fighter Squadron at Atsugi stand next to* Shady Lady*, one of the four P-38Ms which were deployed to the Pacific Theatre shortly after the end of hostilities. Some of the P-38M aircrew had previously flown combat with the 418th in the P-61 Black Widow.*

While most Lightnings were scrapped in the months following the war, a few found their way into civil hands, and a number joined the National Air Races circuit. Inevitably, non-standard modifications were applied to some. Both of these aircraft feature streamlined noses and clipped wingtips, and have had the small sections of tailplane that protrude outboard of the rudders truncated. The aircraft above has a non-standard intake above the engine, while race number '27' has no boom radiators at all.

Among the most interesting of the survey aircraft was Hycon's N91300. This P-38E was the former 'Swordfish' two-seat laminar-flow test aircraft, and it retained its lengthened fuselage and dual controls (albeit with new canopies fitted. The grossly extended nose housed a survey camera and a glazed operator's position.

Following a trials conversion, 75 (some sources say 80) P-38Ls were set aside for M conversion, and work got under way at Lockheed's Dallas Modification Center in March 1945. The first aircraft was lost in an accident, but it had showed considerable promise. Testing in earnest began in July, and soon P-38Ms arrived at Hammer Field, California, the principal West Coast night-fighter training base. Here a cadre of experienced night-fighter crews began working up on the Night Lightning in order to deploy an eight-aircraft group to the war zone for evaluation under combat conditions. The crews who came into contact with the P-38M were very impressed with it as a night-fighter, and it was considered by most to be much better (if a lot less comfortable) than the P-61.

Unbeknown to all but a few, the US was about to drop the two atomic bombs that would bring the Pacific War to an abrupt halt. When the conflict ended, the P-38Ms were still in the US, about to sail for either Iwo Jima or Okinawa, and thus played no part in combat operations. In the event, four Night Lightnings were deployed, initially to Clark in the Philippines and then to Atsugi in Japan, from where they were flown by the 418th Night-Fighter Squadron on general patrols. One of the aircraft was lost in a crash, and in February 1946 the remaining trio was transferred to the 421st NFS at Fukuoka, which flew them for a few months before they were sent back to the US for scrapping. With the war long over and the Twin Mustang about to enter service, any remaining interest in the P-38M evaporated.

Post-war Lightnings

With Japan defeated the USAAF embarked upon a massive demobilisation. Of the wartime fighters, only the P-51 survived in any numbers, and the last P-38Ls were withdrawn in 1949, by which time they had been redesignated as F-38L in the newly created US Air Force. The vast majority of airframes were scrapped, but a fair number got a second lease of life with civilian owners, most of them acquired via the War Assets Administration from the storage site at Kingman, Arizona. Some were used for racing, for which they received various modifications such as clipped wingtips and tailplanes, and more streamlined noses. One aircraft had its tailboom radiators deleted in the quest for greater speed.

Others were converted for the aerial survey role with various extended nose shapes housing large cameras and a prone operator's position. Survey Lightnings operated with companies such as Hycon Aerial Surveys, Mark Hurd Aerial Surveys, Spartan Air Services, Fairchild Aerial Surveys and the Tennessee Valley Authority. They were mainly active in Canada (where the last survey flight was made in September 1961), the US, and in Latin America, but one Mark Hurd aircraft was also ferried to Ethiopia for survey work in 1957/58. At least one aircraft was on the Argentine civil register and another was registered in Venezuela.

In the military field the major recipient was the Aeronautica Militare Italiana, which had been formed in June 1946 following the proclamation of Italy as a republic. To re-equip the fighter force 100 P-38 Lightnings were drawn from USAAF stocks held at Capodichino, and 93 were overhauled by IMM (Aerfer) before issue to the AMI. The remaining seven were held over for spares. The majority of the aircraft were F-5Es and P-38Ls, but a few P-38Hs and P-38Js were also included in the batch. Some were converted to P-38DC standard, with a second cockpit with dual controls added in an extended nose.

Although most survey Lightnings operated with US registrations, some were registered overseas. Above is a Canadian aircraft, operated by Spartan Air Services, one of the stalwarts of the huge Canadian mapping effort in the 1950s. Below is the Argentine-registered LV-HIX.

Post-war survey work

During the war F-5s had been widely used on mapping and survey work, as well as target reconnaissance. Through large swathes of the Pacific region they had been instrumental in producing detail contoured maps to allow accurate military planning. It was no surprise that the Lightning carried on this role after the war. Under civilian ownership, modified Lightnings charted large areas of Canada and Latin America, for example, and assisted with aerial surveys such as that associated with the Tennessee Valley project. Some of the aircraft survived to become 'warbird' display items.

P-38s were issued initially to the 4° Stormo Caccia at Lecce-Galatina, whose 9° and 10° Gruppi converted to the Lightning from the P-39 Airacobra. In early 1948 the 4° Stormo moved to Napoli-Capodichino, where the two gruppi converted to the P-51. The displaced Lightnings were then handed to an interim unit known as the Stormo P-38 at Bari-Palese, which acquired the 28° Gruppo in January 1948 and the 132° Gruppo in April. The Stormo P-38 was officially redesignated as the 3° Stormo Caccia on 1 November 1948. From the mid-1950 the stormo began to transition to the Mustang, although the 28° Gruppo continued to fly P-38s until early 1954.

P-38s also flew with training units, principally the Scuola Volo delle Puglie at Lecce-Galatina, whose main flying unit was initially known as 3° Gruppo Scuola Volo (later 213° Gruppo). This unit was used for advanced training, and operated P-38s until 1955, absorbing some of the aircraft from the 3° Stormo. One other unit to fly the Lightning was the Scuola Caccia Ogni Tempo (all-weather fighter school) at Amendola, which had a pair of dual-control P-38DCs on charge until it was disbanded in 1957. The last Italian Lightnings, operating in a 'hack' role, were not retired until October 1959.

In 1947 a number of P-38Ls were acquired clandestinely in the USA and were sent to Cayo Confites in Cuba to be operated by the Partido Revoluçionario Dominicano, a US- and Cuba-backed group which opposed the dictator Trujillo in the Dominican Republic. Before an amphibious invasion could be launched, with the Lightnings providing air cover, support was withdrawn from the PRD and Cuba confiscated the P-38s. They were flown for some time by the Cuban air force and were soon withdrawn, although one ended up in Honduras.

In September 1947 the US and the Latin American states signed the Inter-American Treaty of Reciprocal Assistance, or the Rio Pact. Among the effects was the release of war-surplus aircraft to re-equip many of the air arms in the region. Bolivia was due to receive the P-38, but none was delivered. Costa Rica operated one aircraft briefly in 1948/49 before passing it on to Guatemala. It was not flown by the FAG and was passed on to Honduras in 1953.

Under the terms of the Rio Pact, the Aviación Militar Hondureña (Fuerza Aérea Hondureña after 1954) received a total of seven P-38Ls to partially equip the service's first combat aircraft unit at Tocontín air base near the capital, Tegucigalpa. P-38L-5s were delivered from 1948, along with five Bell P-63E Kingcobras. From 1956 Vought F4U Corsairs were received to replace the Lightnings and Kingcobras, although the older types were still in use during the border conflict with Nicaragua in early 1957. One P-38 was lost in FAH service, killing Lt Col Hermán Acosta Mejía, who had been appointed as the first Honduran chief of the air force in 1948. Four of the Honduran Lightnings returned to the US in 1960/61, one of which was purchased by the USAF Museum, and is now on display at Dayton (albeit with a P-38J serial).

Around 35 P-38 airframes exist today, of which just a handful is airworthy as flying Lightning numbers have dwindled in recent years through crashes. Included in the current list of flyers is a P-38F – named *Glacier Girl* – with a remarkable history. It, plus five other P-38s and a B-17, force-landed on a glacier in Greenland during a ferry flight to England on 15 July 1942. Between 1990 and 1992 it was recovered from under 260 ft (79 m) of ice by the Greenland Expedition Society and, once returned to the US, restoration to fly began at Middlesboro-Bell County Airport in Kentucky. On 26 October 2002, over 60 years since its last flight, *Glacier Girl* returned to the air again.

David Donald

Most important of the post-war military users was Italy, which acquired large numbers of P-38s to help rebuild its post-war fighter force by equipping one full wing (stormo). A small number were locally converted to the P-38DC dual-control version by adding a second cockpit in an elongated nose section.

***The Lightning was not as common in Latin America as were the P-47 and P-51, Honduras being the main operator. This P-38L is seen in California in 1960 upon its return to the US after service with the FAH. Today it flies with the Lone Star Flight Museum at Galveston, Texas, wearing the colours of 'Mac' MacDonald's* Putt Putt Maru.**

Lightnings continue to grace airshows in the hands of 'warbird' enthusiasts. For many years a very well-known aircraft was Marvin 'Lefty' Gardner's* White Lightning *(below left), but this was badly damaged in a crash in June 2001. An interesting 'warbird' was the Confederate Air Force's N38LL (below) which flew with a survey nose and second cockpit. It crashed in 1994 and is now on ground display.

Gloster Javelin
'Flying Flat Iron'

The Gloster Javelin holds the distinction of being the first and, to date, the only all-British delta wing fighter to enter squadron service, and was also the world's first twin-jet delta. In an era marked by many distinctive and unusual shapes, the 'Ace of Spades', 'Flying Flat Iron' or 'Flying Triangle' was unique but, as a fighter, the Javelin possessed weaknesses and flaws in terms of performance and manoeuvrability that resulted in frequent criticism. However, with the Soviet bomber threat at its height, the Javelin acted as a powerful deterrent, frequently proving itself in exercises to be the finest 'bomber-destroyer' of the era.

Above: The RAF's Javelin era began in early 1956 as deliveries commenced to No. 46 Squadron, based at RAF Odiham. Replacing the obsolescent Meteor and Venom all-weather fighters, the RAF had, at last, an aircraft capable of intercepting the new Soviet bomber force that included the M-3 'Bison' and Tu-95 'Bear' long-range strategic bombers.

Below: As the largest and most complex RAF fighter of the period, the Javelin's unique delta wing and tailplane configuration necessitated a long and sometimes tortuous development period. With the benefits of 'area ruling' yet to be discovered, the considerable width of the Javelin's fuselage is evident here on the first prototype.

In early 1947 two new specifications were issued for RAF fighter aircraft, F.43/46 for a day interceptor and F.44/46 covering a night-fighter. Gloster Aircraft tendered to both, its P.228 night-fighter proposal noticeably derived from the Meteor. Gloster's P.234 day-fighter proposal was, however, of a more radical nature with a delta wing and a V-tail design. Many of Gloster's fighter designs from this period showed some level of Meteor ancestry but as they were refined they gradually took on the features of what would eventually become the Javelin, and by August 1947 the follow-on P.248 and P.250 projects did possess some distinct Javelin characteristics.

Gloster initially concentrated on the F.43/46 day-fighter which called for the destruction of high-speed, high-altitude bombers by an aircraft far superior in performance to the Meteor. A Tender Design Conference was held on 9 April 1947 and the P.234 was declared as the most viable project. Agreement was reached to place an order and Gloster received confirmation to continue design work in June, however, no order was received as the Air Staff altered its requirements and subsequently the

On 26 November 1951 Javelin prototype WD804 made its maiden flight from Gloster's test airfield at Moreton Valence, and immediately the first of a number of design faults was evident. Severe vibration, caused by the interaction of jet efflux and airflow over the rear fuselage, curtailed the flight and remedies were sought. Here the aircraft is seen with one of a number of interim jet pipe fairings tested before the definitive 'pen-nib' configuration was selected. The twin extrusions visible at the rear are oil vent pipe extensions.

Ministry of Supply requested that Red Hawk air-to-air missiles (AAMs) should replace the originally-specified cannon armament. A large air-interception radar was also needed and Gloster responded with the P.259 which, for the first time, was a two-seat design.

The favoured projects within the Air Ministry for the F.44/46 night-fighter were the de Havilland DH.110 and Hawker P.1057, both two-seaters. Hawker, however, had studied single-seat fighters in depth and the Air Ministry began to think in terms allocating Hawker the day-fighter project. With Gloster's P.259 lending itself better to F.44/46, it was decided in January 1948 to switch everything around and order three Hawker F.43/46s (the P.1067, which became the Hunter), three DH.110s and three (later four) Gloster F.44/46s (plus a structural test airframe). Under the SBAC numbering system Gloster's design was eventually called the G.A.5 and, from this point, it was considered purely as a night-fighter.

DH.110 and G.A.5 head-to-head

In mid-1952, when both had begun flight testing, a comparative evaluation was made between the DH.110 and G.A.5, the RAF favouring the latter and, when the first DH.110 crashed at the 1952 Farnborough Show, the Gloster aircraft's future was secure.

F.44/46 was eventually renumbered F.4/48 to accommodate the extensive changes covered by Operational Requirement OR.227. It requested a night-fighter capable of at least 525 kt (973 km/h) top speed at 40,000 ft (12192 m) and was to be fitted with four 30-mm Aden cannon and an AI.Mk 9 radar.

Due to increases in the estimated weight, the decision was taken to adopt more powerful Armstrong Siddeley Sapphire engines instead of the intended Rolls-Royce Avons as the P.280 project.

Contracts were awarded in March 1948 to proceed with mock-up and design work and a general layout, based on the P.280, was stabilised and finally agreed in August. The full armament was to comprise four 30-mm Aden cannon, or four de Havilland Blue Jay AAMs (later called Firestreak) and two cannon. The Mock-Up Conference was held at Bentham on 26 October and the contract to supply four prototypes was placed in April 1949, but in November the total was cut back to two since the aircraft was regarded, at this time, more as an insurance against any setbacks to the DH.110 programme. De Havilland's fighter was the preferable production choice because it was more conventional and easier to build. However, three more prototypes were ordered in March 1951 to ensure that sufficient machines were available for the full flight test programme and CA Release. The Chief Designer for the majority of the Javelin programme was Richard Walker and his Assistant Chief Designer was R.W. Ward. The name Javelin was not selected until the award of a production order in June 1952.

During 1950 it was decided to add ILS automatic approach equipment to the aircraft and armour protection for the crew and, in early 1951, the fuel capacity was increased and additional equipment included, namely the American Hughes E4 Collision Course Fire Control and the APQ-43 search and gun-laying radar (first fitted on the FAW.Mk 2). This period of gestation ultimately produced a twin-engined, two-seat, all-weather fighter of mid-wing delta configuration. The tailplane was placed high on the fin as part of a T-shaped tail unit, completing the Javelin's unique appearance.

In these early days of jet fighter development choosing a delta wing had been seen as a big step when relatively little was known about the flight characteristics of such a planform. In time, however, and with experience gained from research vehicles such as the Boulton Paul P.111 and Avro 707, it became apparent that the delta offered great benefits in high-speed flight, while the large wing volume was useful for carrying more fuel and equipment. Other advantages were a reduction in peak supersonic drag (the rigid structure reduced flutter and helped control by eliminating aileron reversal from wing twisting) and the large wing area cut wing loading which helped manoeuvrability and handling – vital for a fighter. The disadvantages included a poor lift-to-drag ratio in low-speed flight caused by a low aspect ratio, high taper and the large sweepback, and the low loading made the wing very sensitive to gusts when flown at high speed and low level.

Construction

Conventional metal construction dominated on what was a big and impressive aeroplane. The delta wing employed an all-metal stressed-

Here two views of the first prototype illustrate different stages of the development programme. At left the aircraft can been seen at the time of its first flight fitted with the initial jet pipe fairing and the anti-spin parachute fairing on the tailplane. Fences were added to prevent parachute shrouds from jamming the elevator. After initial trials, the fairing and fences were removed (above) as it had been found they added to the buffet problems.

The third Javelin prototype, WT827, was allocated primarily for armament and radar trials, and was the first to be fitted with the four wing-mounted Aden 30-mm cannon armament and AI airborne interception radar. During the radar development trials heavy rain and hail was found to damage the neoprene skin of the radome and a shape change was deemed necessary. A number of radomes were tested, including a short and bulbous 'bull-nose' configuration (left) and a sharper 'needle-nose' configuration (below). During the trials the canopy was modified to incorporate a square window above the rear cockpit. The white fairing, made of Durestos, at the rear of the canopy was for the Gee navigation aerial.

skin structure and a torsion box was created by the main spar, leading-edge ribs and the skins. Five fuel tanks were fitted inside each inboard wing and air brakes were mounted in the top and bottom surfaces a little forward of the trailing edge; slotted landing flaps were unusually located in the wing undersides forward of the airbrakes. Two 30-mm Aden cannon were mounted in each wing outboard of the undercarriage, while the later Javelin marks introduced Firestreak missiles carried on underwing pylons. In due course the Javelin FAW.Mk 7, 8 and 9 aircraft employed in the European theatre usually carried just a single Aden in each wing, however, those deployed in the Far East retained all four guns.

In September 1954 the Air Ministry decided to strengthen the wing structure to take four Firestreaks (at this time still called Blue Jay), but the full installation would not be introduced until much later in the programme.

Adding Blue Jay

On 16 May 1956 Gp Capt. H.N.C. Wheeler, Deputy Director of Operational Requirements, reported the decision to retrofit Blue Jay to Mk 2, 5, 6, 7 and 8 aircraft but, thanks to delays in the missile's development, only the Mk 7, 8 and the later FAW.Mk 9 received the weapon. Design work on Blue Jay integration was finally completed in late December 1955. Finding the correct aerodynamic shape for the launch pylons proved a complex task because they had to be carefully designed to keep drag to a minimum; as a result they were swept back from the missile end.

The landing gear retracted hydraulically, the nose wheel folding beneath the cockpit and the main gears inwards into the inner wing. Power boost was applied to the ailerons on the first prototypes but, from the fourth machine onwards, fully powered ailerons were substituted and adopted for all subsequent aircraft; the flaps and air brakes were hydraulically operated. The first three prototypes had a straight leading edge, which was found to be unsatisfactory for high-speed, high-altitude handling and the second machine, WD808, eventually received a modified slightly 'kinked' leading edge which subsequently appeared on all Javelins from the fourth prototype onwards. Sweepback on the outer panels was reduced; sweep angle at 25 per cent chord was 39.5° on the inner wing and 33.8° for the outer portion.

All-metal stressed-skin was also used in the fuselage and tail unit structure. The fuselage was constructed from four main assemblies, the centre part comprising a longitudinal box-beam structure with large frames forming the engine bays. The Sapphire turbojets fitted snugly into these side fuselage nacelles and were fed by circular air intakes. The two-man crew, pilot and radar operator, were seated in tandem Martin-Baker ejection seats inside a fully air-conditioned cockpit and, after criticism of the original prototype arrangement where a one-piece hood offered only two small holes per side for the radar operator, their canopies were modified from the fifth aircraft onwards with twin sliding hoods. Two more fuel tanks were housed in the forward fuselage giving a total of 12 tanks. The P.280 project had an AI.Mk 17

The loss of WD808

Following the crash of WD804, work was accelerated on the second prototype, WD808, which made its first flight on 21 August 1952. In the spring of 1953 WD808 was involved in tests of the new 'cranked' wing shape when tragedy struck. On 11 June WD808 crashed, close to school playing fields at Flax Bourton, near Bristol, killing test pilot Peter Lawrence. Flying low-speed, aft centre of gravity trials WD808 entered a 'super' or 'deep' stall condition with the nose at a high incidence of 50° and the wing blocking airflow to the tailplane and elevator. The aircraft dropped near vertically from 11,000 ft (3353 m). Despite the aircraft being essentially stable, no workable airflow could be obtained over the tail to allow recovery and the pilot could do nothing to recover the situation. Trying to ensure the aircraft avoided the school children, Lawrence ejected too low for his parachute to fully deploy.

Service experience showed a Javelin fully stalled would sink at a rate of 20,000 ft (6096 m) per minute. Instructions issued after the accident detailed the use of flap with certain combinations of speed to prevent another occurrence.

WD808 is seen here early in its test flying programme when fitted with the original 'straight' wing leading-edge and the spin recovery parachute on the tail. The aircraft was conducting its 67th flight at the time of the accident with test pilot Peter Lawrence having carried out a similar flight profile earlier in the day. Long-term fixes to help prevent the aircraft from entering an unrecoverable stall, including drooped wing leading edges and leading-edge slats, were subsequently trialled but the weight/drag penalties were deemed excessive. Instead, the solution adopted was the fitment of a stall-warning system comprising a hinged metal plate on the upper surface of the wing that was sucked forward by low pressure at the onset of a stall, sounding an alarm in the cockpit.

***Right:* The loss of the first two prototypes was a severe setback and caused substantial delays to the programme. The fourth prototype, WT830, first flew from Brockworth in January 1954 and was the first to be built with the 'kinked' wing and fully-powered ailerons. Designated for aerodynamic and handling tests, the aircraft, after initial trials with Gloster, was sent to the A&AEE at Boscombe Down for official Air Ministry evaluation.**

***Below:* Originally the rear cockpit was covered with a metal canopy with two small windows on either side, as it was believed that the navigator could operate the radar displays better in semi-darkness. This theory was soon rendered obsolete and, from the fifth prototype (WT836) onwards, new twin-Perspex canopies were installed.**

intercept radar in the nose but this equipment was to be replaced on some marks.

The large angular T-tail overshadowed the thick delta wing and shared the same configuration. The tail on the prototypes, Mk 1s and many Mk 2s was of the variable-incidence type with power-assisted hydraulic elevators but, following a decision made in July 1954, this was substituted by an electrically-operated 'all-moving' slab flying tail, one of the first fitted on a British aircraft (a one-piece slab tail had originally been proposed before first flight). Once in service, it was found that high stick forces were needed at high speeds and the fully-powered all-moving version replaced the variable-incidence type from the FAW.Mk 4 onwards to improve handling characteristics at transonic speeds.

The value of a delta tail high on the fin was that it reduced the high angle of attack on approach and touch-down seen on previous delta wing aircraft, a big help when flying in darkness. Like the Panavia Tornado of today, the Javelin could be recognised from a distance by its massive fin, which was built around two fin posts attached to the centre fuselage beam. At 25per cent chord, fin sweepback was 47.6° and tail sweep 42.8°. Light aluminium alloy was employed throughout the whole aircraft with the exception of a few critical steel parts.

Maiden flight

The first metal for the first G.A.5 Javelin prototype was cut in April 1949. This aircraft, WD804, and the second prototype, WD808, were both assembled in sections at the firm's experimental department at Bentham and, by July 1951, WD804 was approaching completion. It was moved by road to Gloster's test airfield at Moreton Valence where it was brought to flying condition. No weapons or radar were fitted to either aircraft, both having an all-metal nose cone with the pitot boom in the centre. WD804 began preliminary ground tests and taxiing a month before its maiden flight on 26 November 1951. In the cockpit for the vital trip was Gloster's Chief Test Pilot Sqn Ldr W.A. 'Bill' Waterton.

The flight was partially successful, but serious rudder buffet was experienced to such a degree that at 174 kt (322 km/h) violent shaking was felt throughout the whole airframe. The problem was jet efflux and airflow mixing together over the rear fuselage and, to stop this, the jet pipe fairing was lengthened. However, the problem was not fully cured until the 'pen-nib' fairing was introduced, a feature on all marks up to the FAW.Mk 7, which introduced a different engine. Flying continued until 29 June 1952 when flutter resulted in both elevators being lost during a high-speed run at 3,000 ft (914 m). With great skill, Waterton managed to land WD804 at Boscombe Down, but the aircraft was wrecked when the undercarriage collapsed. Fortunately, Waterton was uninjured and, for saving the aircraft and recording equipment so bravely, he received the George Medal. A replacement was ordered immediately, serialled XD158 and known as the sixth prototype – this becoming the FAW.Mk 2 trials aircraft.

WD808 made its first flight on 20 August but spent the rest of the year being prepared for flight resonance tests to investigate the flutter which had caused WD804's loss. The Javelin was the first British aeroplane fitted with a resonance

One of the main improvements during the test phase was the introduction of the 'kinked' wing and modified wingtips (right), replacing the original planform (above). The new wing significantly increased lift at the wing tips by reducing span-wise migration of airflow, prevented 'tip-stalling' and increased manoeuvrability at higher altitudes.

Above left: The first production Javelin FAW.Mk 1 takes off on its maiden flight on 29 July 1954. After appearing at the SBAC Farnborough airshow in September of that year the aircraft was used for armament and instrumentation trials before being retired to No. 4 School of Technical Training at RAF St Athan in February 1958.

Above: Incorporating the modifications introduced by the prototypes, including the 'kinked' wing and all-Perspex canopy, a total of 40 FAW.Mk 1s was produced at Hucclecote. XA563 and '564 are seen here in 1955 demonstrating the shorter 'pointed' nose radomes fitted to a number of Mk 1s. XA564 was earmarked for conversion with Rolls-Royce Olympus engines for the 'Thin Wing' supersonic Javelin project, but cancellation of the project during the conversion process left the aircraft with no purpose and it was retired to the SoTT at RAF Locking with only 2 hours 35 minutes flying time.

Left: With research into the best jet pipe arrangement continuing, the FAW.Mk 1s were initially fitted with this extended fairing between the jet pipes. XA550 was retained by Glosters at Moreton Valence for ongoing trials between 1955-57.

system and this flight testing began in January 1953 covering speeds up to 520 kt (964 km/h). By May WD808 had received the extended chord wing tips (cranked wing) to help cure a tendency for tip stalling, which made the fighter tighten into a turn, plus increased elevator boost and a stiffened fin, tailplane and elevators, all designed to improve the aircraft's handling. The centre section spar was modified to clear the aircraft to 530 kt (982 km/h) maximum speed and when flight testing resumed on 23 May the preliminary trials indicated that handling had improved. Javelin development was, however, to suffer a major setback when WD808 crashed on a test flight on 11 June 1953, killing the pilot. This second second crash pushed the Controller Aircraft (CA) Release date back to February 1955.

On 7 March 1953 the third prototype, WT827, joined the flying programme and was the first to carry cannon and radar. Following experiments with both blunt bull-nose and pointed radomes, after rain erosion had damaged the original version, the pointed format became standard.

Trials also revealed that vortices entering the Aden's barrels were heating the cannon rounds, with the result that the leading-edge blast tubes had to be extended very slightly. During March one prototype was test flown by two USAF pilots who criticised the small quantity of internal fuel and WT827 was quickly fitted with two 250-Imp gal (1137 litre) ventral tanks, which were to became standard. The fourth machine, WT830, was unequipped but it did have fully-powered ailerons and the other improvements previously incorporated into WD808, making its first flight on 14 January 1954. This aircraft became the aerodynamic test airframe and, after completing the Contractor's handling trials, it was sent to the Aircraft and Armament Experimental Establishment (A&AEE), Boscombe Down, for official evaluation.

Last of the prototypes

The last prototype, WT836, became airborne on 20 July in the hands of 'Dicky' Martin and was used to prove some of the engineering aspects. Wg Cdr R.F. Martin succeeded Waterton as Chief Test Pilot in March 1954 and spent a great deal of time developing the technique employed for Javelin spin recovery. However, by June 1955 Gloster and the Royal Aircraft Establishment (RAE) at Farnborough still could not agree as to the correct recovery action for the aircraft; the spin direction in a Javelin was unpredictable and often could be reversed during the spin. In October, Gloster and A&AEE did finally agree that the recovery technique required full in-spin aileron to be applied when the nose dropped, followed by moving the stick fully forward and holding until the Javelin finished a nose-down pitch.

Above: The revised cockpit arrangement is clearly seen on XA550 during a test flight over Gloucester. The opportunity was taken to improve the pilot's field-of-view by removing the cockpit frames of the early prototypes. In addition, a more bulged contour was introduced to allow the crew sufficient room to reach ejection seat blind-handles while wearing the new 'bone dome' helmets then being introduced into service.

Right: XA549 introduced the new 'pen-nib' jet pipe fairing, which finally cured the buffeting problems. The aircraft was also used for radar trials before finally entering RAF service with No. 87 Squadron in November 1959.

Left: This fine colour shot of FAW.Mk 1 XA563 was taken in 1957 during A&AEE trials at Martlesham Heath. The aircraft was at this time still restricted in the manoeuvres it was permitted to perform due to the dangers of 'tail-blanking' and entering a deep stall.

Below: No. 46 Squadron at RAF Odiham was allocated the honour of introducing the Javelin FAW.Mk 1 into front-line service. The first aircraft were received in February 1956 and by late 1956, when the unit conducted this 16-ship formation, the squadron was at full strength.

A production order covering 200 Javelins for the Royal Air Force was placed on 7 June 1952. Despite growing interest from some European air forces and in particular Belgium, it appears that the aircraft was never close to securing an export order.

As predicted, the two original prototypes proved far from sufficient to deal with all the development flying of this large and complex aircraft and, as a result, by 1 March 1956 no less than 21 FAW.Mk 1 aircraft were allocated to trials work. This figure, however, did include aircraft sent to Armstrong Siddeley, Rolls-Royce and Bristol Engines for Sapphire, Avon and Olympus engine development.

Along with the de Havilland DH.110, Hawker Hunter and Supermarine Swift, Gloster's Javelin became one of the new fighter aircraft capable of making a supersonic bang in a dive, but there were comments within the industry that the Javelin could not go supersonic. In addition, former Chief Test Pilot Bill Waterton had become the air correspondent for the *Daily Express* newspaper and an article written by him strongly criticised the aircraft. Consequently, and to prove a point, during the night of 4/5 July 1954 Martin flew a Javelin prototype right across South London (allegedly where Waterton lived) projecting a sonic boom over the entire area.

Flight testing

By 1954 the Javelin's slow-speed characteristics were causing a lot of anxiety. On 21 October the third production aircraft, XA546, crashed into the Bristol Channel while conducting spinning tests, emphasising the importance of the low-speed investigation, which, from then on, had priority over all other flight test work (the pilot was killed). By February 1955, instability at the stall, especially with flaps down, was seen as a fundamental feature of the design. On the landing approach a high rate of descent warned the pilot of an impending stall but at high altitude there were no landmarks available to judge the rate of descent. The aircraft had shown a self-stalling tendency and a stall on approach would probably be fatal; the possibility that a superstall might follow made little difference. The solution was the introduction of a stall warning device.

Other aerodynamic problems identified at this time included some disturbing features at high Mach numbers – nose-down pitch at 0.88 Indicated Mach number (IMN) was followed by slight buffet at 0.90, rocking at 0.92, severe buffet at 0.94 and yaw at 0.97; after this the aircraft passed smoothly to Mach 1.05 IMN. It was considered that all these effects were due to shock-induced boundary separation, although at the time the manufacturers were concentrating on improving the flow over the wing and tail. There was a likelihood that the buffet and yaw would involve a serious risk of structural failure if the aircraft stayed within this speed band for a sufficient period.

In September 1956 the All Weather Development Squadron at Central Fighter Establishment, RAF West Raynham, completed its tactical trials of the FAW.Mk 1. The final report stated that the fighter's controls were light and well balanced at low and medium speeds but, as speed increased beyond 450 kt (834 km/h) IAS, they became progressively stiffer until at the maximum permissible IAS of 535 kt (991 km/h) the rudder became near uncontrollable and the elevators were heavy. Adequate manoeuvrability was experienced up to 40,000 ft (12192 m) but above this height the speed could not be maintained in turns at large angles of bank without sacrificing height. For example, at 45,000 ft (13716 m) and 0.91 IMN, a sustained turn at 30° of bank was the maximum possible.

Limiting speeds and Mach numbers were; above 20,000 ft (6096 m) Mach 0.93 or the onset of buffet, Mach 0.91 between 7,000 ft (2134 m) and 20,000 ft (6096 m), and 535 kt IAS below 7,000 ft (2134 m). These figures could be reached and exceeded in level flight up to 35,000 ft (10668 m) and, between 35,000 ft (10668 m) and

No. 46 Squadron's Javelins received the unit's red arrow insignia either on the nose roundel or high on the tail fin. The pilots conducted a conversion course consisting of seven flights by both day and night, which covered general handling, blind landings, combat manoeuvring to the point of stall at high altitude, asymmetric flight (including approaches) and day and night interception.

Above: Javelin FAW.Mk 1 XA549 is lowered onto supports from a lifting crane at RAF St Athan. The aircraft had served with the second of two squadrons to operate the Mk 1 – No. 87 Sqn – and still wears the unit's dagger insignia on the tail fin.

Left: FAW.Mk 1 XA552 demonstrates the 'bosom' or 'Sabrina' underfuselage auxiliary tanks, which had first been tested, following criticism of the Javelin's range, on prototype WT827. XA552 was later converted as a testbed for the Gyron Junior engine.

40,000 ft (12192 m), Mach 0.93 could be achieved. Above that, Mach 0.92 was the best possible. Acceleration up to 35,000 ft (10668 m) was good but fell away above that height until at 48,000 ft (14630 m) it was impossible to accelerate from 0.71 IMN in level flight. This would make it necessary to maintain a high speed (at least Mach 0.85) until late in an interception.

A report on the Mk 4 followed in March 1957. This aircraft could intercept targets at Mach 0.85 and 45,000 ft (13716 m), but was dependent on the amount of early warning given and the available fuel in bad weather. In addition, this was achieved 'in spite of, rather than because of, the performance of the AI.Mk 17 radar'. The range of the equipment was such that a successful interception depended initially on the Ground Control Interception controller being absolutely precise in positioning the fighter; improvements in this area were recommended.

Assessing the Mk 4

CFE found that the FAW.Mk 4 was no more manoeuvrable than the Mk 1 and its heavier elevator and apparently heavier ailerons made it less pleasant and more tiring to fly, except above Mach 0.92 when the all-flying tail was a definite improvement. The least satisfactory manoeuvrability characteristic was the rapidity with which speed dropped off in turns and when the nose was well above the horizon. On the Mk 4 the limiting speed above 35,000 ft (10668 m) was Mach 1.04 (achieved in a dive). Maximum level speed at high altitude was actually less than the Mk 1 – Mach 0.88 to 0.89 at 48,000 ft (14630 m), though Mach 0.92 at 40,000 ft (12192 m) was possible. The acceleration in a dive was described as rapid.

A&AEE Boscombe Down tested a Mk 1 from July 1955 and described the type as generally easy to fly but deficient in several important areas – strong pro-spin tendencies at the stall with inadequate stall warning, inadequate longitudinal control and severely restricted lateral control at high speeds, poor dynamic directional stability especially at high altitude and the presence of buffet in level flight above Mach 0.90. It was hoped that the development of the all-flying tail, yaw damper, stall warning etc., would cure most of these problems. However, the Javelin was praised for its effective airbrakes, good asymmetric characteristics and lack of engine problems. The Mk 1 theoretically met the requirement for intercepting bombers flying at 40,000 ft (12192 m) and Mach 0.80, but by the standards of the day its performance and weapon system were felt to be inadequate. A restricted CA Release was given purely to enable early Service experience to be obtained on the small number of Mk 1s available.

The FAW.Mk 4 was test flown by A&AEE from June 1956, the all-flying tail, stall warning system and vortex generators all new from the Mk 1. Improved longitudinal control was appreciable under most conditions, but disappointingly small at high Mach numbers; in most other respects the FAW.Mk 4 was similar to the Mk 1. The handling characteristics of FAW.Mk 2, 5 and 6 aircraft were found to be little different from the Mk 4.

A&AEE got its hands on an FAW.Mk 7 in August 1957, the principle changes being more powerful Sa.7 engines, a yaw damper and full power rudder. Despite a 15 per cent increase in take-off weight over the Mk 4, altitude capability without external stores was about 5,000 ft (1524 m) higher than the earlier model and just sufficient to attain 50,000 ft (15240 m). At that height however, there was very little margin of thrust available for acceleration, and buffet-free *g* levels for manoeuvring were very small; full power was essential for most of the time at this altitude. The yaw damper gave great improvement to lateral and directional stability compared to earlier marks, while the behaviour with Firestreak missiles aboard was in all respects satisfactory.

The reheated marks of Javelin showed an improved performance in some areas above about 10,000 ft (3048 m) but their maximum

Javelin T.Mk 3 dual-control trainer

Gloster first considered a dedicated dual-control trainer under Specification F.4/48 in June 1950 as the P.319, rejecting the side-by-side configuration of the Hawker Hunter trainer to ensure minimal structural alteration. The project was initially rejected, but the need for such an aircraft was soon recognised and a prototype, WT841 (foreground), was ordered in 1951. Specification T.118D was raised to cover the aircraft which had Sapphire Sa.6s and was based on the FAW.Mk 6. WT841 was designed and built using Gloster components by a subsidiary firm, Air Service Training Ltd of Hamble, and flew on 20 August 1956. Some delay was experienced in getting the aircraft ready; for instance by July 1952 the rear fuselage that was largely common to the fighter prototypes was well advanced but the front fuselage had progressed little and was about 12 months late.

The T.Mk 3 lacked the nose radar but had the all-moving tail while the guns were retained for instructional purposes. Compared to the standard fighter variants its fuselage was 3.67 ft (1.12 m) longer and it had an enlarged rear canopy. Removing the radar moved the centre of gravity rearwards and the fuselage extension acted as a counter, also proving useful in allowing two additional 50-Imp gal (227-litre) fuel tanks to be added. The first of 22 production machines (XH390) made its maiden flight on 6 January 1958.

Below: The FAW.Mk 2 equipped No. 89 Squadron (later renumbered as No. 85 Squadron) at RAF Stradishall. The only other RAF squadron to operate the variant was No. 46 Sqn (seen here), which relinquished its Mk 1s for Mk 2s in the autumn of 1957.

Below right: Built as a standard FAW.Mk 2, XA778 was retained by Gloster and converted with Sapphire Sa.7 engines to act as the prototype for the FAW.Mk 7. It is seen here post conversion at the 1956 SBAC show.

Javelin FAW.Mk 2

Making its maiden flight shortly after the FAW.Mk 4, the Mk 2 suffered a number of development problems and, after protected testing, only 30 were ever produced. The Mk 2 was in essence the basic FAW.Mk 1 except for some changes in equipment, the most important being a different radar. In March 1952 it was agreed that the American Westinghouse APQ-43 radar would be introduced into production as soon as possible without delaying the delivery of AI.17-equipped machines. The sixth prototype, XD158, was selected for the trial installation, in effect the basis of a new mark, and its completion was thus delayed because of the necessary alterations; it flew for the first time, with a metal nose, on 31 October 1955.

In British service APQ-43 was designated AI.Mk 22 and required a larger radome, which actually pulled out and hinged sideways instead of having to be removed, a feature which allowed easier and quicker access. Comparing the two systems, AI.17 became reasonably reliable after the first year or so of service. Its range was about 20 miles (32 km) at height, more if there was a big target, but, depending on the height and level of ground returns, this figure could drop to as low as 5 to 7 miles (8 to 11 km) at low level and against fighter-sized targets. AI.22 was supposed to be more refined but suffered many problems and low serviceability during its early career. This improved with experience but the 'lock-on' characteristics would not always perform 'as-advertised' and the variant was more often outperformed by AI.17-equipped Javelins during interception exercises.

speed was slightly less than the early versions. The FAW.Mk 8 and 9 could reach 610 kt (1130 km/h) or Mach 0.92 at sea level compared to 616 kt (1141 km/h) or Mach 0.925 for the Mk 1, 2 and 7. In fact, the reheat system required a re-design. Reheat could be engaged at sea level but the Sapphire Sa.7R was fed by a fuel pump which had a constant output and such action below about 8,000 to 10,000 ft (2438 to 3048 m) cut the quantity of fuel available to the combustion chamber and hence reduced the basic cold thrust of the engine. Above this height the main engine did not demand the same level of fuel and when the pump output exceeded demand, spare fuel became available to light the reheat.

Javelin variants

The development of the Javelin was to become a protracted affair and resulted in no less than nine variants of the type covering just 428 production aircraft.

The initial production version, designated FAW.Mk 1, had the 8,300-lb (36.9-kN) thrust Sapphire Sa.6 engines and the AI.Mk.17 S-band search, lock-on and blind firing radar. XA544 was the first example of 40 Mk 1s, all built by Gloster's Hucclecote factory, and was flown by Dicky Martin on 22 July 1954, just two days after the fifth prototype had become airborne. CA Release was given to the FAW.Mk 1 on 30 November 1955 but with some restrictions – stalling and looping were prohibited until the spin recovery technique was confirmed while the upper and lower speed limits were 540 kt (1001 km/h) and 125 kt (232 km/h). As a result, when deliveries to RAF Fighter Command began in February 1956 the Service received its first fighter to be prohibited from stalling. An announcement made in October 1954 confirmed that some Javelins would be paid for by the USAF under the US offshore procurement policy, all of them to serve with the RAF.

Of the Mk 1s retained for tests, XA545 became the first aircraft to examine the all-moving flying tail for the FAW.Mk 4, XA546 tested the Gee navigation aid and XA547 became the first Javelin to try the Firestreak pylons, albeit with dummy weapons. XA548 had its wing fitted with slats and a drooped leading edge to perform some prolonged spinning trials, and XA549 was used to sample various radar installations. XA552 found employment as an engine test bed for the 10,000-lb (44.4-kN) thrust de Havilland Gyron Junior, in preparation for the Bristol 188 supersonic research aircraft which was to use this engine. After its conversion by Napier, XA552 was painted Royal Blue with a white trim and a gold 'Gyron Junior' logo on the nose. XA560 was the first to get reheated Sapphires, flying with 11,000-lb (48.9-kN) Sa.7s on 30 September 1955, while XA562 became a Rolls-Royce Avon RA.24R test bed.

The first production FAW.Mk 2, XA768, flew on 25 April 1956 and 30 were built at Hucclecote on a near parallel production line to the Mk 1. Some were used for experimental work (in fact the Javelin was to achieve an impressive test aircraft record), XA769 together with the prototype being employed on radar trials while XA771 went to work on full-scale missile trials. In addition, XA778 was used by A&AEE until 1968 as a calibrated 'pacer' aircraft for which it was painted bright fluorescent orange and fitted with an FAW.Mk 7 type non-reheat Sa.7 Sapphire and FAW.Mk 8 control system.

A total of 22 dual-control T.Mk 3 trainers was produced for the RAF. The version served with No. 228 OCU as well as in small numbers with the majority of operational squadrons.

The FAW.Mk 4 was basically an FAW.Mk 1 with Sa.6 engines, AI.17 radar and the new all-flying tail. The first example, XA629, was

Although the Javelin's reportedly docile handling and the reliability of the Sapphire engines bestowed the aircraft with a relatively good safety record for a jet fighter of the period, the addition of the dual-control T.Mk 3 was seen as a vital tool during initial conversion training with No. 228 Operational Conversion Unit (OCU), based at RAF Leeming. From mid-1959 one or two T.Mk 3s were allocated to each operational squadron for continuation and refresher training as seen here in the colours of No. 64 Squadron (above) and No. 60 Sqn (left), both when based in the Far East during the Indonesian confrontation.

***Left:** The Javelin Mk 4 was introduced into RAF service by No. 141 Squadron in February 1957 after an 18-month test and evaluation programme. XA644 was one of a number of Mk 4s retained by Gloster for FAW.Mk 7 development work, before being destroyed in a collision with a Hawker Hunter near Moreton Valence on 24 August 1956, killing Gloster test pilot Brian Smith and his navigator.*

***Top:** A characteristic of the Javelin Mks 1-7 was the vast amount of smoke, pungent smell and ear-splitting squeal made by the large cordite-filled starting cartridges.*

***Middle:** FAW.Mk 4 XA631 sits on a compass swing pan in the winter on 1955-56, while undertaking radio and instrumentation trials with the A&AEE.*

***Above:** A total of eight squadrons eventually operated the FAW.Mk 4 including No. 11 Sqn (seen here), before Firestreak-capable variants replaced the last examples in 1959.*

considered to be the prototype and flew on 19 September 1955 after being pushed up the production line to accelerate the programme (this step followed the tests undertaken with XA545). A total of 50 Mk 4s was built, with production shared between Gloster and Armstrong Whitworth, Baginton. The first series aircraft flew on 27 February 1956 and CA Release was achieved in December of that year.

One modification made to XA629 was the fitting of an audible stall warning apparatus, triggered by electro-mechanical vanes fixed on the wings. Later, on 28 March 1956, it was flown with some large streamlined fairings known as Küchemann 'carrots', initially on the wing trailing edges just inboard of the ailerons but later with additional smaller versions on the wingtip trailing edges. Their role was to reduce the effect of shock waves and cut airflow separation when flying at high subsonic speeds, so extending the buffet boundary. This concept was rejected when it was found that two rows of vortex generators placed along the outer leading edge, plus a thickening of the trailing edge, were just as effective. This solution re-energised the boundary layer and was also a cheaper modification, so was utilised on all Javelins from the Mk 4 onwards.

The usual quota of FAW.Mk 4s spent time on experimental duties, including XA630 which cleared the pen-nib jet pipe arrangement, and XA720, XA721 and XA723 sent to Canada for cold weather tests. XA763 and XA764 went to the Central Fighter Establishment (CFE).

The FAW.Mk 5 and Mk 6 were based on the Mk 1 and Mk 2, respectively, the most significant difference being an increased internal fuel capacity.

The next significant variant was designated FAW.Mk 7. Based on the Mk 5, the Mk 7 was the first Javelin to receive uprated engines, the 11,000-lb (48.9-kN) thrust Sapphire Sa.7, plus a revised nozzle and extended rear fuselage to help reduce drag. The AI.17 and four Aden were retained but, for the first time, the four Firestreak AAMs or alternatively four 100-Imp gal (455-litre)

Supersonic development – the Thin Wing Javelin

The Gloster G.A.6 'Thin Wing' or 'Super' Javelin fighter programme began in July 1953 as the P.356 project. The intention was to attach a new wing and new engines to the original Javelin fuselage. Alternative weapon loads comprised two massive Red Dean AAMs, four Firestreaks or two of each, plus four Aden cannon. An AI.18 radar was to be fitted and power was to come from two 16,000-lb (71.1-kN) Bristol Olympus 6s. The covering specification was F.153D and one prototype, XG336, was ordered in March 1954 with Sa.7 Sapphires. A development batch of 18 aircraft received approval in January 1955 (serials XJ836 to XJ842 and XJ877 to XJ887) and production was expected to start in 1958. (Note:- XG336 was also allocated to an English Electric F.Mk 1 Lightning.)

However, because of the need to accommodate the expected improvements in the performance of enemy bombers, the 'Thin Wing' Javelin concept was eventually switched from being a standard Javelin with new wings to an all-new aircraft that looked similar to the original. In late 1956 it was realised that on service entry (in 1961) the aircraft's maximum speed of Mach 1.01 at 40,000 ft (12192 m) would be insufficient to deal with a new generation of bombers flying at Mach 2, so the programme was terminated on 31 May 1956. At that stage XG336, under construction in the Gloster Experimental Shop, was apparently 60 per cent complete. Its planned flight date was December 1958.

Increasing the range with the FAW.Mk 5 and 6

Again based essentially on the FAW.Mk 1 with AI.17 radar and the all-flying tail, the FAW.Mk 5 (XA652 of 228 OCU; above) incorporated a modified wing that externally was no different to the Mk 4's but which carried a total of 125 Imp gal (568 litres) of additional internal fuel in an attempt to counter one consistent Javelin criticism, lack of range. It was also the first mark with provision for the four Firestreak missiles, but the weapons were never carried. XA641 was the first of 64 built, production split between Gloster and Armstrong Whitworth, and flew on 26 July 1956.

The new wing and increased fuel that turned the Mk 1 into the Mk 5 was also fitted to the Mk 2 to produce the FAW.Mk 6 (formation of 85 Squadron examples; right). The AI.22 radar and standard Sa.6 Sapphire engines were retained. The 'prototype' XA815 flew on 15 January 1957 and 33 examples were manufactured by Gloster at Hucclecote.

drop tanks could be attached to the underwing pylons. Production was in fact underway before the missile installation was ready and nearly 30 machines had to be retrofitted. Many improvements were also made to the flight-control system including a yaw stabiliser on the fully-powered hydraulically-operated rudder.

The FAW.Mk 7 was to become the most numerous Javelin with 142 built split between the two production lines. The first example flew on 9 November 1956, although CA Release was to take just over another year. A good number of Mk 7s were held back for trials and research including XH706 which assessed handling and the new auto-stabiliser, XH707 which tested the reheated Sapphire Sa.7R in the hands of Armstrong Siddeley, and XH713 which undertook tropical testing in Bahrain during August 1957. XH754 was used by RAE Farnborough from January 1960 for rain dispersal trials as part of the TSR.2 bomber programme, which included a detachment to Singapore in March 1961, and XH756 and XH757 were used for missile trials. Before production was completed the reheated FAW.Mk 8 appeared, which resulted in over 80 Mk 7s passing into storage at 5 MU, Kemble direct from manufacture; 116 Mk 7s were eventually converted into FAW.Mk 9s.

The last new-build Javelin, the FAW. Mk 8, was an FAW.Mk 6 fitted with the 11,000-lb (48.9-kN) Sa.7, but now modified with limited reheat as the Sa.7R for a total thrust of 12,300 lb (54.7 kN). In fact the reheat was only effective above a certain height, officially 20,000 ft (6096 m) though in practice about 8,000 to 10,000 ft (2438 to 3048 m), but it helped to address the FAW.Mk 7's weakness of poor manoeuvrability at height. The Mk 8 had Firestreak, while other new aerodynamic improvements made to both Mk 8 and 9 aircraft included a drooped leading edge in the outer mainplane panels, a pitch stabiliser incorporated in the aileron system and an auto-stabilised rudder.

All 47 Mk 8s were built at Hucclecote, the first of these (XH966) flying on 9 May 1958. XJ165, which flew on 8 August 1960, was to be the last new-build Javelin and, in fact, the last all-new aircraft to be built by Gloster. Eight days later it left to join No. 41 Squadron but FAW.Mk 9 conversion deliveries did continue into 1961. Thirteen Mk 8s (XJ166 to XJ178) were cancelled in August 1960 during final assembly, the machines being cannibalised for spares and then broken up after the decision was made to convert stored Mk 7s to the same standard. XJ125 was used for Sapphire Sa.7R development.

In January 1957 it was agreed that no marks earlier than the FAW.Mk 7 should be fitted with reheat, but the equipping of 116 FAW.Mk 7s with Sa.7R powerplants, to produce the FAW.Mk 9, went ahead and proved to be a big undertaking, the work including adding the drooped leading edge plus variable-area nozzles. XH715 was the last aircraft to be converted, the task being completed in December 1961. One Mk 7 upgraded to Mk 9 standard was XH897 and in January 1967 this machine joined A&AEE to serve on airborne calibration work, for which it was painted in a very attractive red and while livery.

Javelin in service

The first Javelin deliveries were made in February 1956 and the first squadrons to equip with the FAW.Mk 1 were Nos 46 and 87 at Odiham and Brüggen (West Germany), respectively. No. 46 received its aircraft direct from No. 23 Maintenance Unit (MU) at Aldergrove and, under orders, it clocked up 1,000 flight hours in eight weeks so that any problems were cleared as quickly as possible. No. 87 Squadron began to receive its machines from August 1957, the great majority of which had previously seen service with No. 46.

The majority of the Mk 2s also went to No. 46 Sqn, but a few also found their way to No. 89 (later No. 85) Sqn.

The Javelin was never to experience real fighter-to-fighter combat so its capabilities were never tested to the ultimate. However, it did get the opportunity to participate in numerous exercises and was also able to exert an influence in several potential areas of conflict.

The first chance for the Javelin to prove itself

Gyron Junior engine testbed

Javelin FAW.Mk 1 XA552 was removed from test duties with Gloster early in its flying career for conversion with two 10,000-lb (44.4-kN) de Havilland Gyron Junior DGJ.10 engines, then under development for the Bristol 188 research aircraft programme. The rear fuselage of the aircraft was modified to accommodate the larger diameter engines and the associated reheat nozzles. In May 1963 XA552 completed its tasking and the engines were removed and used as spares for the Bristol 188, XA552 being retired from use after only 85 flying hours.

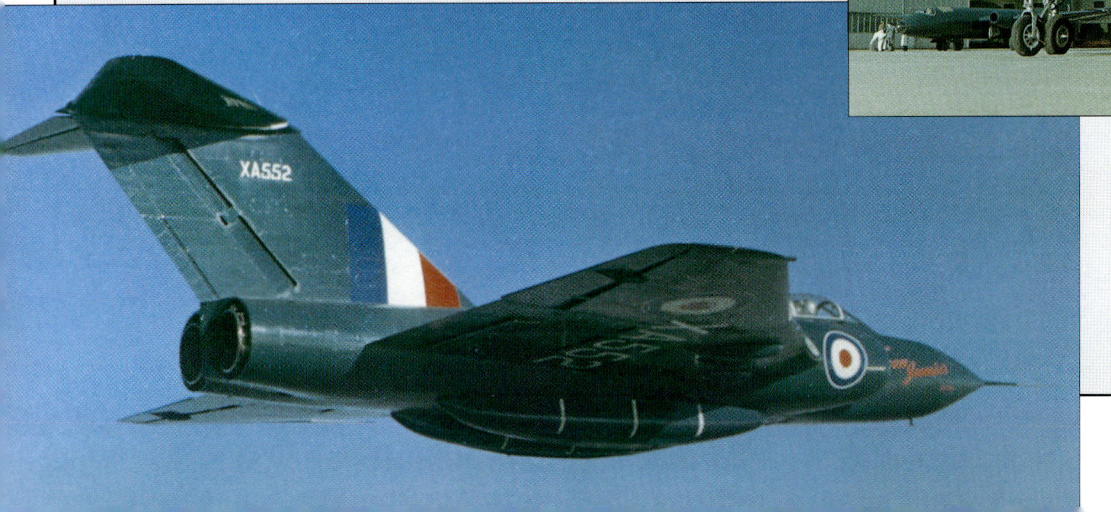

The Gyron Junior testbed flew both with (left) and without (above) the standard 'Sabrina' auxiliary fuel tanks beneath the fuselage. The final stage of the test programme from March-May 1963 involved test flights by Rolls-Royce from its base at Filton, near Bristol, accompanied by a Canberra chase aircraft (above). The Javelin's good handling was retained despite the modification of the rear fuselage, although the high fuel consumption rates when the reheat was engaged ensured most flights were of short duration.

The introduction of the Javelin Mk 7 finally provided the RAF with a fighter capable of successfully intercepting and destroying a wave of incoming high-altitude jet bombers. The increased thrust of the Sa.7 engines boosted the rate of climb, but the most important aspect was the belated introduction of the Firestreak missile, following trials by XH756 of the AWDS (below). Up to four missiles could be carried on wing pylons, as shown on this pair of No. 23 Squadron Mk 7s (right), the second of four front-line units to operate the variant.

XH783 was one of the FAW.Mk 7s assigned to the Guided Weapons Test Squadron (GWTS) at RAF Valley. From 1959 to 1962 the GWTS helped develop the tactics that made the later versions of the Javelin such potent interceptors.

was an annual UK air exercise held in October 1955 called Exercise Beware: two test Mk 1 aircraft (XA554 and XA559) took part flying from RAF Coltishall. The 'enemy' included English Electric Canberra bombers, 10 examples of which were successfully intercepted and 'destroyed' 100 miles (161 km) out to sea, but the bag also included some USAF North American F-100 Super Sabres that happened to fly into the area.

The first unit to receive the FAW.Mk 4 was No. 141 Squadron in February 1957 and the variant was to serve with seven other squadrons. By now the Javelin was serving in the 'immediate readiness' role (code-named Fabulous) where the fighters were parked alongside the runway ready to fly with the aircrew in their seats.

The first recipients of the Mk 5 were the Central Fighter Establishment and No. 151 Squadron at Leuchars, the aircraft arriving from May 1957; another six units operated this mark plus No. 228 OCU. Throughout the Javelin's career the movements of individual airframes between units was extensive, a situation that was partly attributable to the large number of different marks produced; for example, No. 23 Squadron's only FAW.Mk 5, together with the majority of its Mk 4s, were transferred to No. 72 Squadron when No. 23 re-equipped with the FAW.Mk 7. The first Mk 6s joined No. 89 Squadron, the first of three to equip with this version, in October 1957.

Some initial Mk 7 deliveries were made to A&AEE through 1957, before starting in earnest to No. 33 Squadron in July 1958. Three more squadrons received the Mk 7 together with the Guided Weapons Development Squadron (which on 1 January 1959 became No. 1 Guided Weapons Trials Squadron and was disbanded on 1 June 1962). One Mk 7 was sent complete with Firestreak AAMs to the Paris Air Show at Le Bourget where, being the only aircraft present fitted with missiles, it generated

de Havilland Firestreak

Despite Javelin Mks 1-6 being only equipped with four 30-mm cannon, guided weapons were always envisaged as the preferred option. Originally known as the Blue Jay, the Firestreak was at the cutting edge of missile technology for the period and incorporated an infra-red seeker allowing successful interceptions from the rear quadrant. Being a relatively large weapon much work was done to reduce drag of both the missile and the pylon, resulting in only a minimal reduction in the aircraft's overall performance at altitude.

Right: From July 1956 FAW.Mk 4 XA632 was used extensively in the Firestreak development programme using dummy missiles. It conducted numerous flights to assess how various combinations of missiles and pylon shapes affected handling and performance.

Left: The rocket-powered Firestreak was stabilised by four large 'wings' on the missile centre section and controlled by small flying surfaces at the rear which acted on inputs from the infra-red seeker via the guidance electronics.

Below: A No. 23 Squadron FAW.Mk 9 launches a Firestreak during an armament practice camp (APC) at RAF Valley. All operational units rotated through Valley to practise interception and missile firing techniques, using both dummy and live weapons against drone targets.

'Pacer' Javelin

One of the more conspicuous Javelins was FAW.Mk 9 XH897. After front-line service with No. 5 Squadron, the aircraft was seconded to Bristol Siddeley engines for the investigation of engine blade vibration problems encountered by the Javelin during sustained low-level operations.

Subsequently, in April 1968 the aircraft was transferred to 'A' Flight of the A&AEE at Boscombe Down. Here, painted in a dramatic red and white livery, the aircraft was used for position error (PE) or Pacer correction duties, and also as a photo-chase aircraft. Early flights included the calibration of the Aberporth range for supersonic PE pacing measurements for the Concorde programme and conducting photo-chase sorties with the Shorts Canberra PR.Mk 9, then under evaluation by the MoD. Further Pacer calibration duties with Concorde 001 were conducted in the early 1970s and, towards the end of its career, the aircraft was used for PE and chase duties in the early stages of the MRCA (Panavia Tornado) programme. As the last airworthy Javelin, XH897 was retired to the Imperial War Museum at Duxford in 1977.

great interest. To support the new marks and their guided weapons, missile 'Sites' had to be established, which had special support equipment available; the first of these to become operational was at RAF Coltishall.

An example of the support given by Coltishall before other Sites were available was the airlifting of 52 missiles by No. 85 Sqn FAW.Mk 8 aircraft to arm No. 41 Sqn at Laarbruch on 25 September 1961. Firestreak was 'partially released' to the Service in August 1959, but it was not until 2 June 1960 that Fighter Command fired its first missile, a No. 23 Sqn Mk 7 being the lucky aircraft. No. 1 GWTS actually received its first Mk 7 on 1 January 1959 and the first Firestreak test firing destroyed a Gloster Meteor target drone on 21 December of that year.

The first deliveries of Mk 8s were completed in the spring of 1959 to the A&AEE for official handling and performance assessment, but the first squadron to get this version, No. 41 at Wattisham, did not receive its aircraft until November. Examples also went to No. 85 Squadron and some were retained by the manufacturer for auto-stabiliser testing. The final Javelin, the FAW.Mk 9, joined No. 25 Squadron first from December 1959 and went on to equip another seven squadrons.

While serving with No. 228 OCU FAW.Mk 9, XH898, was uniquely finished in non-standard natural metal. However, the standard colour scheme throughout the Javelin's operational career was dark green and dark sea grey upperworks and medium sea grey for the fuselage underparts and lower wing face. Most squadron aircraft would also be decorated in the respective unit's individual markings, but a Mk 1 used by the College of Aeronautics at Cranfield in 1960, XA568, received some extra Day-Glo patches around the engine intakes, wing tips and fin. In 1957 Javelin Mk 4 XA723 undertook winterisation trials at RCAF Namao in Canada, for which it received additional large patches of Day-Glo for identity purposes on the wing tips, fin and rear fuselage.

Another non-standard colour scheme was sported by Mk 7 XH754, which flew with the Royal Aircraft Establishment at Farnborough from 1964 onwards with white uppersurfaces, Day-Glo on the nose and the rear fuselage, and an underside painted in black and yellow target-towing stripes.

Javelin training

No. 228 OCU was disbanded in September 1961, but was reformed in June 1965 at Leuchars in order to supply the extra pilots which were deemed necessary for the growing confrontation between Indonesia and Malaya – this represented a change of policy since it had been planned to phase out the Javelin shortly afterwards. When the unit disbanded in November 1966 it had become the last regular operator of the Javelin to be based in Britain.

The one potentially serious situation to involve the Javelin was the Unilateral (one-sided) Declaration of Independence (UDI) made from Britain in November 1965 by the minority ruling party in Rhodesia. As a result, No. 29 Squadron's Javelin Mk 9s were

Another brightly-painted Javelin was FAW.Mk 2 XA778, which had previously been converted as the prototype Mk 7. The aircraft preceded XH897 as Boscombe Down's Pressure Error Correction, or Pacer aircraft, flying alongside all new aircraft types to measure how the pitot and static pressure sources on the new type varied from the 'true' reading provided by the Javelin.

Above: Only 47 of the last production variant of the Javelin, the FAW.Mk 8, were built, the procurement being cut in favour of conversions of Mk 7s to Mk 9 standard. XJ125 was retained by Gloster and used as a testbed for the Sa.7R reheated Sapphire engine.

Below: No. 41 Squadron was one of five squadrons to operate the FAW.Mk 8 (alongside Nos 23, 64, 72 and 85 Sqns), before the type was superseded by the English Electric Lightning. Here one of the squadron's aircraft formates with a pair of Lightning F.Mk 1As of Nos 111 (foreground) and 56 Squadrons in early 1962.

temporarily moved to Ndola and Lusaka in the neighbouring country of Zambia, from where they carried out border patrol flights well into 1966. Moving to a new country presented many problems, not least an army of termites that marched up the main undercarriage leg of one Javelin in order to eat the solid rocket fuel inside a Firestreak missile. Had a full-scale conflict broken out, opposition for these fighters would have been British-built Rhodesian Hawker Hunters FGA.Mk 9s delivered in 1963, but fortunately this never materialised.

For several years into 1967, No. 29 was based in Cyprus and here some effort was put into what was essentially a combat situation. In the mid-1960s Turkey began to rattle its sabres over its claims towards the island and numerous Javelin flights were made to monitor and buzz Turkish submarines or chase away Turkish Air Force aircraft attempting to break into Cypriot airspace. Even Egyptian aircraft tried to test the Javelin's response by approaching the island, but recovered to home ground the instant the missile-armed fighters were scrambled. On several occasions No. 29 Squadron's Javelins also intercepted Soviet Tupolev Tu-16 Badger bombers on route to Egypt.

Most Javelins were deployed in the Cold War Air Defence role in squadrons split between Britain and stations in West Germany such as Brüggen and Geilenkirchen. For example, the FAW.Mk 4s of Nos 3, 11, 87 and 96 Squadrons all operated in Germany while No. 23, 41, 72 and 141 were based in the UK; the FAW.Mk 5s of Nos 5, 11 and 87 saw service in Germany but Nos 41, 72 and 151 stayed in the UK. The majority of Mk 6, 7 and 8 aircraft were based in the UK but the FAW.Mk 9s of Nos 5 and 11 Squadrons were deployed in West Germany.

Increasing the range

A problem suffered by early Javelins was a lack of range and this was not solved until the large external 'bosom' tanks were introduced which were fixed side-by-side in the ventral position beneath the fuselage. In October 1957 an inadvertent release of these tanks resulted in a restriction being imposed to prohibit their carriage until specially tested release slips were fitted to service aircraft. These tanks gave a much needed boost to the aircraft's range, but getting aircraft out to potential trouble spots quickly brought another important addition to the Javelin's capability, in-flight refuelling. This became a very important facility after a number of overseas RAF bases and other staging points were closed.

Airwork Ltd equipped Mk 4 XA634 with a wing-mounted probe and this was first tested with a Canberra tanker in 1957, although these hook-up trials were 'dry' and no fuel was passed to the fighter. Two more Javelins, Mk 7 XH780 and Mk 9 XH965, were used for further tests in 1959 and 1960, respectively, that introduced a huge 20-ft (6.10-m) long fuselage probe mounted on the starboard side of the cockpit; this proved a success and was eventually fitted to over 40 Mk 9 aircraft which were re-designated FAW.Mk 9(F/R)s. The first squadron equipped with in-flight refuellable aircraft was No. 23 which dispatched four Mk 9s on a non-stop proving flight to Singapore, demonstrating at the same time how Britain could now swiftly re-enforce its overseas bases; Valiant tankers topped up these Javelins on the way. With the probe in position, the fighter's speed was limited to 450 kt (834 km/h).

During the 1960s the Javelin was progressively replaced in the air defence role by the supersonic English Electric Lightning, the final fully operational UK squadron being No. 23 at Leuchars, which relinquished its Javelin Mk 9s

Above: The advent of the reheated Sapphire engine in the FAW.Mk 8 provided an additional 12 per cent of thrust at altitude (the reheat was not effective at low altitude), allowing the aircraft to reach 40,000 ft (12192 m) in four minutes and 50,000 ft (15240 m) in some nine minutes. The additional thrust also endowed the aircraft with much improved manoeuvrability at altitude. No. 85 Squadron (as seen here) performed particularly well with the variant during exercises, regularly intercepting 'enemy' Vulcans and Victors long before they reached their 'targets'.

Right: XJ128 is seen here being prepared for take-off as the very last Javelin to leave Moreton Valence in 1961. Its flying career was, however, short, the aircraft being burnt out in a starting accident at West Raynham in July 1962.

Gloster Javelin FAW.Mk 9

No. 11 Squadron, Royal Air Force RAF Geilenkirchen, West Germany May 1964

Flying surfaces
Unusually for delta-winged aircraft the Javelin incorporated a tailplane instead of elevons on the trailing edge of the wing for pitch control. The aircraft therefore had conventional hydraulically-powered controls comprising elevators, ailerons, rudder and flaps. The early variable-incidence electrically operated tailplane with hydraulically-boosted elevators was replaced from the Mk 4 with a full all-moving tailplane – the elevators acting only as anti-balance tabs.

Cartridge start
On the Javelin Mks 1-7 the engines were started using two 5-in (12.7-cm) cordite-filled cartridges. On firing cordite gas fed a turbine starter which accelerated the engine to 2,000 rpm with an associated high-pitch 'squeal' and clouds of dense black smoke. At this point the starter was designed to disengage, however, on a number of occasions this failed leading to disintegration and sometimes catastrophic fire. The problems led to the introduction of a liquid fuel cartridge engine starting system on the Mks 8 and 9, although problems still persisted and ground fire on start-up remained a real danger throughout the Javelin's service life.

Airbrakes
One of the Javelin's most novel features was the powerful slotted-plate type airbrakes fitted to the top and bottom surfaces of the inner wing. The airbrakes not only allowed rapid deceleration, but also allowed rapid and near-vertical descents from cruising altitude into the circuit pattern, saving both fuel and turnaround time. Techniques for using the airbrakes in combat to outmanoeuvre enemy fighters were also developed.

Landing gear
The nose and main undercarriage units were each fitted with Dowty liquid-spring shock absorbers, topped up automatically in flight from the Javelin's hydraulic system. An anti-skid braking system, utilising Dunlop disc brakes, allowed the aircraft to rapidly stop on a wet surface and the castored nose-wheel was fitted with a self-centring device. Steering was achieved using toe-mounted brakes on the rudder pedals, a system that led to a number of early accidents as pilots adjusted from the hand-lever brake on night-fighter versions of the Gloster Meteor.

FAW.Mk 9 specification

Powerplant: Two Armstrong Siddeley Sapphire Sa.7R turbojets rated at 11,000 lb (48.9 kN) dry, and 12,300 lb (54.7kN) with limited reheat at altitude.
Span: 52 ft 0 in (15.85 m)
Length (overall): 56 ft 9 in (17.30 m)
Wing area: 927 sq ft (86.2 m^2).
Max. level speed: 702 mph (1130 km/h) at sea level, 616 mph (991 km/h) at 45,000 ft (13716 m)
Max. dive speed: Mach 1.08
Rate of climb: 7.24 minutes to 40,000 ft (12192 m), 9.25 minutes to 50,000 ft (15240 m)
Service ceiling: 54,000 ft (16459 m)
All-up-weight with ventral tanks: 43,165 lb (19580 kg)
Armament: Two or four 30-mm Aden cannon, plus up to four Firestreak AAMs

Radar
The FAW.Mk 9 was fitted with the AI.17 wide-scan radar, which proved superior in service to the US-built AI.22 unit fitted to the Mk 8. The two-crew configuration of the Javelin certainly reduced the workload and aided successful radar interception under conditions of heavy enemy jamming.

Above: In late 1965 No. 29 Squadron deployed 10 FAW.Mk 9Rs to Ndola, Zambia, in response to Rhodesia's unilateral declaration of independence (one of the squadron's aircraft is seen here landing at Ndola). The non-stop ferry flight from Akrotiri, Cyprus, with four underwing fuel tanks in addition to the two 'bosom' tanks, caused outrage in Egypt as the aircraft transitted its airspace. Operating in basic conditions, the Javelin fulfilled its peacekeeping role well for the duration of the 10-month deployment.

Above: Combining the afterburning Sapphire engine, developed for the FAW.Mk 8, with the superior AI.17 radar, the FAW.Mk 9 – especially in Mk 9R form – was undoubtedly the most capable of all Javelin variants. Here a mixture of No. 5 and No. 11 Squadron FAW.Mk 9s are seen during an exercise in West Germany in 1964. The two squadrons provided all-weather fighter defence on the central front throughout the first half of the 1960s, operating from Laarbruch and Geilenkirchen, respectively.

Left: A No. 25 Squadron FAW.Mk 9 is seen with a full load-out of four Firestreaks during a training sortie in 1961. The squadron had previously become the first operator of the final variant, receiving its initial aircraft in December 1959.

for Lightnings in September 1964. Those Javelins based in Germany lasted a while longer and No. 11 Squadron at Geilenkirchen did not give its aircraft up until January 1966. Cyprus-based No. 29 Sqn flew Javelins until May 1967 and further operations were needed when Nos 60 and 64 Squadrons were moved to Malaya to guard that country's airspace against Indonesia.

Conflict in Asia

In April 1963 some Indonesian guerrilla groups started to make raids into Sarawak and two No. 60 Squadron FAW.Mk 9s were among the first aircraft to be dispatched to Labuan and Kuching in response. The rest of No. 60 Squadron followed and in due course No. 64 Squadron was also sent to the area. No. 64 actually began its operations by deploying six aircraft from RAF Binbrook in Lincolnshire on a rotation basis before it disbanded to reform at RAF Tengah, Singapore, in April 1965. In 1964 an escalation of the conflict saw regular Indonesian forces become involved and the Javelins, plus Hawker Hunters and Fleet Air Arm DH Sea Vixens, which were also present, began low level patrols along Indonesia's border. During September, one No. 60 Squadron Javelin forced an Indonesian Lockheed C-130 Hercules transport to crash while taking avoiding action.

In due course an Air Defence Identification Zone was established over West Malaysia where all unidentified intruders were intercepted including, on one occasion, a Soviet Tupolev Tu-16 bomber by a No. 60 Squadron aircraft. Eventually a peace agreement was signed and the security forces began to leave. No. 64 Squadron disbanded at Tengah in June 1967, but No. 60 continued to operate until 30 April 1968 when it disbanded as the last Javelin squadron. These overseas operations probably extended the Javelin's service career by a year.

The Gloster Javelin eventually completed 12 years of squadron service with the RAF, although unfortunately many airframes actually survived for only a relatively short life. On withdrawal, many aircraft were passed to Maintenance Units before being Struck Off Charge and scrapped, but some did find further use as Ground Instructional Airframes at a variety of RAF stations. In addition a quantity were also employed as gate guards. The Javelin has been well served in preservation with no less than 10 surviving worldwide in a relatively complete state (eight in Britain and two abroad).

Service assessment

Brian Carroll completed over 1,000 hours on Mk 5 and Mk 9 Javelins with Nos 151 and 29 Squadrons. He felt that it was a very comfortable aircraft to fly with a big cockpit, well displayed instruments and good all-round visibility, essential in a fighter, and in combat situations in Cyprus he always had complete confidence in the aircraft.

Javelin FAW.Mk 9R

From early in its career the Javelin was criticised for a perceived lack of range. Prototype WT827 and Mk 1 XA634 were modified by Airwork Ltd with an inflight refuelling probe on the starboard and port wings, respectively. These trials were, however, unsuccessful and it was not until the introduction of the huge nose mounted probe on the FAW.Mk 9(F/R) (right) that air-to-air refuelling became possible. Following trial work refuelling from both Valiant and Canberra (above) tankers, the majority of FAW.Mk 9s were modified to (F/R) standard. They were later then further modified to FAW.Mk 9R standard with the capability of carrying four 230-Imp gal (1046-litre) underwing drop tanks.

Left: No. 228 Operational Conversion Unit played a vital role in the combat readiness of the operational squadrons. As well as operating the T.Mk 3 dual-control trainer, the OCU also acquired small numbers of each operational variant as the version entered service. The unit also encompassed the Javelin Mobile Training Unit, manned by experienced pilots, which assisted squadron conversion to the type at its home base. No. 228 OCU FAW.Mk 9 XH898 was unique as the only Javelin to wear a bare metal scheme and was flown by Sqn Ldr George H. Beaton – hence the non-standard tail code.

There were, however, some weaknesses that were far too restrictive. Above 20,000 ft (6096 m) and Mach 0.95, pilots were limited to 3-*g* manoeuvres if the ventral tanks had any fuel in them and only 4 *g* when empty, which was hardly sufficient for a combat situation. The power controls were such that it was easy to over-cook things with the result that a good deal of time was taken up trying to stay within limits rather than concentrating on the opposition. The aircraft's size and radar signature were large, which negated any chance of surprise, but the efficiency of the airbrakes was legendary and allowed for quite spectacular tactical recoveries to be made. A descent from a high level of around 36,000 ft (10973 m) could begin only 10 to 12 miles (16 km to 19 km) from the runway threshold yet still allow a straight run in to break from about one mile (1.6 km), holding high speed all the way.

No. 60 Squadron's FAW.Mk 9Rs were heavily involved in air operations during the 'Confrontation' against Indonesia. At one time the largest squadron in the RAF and based at Tengah, Singapore, the unit flew lengthy border patrols at altitudes as low as 50 ft (15 m), placing great demands on the crew. Aircraft servicablity was generally good although a lack of equipment often led to novel maintenance procedures (right).

Positive performance

Javelin plus points were two engines and good single engine performance, fully powered well balanced controls with duplicated hydraulics, a good armament of cannon and four missiles with radar (it was a very stable gun platform), reasonable endurance, wheels-up landings allowed in emergencies (ventral tanks jettisoned first) and excellent airbrakes giving *g* retardation and extremely effective aerodynamic braking.

The negatives were a shortage of thrust (reheat was only useable when the fuel pump output exceeded normal engine demands), a complex fuel system where it was possible to cross transfer in such a way as to starve both engines, restricted flight envelope with looping, stalling and spinning prohibited (unreal for a fighter), low-level speed restricted to 535 kt (991 km/h), *g* limits too low (5.5 to as little as 3 in some flight conditions), no spin recovery chute and inverted flight time limited to 15 secs.

Against the Hunter in one versus one and two versus two combat, the Javelin came away with honours even. The effective slow down from the barn door brakes could force a fly-through, while the head-up collimator displayed through the gunsight proved invaluable when a target attempted evasion by entering cloud; the Javelin could match a Hunter above 20,000 ft (6096 m) because at high level it could out-turn it.

The Javelin's great naval rival, the de Havilland Sea Vixen, could usually come out best at most heights, since the RAF machine was minimum speed limited and the Sea Vixen could scissor around at slow speed, but the Javelin's big advantage here was its cannon armament whereas the Vixen had none.

Nevertheless the Javelin was a good enough fighter to do the job required of it. Restrictions in the flight envelope did not help it to be a dogfighter, but as a bomber interceptor this was not so big a problem. Firestreaks could be fired at far greater ranges than the Aden cannon and, controlled effectively, the Javelin could intercept any V-bomber with or without the need for AAMs; fortunately, it never had to fire its weapons in anger and it appears no effort was made to fit the improved Red Top AAM.

The loudest criticism made of the Javelin concerned its flying capabilities, but once the aircraft's limitations had been established it was possible to use the aircraft in a way that prevented it from going beyond these limits. Thanks to the missile armament, it possessed a quite formidable combat potential which was available in all weathers. Many within the RAF were happy with the Javelin because it was just what they needed to knock down Soviet bombers, but there were others who did feel that the DH.110 would have been a better choice.

When silhouetted against the sky the Javelin looked like an Ace of Spades and the type was also known as the 'Harmonious Drag Master' from the weird whine of its two Sapphire engines coupled with the organ-like sound created by air flowing over the gun blast-tubes protruding slightly ahead of the wing leading edge. The aircraft left an indelible impression on both ears and eyes – it was a radical design but the T-tail delta format was never used again.

Tony Buttler

Although only in front-line service for a dozen years, the Javelin represented a big stride forward for the RAF in terms of mission capability. At the time of its introduction obsolete fighters could not hope to intercept the new generation of Soviet bombers, however, the Javelin remedied this and, with crews primed in the cockpit at Quick Readiness Alert 24-hours a day, provided a powerful deterrent to any aggression. A No. 33 Squadron Javelin FAW.Mk 9 is seen here shortly before the unit disbanded in 1962, as the RAF was re-equipping with the new-generation Mach 2-capable English Electric Lightning.

Javelin operators

Above: A quartet of No. 25 Squadron Javelin Mk 9s is seen when operating from the unit's home base at RAF Waterbeach.

Royal Air Force

No. 3 Squadron

No. 3 was re-formed out of No. 96 Squadron at Geilenkirchen in Germany on 21 January 1959 and operated FAW.Mk 4s. As such it formed part of RAF Germany's all-weather force until disbanded on 31 December 1960 to re-form with Canberras.

No. 5 Squadron

Formerly No. 68 Squadron, this unit was based at Laarbruch in Germany and its Gloster NF.Mk 11 Meteor night-fighters were replaced by Javelin FAW.Mk 5s in January 1960. They were succeeded by Mk 9s in November 1962, the unit moving to Geilenkirchen a month later. In October 1965, when the squadron lost its Javelins for Lightnings, it moved to Binbrook in Lincolnshire.

No. 11 Squadron

Another renumbered Meteor NF.Mk 11 unit (formerly No. 256 Sqn), No. 11 Sqn received Javelin Mk 4s from October 1959 when based at Geilenkirchen, West Germany. From March 1962 the Mk 4s were gradually replaced by Mk 5s. Ex-No. 25 Squadron FAW.Mk 9s were received from December 1962, and this final variant stayed until No. 11 Squadron disbanded in January 1966.

FAW.Mk 9 XH776 joined No. 11 Sqn in late 1960 and continued to serve with the unit until it was disbanded in 1966. After a brief period of service with No. 29 Sqn the aircraft was scrapped in 1968.

No. 25 Squadron

Originally operating as No. 153 Sqn and based at Waterbeach, this unit received Mk 7 Javelins to replace Meteor night-fighters in March 1959, then Mk 9s from December 1959. In October 1961 it moved to Leuchars before disbanding in November 1962 to become the first Bloodhound surface-to-air missile unit.

Left: No. 29 Squadron's FAW.Mk 9s were based abroad for much of the 1960s and it was the first unit to operate the inflight refuelling-capable FAW.Mk 9R variant.

No. 29 Squadron

A much travelled Javelin unit, No. 29 Squadron received its first FAW.Mk 6s to replace Meteor NF.Mk 11s in November 1957, while flying out of Acklington. The switch to FAW.Mk 9s began in April 1961, nearly three years after moving to RAF Leuchars. On 28 February 1963 No. 29 moved to Nicosia, Cyprus, as the all-weather element of the Near East Air Force. The next move was to Akrotiri in March 1964, however, for nine months from December 1965 was detached to Ndola (Zambia) during the 'Rhodesia Crisis'. No. 29 subsequently remained at Akrotiri until May 1967 when it moved to Wattisham to re-equip with the Lightning.

No. 41 Squadron

Formed on 1 February 1958 at Coltishall when the 'old' No. 41 with Hawker Hunter F.Mk 5s was replaced by renumbering No. 141 Squadron, then operating Javelin Mk 4s. FAW.Mk 5 aircraft were added in August 1958 after a move to Wattisham and both marks were replaced by FAW.Mk 8s from November 1959. The squadron disbanded in December 1963.

Above: No. 46 Squadron pioneered the service entry of the Javelin, when based at RAF Odiham, Hampshire.

No. 33 Squadron

Formerly No. 264 Squadron, No. 33's Meteor NF.Mk 14s were replaced by Javelin Mk 7s in July 1958 while it was based at Leeming. A move to Middleton St George followed two months later and Mk 9 aircraft began to arrive in October 1960. No. 33 disbanded on 17 November 1962.

No. 46 Squadron

This became the first operational Javelin squadron when FAW.Mk 1s replaced its Meteor night-fighters in March 1956, while the unit was based at RAF Odiham. It converted to the FAW.Mk 2 in August 1957, the last Mk 1s leaving in November to join No. 87 Squadron. In May 1958 they were replaced by the FAW.Mk 6, which served until disbandment in June 1961, the unit having moved to RAF Waterbeach two years previously.

No. 23 Squadron

DH Venom NF.Mk 3 night-fighters were replaced by Javelin Mk 4s in April 1957 while No. 23 Sqn was based at Horsham St Faith. Soon afterwards the squadron moved to RAF Coltishall, with FAW.Mk 7s arriving from April 1959 and then Mk 9s from April 1960. These served until replaced by Lightnings in September 1964, No. 23 having moved to Leuchars in March 1963. The unit was a leader in long-distance in-flight refuelling and, as such, became the first Fighter Command squadron to deploy by air to Singapore.

Below: A line-up of No. 23 Sqn FAW.Mk 4s, wearing the unit's famous eagle badge on the tail fin, at RAF Coltishall in 1958.

Above: FAW.Mk 9 XH758 served with No. 33 Sqn from March 1961 until the unit disbanded in November 1962

No. 60 Squadron

Javelin FAW.Mk 9s were operated by this unit from July 1961 until 30 April 1968 when it disbanded. The whole of this period was spent at Tengah, Singapore, and No. 60 was to be the last front-line operator of the type. This service included operations during the 'Indonesian Confrontation'.

Baking in the hot midday sun, a line-up of No. 60 Sqn FAW.Mk 9s is seen here while operating from Tengah, Singapore, during the confrontation with Indonesia.

In September 1958 No. 64 Squadron finally relinquished its obsolete Meteor night-fighters for Javelin FAW.Mk 7s. Initially, the squadron was equipped with the four-gun, non-Firestreak capable version as seen here.

No. 64 Squadron

Another long-term operator, No. 64 Squadron, used the Javelin FAW.Mk 7 from September 1958 until June 1960 when Mk 9s arrived; these stayed until disbandment in June 1967. No. 64 was based at Duxford until July 1961, then Waterbeach until August 1962 when the unit moved to Binbrook. In 1964 the squadron was re-deployed to Tengah to provide all-weather defence for Singapore. These final duties actually began with detachments but the whole squadron finally moved on 1 July 1965.

No. 72 Squadron

This unit operated Javelins for a little over two years, receiving Mk 4s in April 1959 and Mk 5s three months later, keeping both until it disbanded on 30 June 1961 to eventually become a helicopter unit. On 28 June 1959 No. 72 moved from Church Fenton to Leconfield, where it stayed until losing its fixed-wing aircraft.

No. 85 Squadron

On 30 November 1958 No. 89 Squadron, based at RAF Stradishall and already a Javelin operator with FAW.Mk 2s and Mk 6s, was renumbered No. 85 Squadron. As such it retained these variants until March and June 1960, respectively, when they were succeeded by FAW.Mk 8s, the first example arriving in March. In June 1959 No. 85 had moved to West Malling and then in September 1960 to West Raynham, were it stayed until disbanded on 31 March 1963.

No. 87 Squadron

Based at Brüggen, West Germany, No. 87's Meteor NF.Mk 11s were replaced by Javelin Mk 1s from No. 46 Squadron from September 1957. FAW.Mk 4 aircraft began to join from December 1959 and both versions stayed until disbandment on 3 January 1961. A few FAW.Mk 5 aircraft were also on strength between September 1958 and November 1960, the squadron remaining at Brüggen.

The longest serving FAW.Mk 5 operator was No. 151 Squadron, which operated the variant between 1957-61.

No. 96 Squadron

No. 96 Squadron operated FAW.Mk 4s out of Geilenkirchen from September 1958 until January 1959, when it re-formed as No. 3 Squadron.

No. 141 Squadron

Operating from Horsham St Faith with Venom NF.Mk 3s, this unit received Javelin Mk 4s in February 1957 and moved to RAF Coltishall three months later. On 1 February 1958 it was renumbered as No. 41 Squadron.

Above: No. 141 Squadron's time as a Javelin operator (using Mk 4s) lasted just 11 months from Feb. 1957 to Jan. 1958.

No. 89 Squadron

Javelin FAW.Mk 2 and Mk 6 variants replaced No. 89 Squadron's de Havilland Venom NF.Mk 3s at RAF Stradishall in October 1957. They remained until the unit was renumbered as No. 85 Squadron in November 1958.

No. 151 Squadron

This unit's Venom NF 3s were replaced by Javelin Mk 5s in June 1957 and, after a brief period at Turnhouse, it flew this variant out of RAF Leuchars until it disbanded on 19 September 1961.

No. 228 OCU

The Javelin Operational Conversion Unit at Leeming used T.Mk 3 and FAW.Mk 4, 5, 7 and 9 aircraft. It was disbanded on 15 September 1961 because it was considered that sufficient aircrew were now fully trained to see the fighter through to its retirement. The task given to Nos 60 and 64 Squadrons to defend Malayan airspace, however, brought the need for additional aircrew and No. 228 OCU was re-formed on 1 May 1965 at Leuchars with Mk 3s and Mk 9s. Later it was given the title of No. 11 (Reserve) Squadron before disbanding in November 1966. T.Mk 3s were also operated by Nos 3, 5, 11, 23, 25, 29, 33, 41, 46, 60, 64, 72, 85, 87 and 151 Squadrons and by the Fighter Command Instrument Rating Squadron.

Left: In the early 1960s No. 85 Sqn utilised the reheat-equipped Mk 8 version of the Javelin. XJ121 served solely with the unit, before being scrapped in 1964.

Above: In 1966 228 OCU was allocated the shadow identity of No. 11 Squadron, as demonstrated by XH958 wearing No. 11's distinctive insignia on the tail.

Other operators

Other Javelin operators included the Central Fighter Establishment at West Raynham, plus several CFE sub-units including the All-Weather Wing (Mk 1s), the All-Weather Development Squadron which initially used Mk 1s at Coltishall and eventually employed Mk 2, 4, 5 and 7 aircraft, and the Air Fighting Development Squadron which used most of the marks. The All-Weather Fighter Leaders School used some Mk 4s and 5s, the College of Aeronautics at Cranfield used the Mk 1 and No. 1 Guided Weapons Trials Squadron made use of half a dozen FAW.Mk 7s for Firestreak missile trials out of Valley.

Left: The All-Weather Development Squadron (AWDS) used three FAW.Mk 7s from the spring of 1958, to help develop effective operational use of the Firestreak missile – as carried here by XH756.

Above: The Aeroplane and Armament Evaluation Establishment (A&AEE), based at Boscombe Down, operated all variants of the Javelin at different times. XH966, the first Mk 8, was used from 1958-62.

INDEX

Page numbers in **bold** refer to illustration captions.

Picture acknowledgments

Front cover: Peter Liander, via Mike Stroud, Frank Smith. **4:** Lockheed Martin via Tom Kaminski, BAE Systems. **5:** Lockheed Martin via Tom Kaminski (two), Boeing via Tom Kaminski. **6:** via Tom Kaminski (two). **7:** Alexander Mladenov, David Donald. **8:** Bell via Tom Kaminski (two). **9:** Raytheon (two). **10:** AgustaWestland, Andrew Cline. **11:** US Navy (four). **12:** Sven-H. Tank (two). **13:** Eric Katerberg (four), Andrew Cline. **14:** Jim Winchester (two), Peter Liander. **15:** Marcus Fülber/Jens Schymura (three), Chris Ryan. **16:** Georg Mader (four). **17:** Georg Mader, Andrew Cline. **18:** Andrew Cline (three). **19:** Boeing (four). **20-21:** Lockheed Martin via Chris Pocock. **22:** Chris Pocock (three). **23:** NASA (two), Lockheed Martin. **24-25:** Lockheed Martin via Chris Pocock. **26:** Lockheed Martin via Chris Pocock, USAF. **27:** USAF (two), via Chris Pocock. **28:** Cees-Jan van der Ende (three). **29:** Frank Smith (two), Cees-Jan van der Ende. **30:** Cees-Jan van der Ende (four). **31:** Cees-Jan van der Ende (two), Frank Smith. **32:** Cees-Jan van der Ende (two), Frank Smith. **33:** Cees-Jan van der Ende (four). **34:** Frank Smith, Cees-Jan van der Ende (two). **35:** Cees-Jan van der Ende (four), Frank Smith. **36:** Hugo Mambour, Marcus Fülber. **37:** Hugo Mambour (two). **39:** Piotr Butwoski, Hugo Mambour (two). **40:** Marcus Fülber. **41:** Piotr Butowski (two), Hugo Mambour. **42:** Hugo Mambour. **43:** Marcus Fülber, Hugo Mambour. **44:** Daniel Karlsson. **45:** David Donald (DD) (four). **46:** Peter Liander, Saab via DD, David Donald. **47:** David Donald, Luigino Caliaro. **48:** Saab via DD, Peter Liander (two). **49:** Peter Liander (two). **50:** Åke Anderson/FMV via DD, Daniel Karlsson, Peter Liander. **51:** Peter Liander (two), Pia Ericson/FMV via DD. **52:** Erland Johansson/FMV via DD, F4 via DD (two), Peter Liander. **53:** David Donald (two). **54:** Peter Liander (three). **55:** David Donald, Peter Liander. **56:** David Donald, Peter Liander, Luigino Caliaro. **57:** Luigino Caliaro, Marnix Sap/MIAS (two). **58:** Jerry Pousette/FMV via DD, David Donald, Daniel Karlsson. **59:** Peter Liander, Luigino Caliaro. **60:** Daniel Karlsson, David Donald. **61:** Luigino Caliaro, David Donald (three). **62:** David Donald (three), Daniel Karlsson. **63:** David Donald (three). **64:** Luigino Caliaro, David Donald (five). **65:** Volvo Aero via DD. **66:** David Donald (three). **67:** Pia Ericson/FMV via DD, David Donald (two), Luigino Caliaro, Jim Winchester. **68:** Ericsson Microwave via DD (two), David Donald (five). **69:** Luigino Caliaro, Peter Liander. **70:** Luigino Caliaro (two), David Donald. **71:** Ulf Hugo (two), David Donald (two). **72:** Daniel Karlsson (two), Saab via DD. **73:** Peter Liander, David Donald, Luigino Caliaro. **74:** Saab via DD, David Donald. **75:** David Donald, Peter Liander. **76:** US Navy via Tom Kaminski (two). **77:** US Navy via Tom Kaminski (three). **78:** US Navy via Tom Kaminski (two), Mike Anselmo/AIR. **79:** US Navy via Tom Kaminski (three). **80:** US Navy via Tom Kaminski (three). **81:** US Navy (two), Sikorsky via Tom Kaminski, Ray Rivard via Tom Kaminski. **82-83:** US Navy via Tom Kaminski. **84:** Heinz Berger, Aerospace. **85:** David Willis, Swedish Air Force, Aerospace. **86:** Alexander Mladenov (six). **87:** Swedish Air Force. **88:** Aerospace, Alexander Mladenov. **89:** Heinz Berger, Alexander Mladenov. **90:** Emiel Sloot, Alexander Mladenov. **91:** MATRA/CEV, Alexander Mladenov (two). **92:** US Navy via TRH, Alexander Mladenov. **93:** Alexander Mladenov, Aerospace. **94:** Mikoyan via Zaur Eylanbekov (four). **95:** Terry Panopalis, David Willis, Mikoyan via Zaur Eylanbekov. **96:** Mikoyan via Zaur Eylanbekov, Aerospace (two). **97:** Alexander Mladenov, Heinz Berger, Aerospace (four), Emiel Sloot, Chris Ryan. **98:** Mikoyan via Zaur Eylanbekov, Terry Panopalis, Aerospace. **99:** Alexander Mladenov, Heinz Berger, Emiel Sloot. **100:** Terry Panopalis, Hugo Mambour, David Willis. **101:** Alexander Mladenov, Piotr Butowski. **102:** Col José Navarro via Santiago Rivas (SR), Museo Nacional de Aviación/El Salvador via SR. **103:** Museo del Aire/Honduras via SR, Museo Nacional de Aviación/El Salvador via SR, via Santiago Rivas and Juan Carlos Cicalesi. **104:** Douglas Cornejo Archive via SR, Museo Nacional de Aviación/El Salvador via SR (two), Dan Hagedorn via SR. **105:** via Santiago Rivas and Juan Carlos Cicalesi, Archie Baldocci via SR, Museo del Aire/Honduras via SR, Marco Lavagnino Archive via SR (two). **106:** via Santiago Rivas and Juan Carlos Cicalesi (two). **107:** Museo Nacional de Aviación/El Salvador via SR (two). **108:** Museo Nacional de Aviación/El Salvador via SR (two), Rob Mears via SR. **109:** Museo del Aire/Honduras via SR (two), Amado Aguilez Ferrari via SR, R. Erazo via SR. **110:** Rob Mears via SR. **111:** Museo Nacional de Aviación/El Salvador via SR, Douglas Cornejo Archive via SR. **112:** Dassault, Dassault via Alexis Rocher. **113:** François Robineau/Dassault Aviaplans via Alexis Rocher, Dassault via Alexis Rocher. **114-116:** Dassault via Alexis Rocher. **117:** Dassault via Alexis Rocher (three), Dassault. **118:** B. Marembert (five). **119:** Dassault via Alexis Rocher, Dassault. **120:** François Robineau/Dassault Aviaplans via Alexis Rocher, Dassault via Alexis Rocher (three). **121-122:** Dassault via Alexis Rocher. **123:** Dassault via Alexis Rocher, Dassault, François Robineau/Dassault Aviaplans via Alexis Rocher. **124-125:** Lockheed. **126:** Lockheed, USAF. **127:** Aerospace, Lockheed, USAF. **128:** Aerospace (three), Lockheed. **129:** Lockheed, Aerospace (four). **130:** Lockheed, Aerospace, USAF (two). **131:** Lockheed (two), Aerospace. **132:** Lockheed (three). **133:** Lockheed via Warren Thompson (two), Lockheed. **134:** Lockheed (three). **135:** Lockheed (two), Aerospace. **136:** USAF, Aerospace (two). **137:** Aerospace. **138:** Harley Barnhart via Warren Thompson, Walt Zurney via Warren Thompson (two). IWM via TRH. **139:** RAF Museum via TRH, Aerospace, Arthur Rowley via Warren Thompson. **140:** Arthur Rowley via Warren Thompson, IWM, J.W. Phegley via Warren Thompson. **145:** Lockheed (two), USAF, Aerospace. **146:** Ken Jorgensen via Warren Thompson, Aerospace (two), USAF. **147:** Lockheed, RAAF. **148:** Charles Van Bibber via Warren Thompson (two), TRH. **149:** USAF, William Fowkes via Warren Thompson, TRH, Aerospace. **150:** Aerospace, USAF via TRH. **151:** Aerospace (three), Paul Giguere via Warren Thompson. **152:** USAF via Warren Thompson (two). **153:** USAF via Warren Thompson (two), David Hopwood via Warren Thompson. **154:** Aerospace (five). **155:** Aerospace (two), Peter R. March (two). **156:** Aerospace, via Mike Stroud. **157:** via Mike Stroud (two), Tony Buttler Collection. **158:** via Mike Stroud (two), Tony Buttler Collection. **159:** via Mike Stroud (three), Gloster. **160:** via Mike Stroud (two), Tony Buttler Collection, Gloster (two). **161:** Tony Buttler Collection (two), via Mike Stroud. **162:** Tony Buttler Collection, via Mike Stroud (two). **163:** Peter R. March, via Mike Stroud (two), via Terry Panopalis. **164:** via Mike Stroud, Gloster, Tony Buttler Collection (two), Peter R. March. **165:** Peter R. March, via Mike Stroud, via Peter R. March (two). **166:** Tony Buttler Collection (three), Gloster, via Mike Stroud (two). **167:** via Terry Panopalis (two), Tony Buttler Collection. **168:** via Peter R. March, Aerospace, Tony Buttler Collection (two). **170:** Tony Buttler Collection (two), via Peter R. March, via Terry Panopalis (two). **171:** Chris Ryan, via Mike Stroud (two), Tony Buttler Collection. **172:** via Mike Stroud (three), Tony Buttler Collection (two), Chris Ryan, via Terry Panopalis. **173:** Tony Buttler Collection (three), via Terry Panopalis, Aerospace, via Mike Stroud, Chris Ryan.